Descriptive Electronics

Descriptive Electronics

Ralph R. Huffsey
Charlotte Aull Huffsey
Missouri Valley College

Holt, Rinehart and Winston, Inc.
New York Chicago San Francisco Atlanta
Dallas Montreal Toronto London Sydney

Library of Congress Catalog Card Number: 76-97859

SBN: 03-081623-8

Printed in the United States of America

9 8 7 6 5 4 3 2 1

To Holly

Preface

Our intention in writing this book was to explain the function of representative electronic circuits without the use of extensive mathematics. Believing that an understanding of how electronic circuits work need not be limited to the physicist and electronics engineer, we hoped to make the area more accessible to other interested persons. The text thus requires only a knowledge of basic high school algebra. However, we have attempted to make the explanations logically sound and as complete as possible without the use of more sophisticated mathematical language.

We had observed a need for background in basic electronic circuits on the part of students whose major interest is chemistry, biology, psychology, or premedicine. To meet this need, we have offered a one-semester course in electronics for nonphysics majors. The text of this book grew out of this course. The level of the book is such that it can be read by interested high school students as well as being suitable for college courses of the type mentioned.

To the reader may we suggest reading the book a section or two at a time. You may find it helpful to write down some of the ideas or discuss them with others.

Try to carry out as many of the suggested experiments as are feasible for you. They will help you to better understand the ideas and to maintain enthusiasm for the subject.

We would like to extend our thanks to John M. Doyle, Electronic Aids, Inc., and Prof. James G. Brazee, Dutchess Community College, for their suggestions after reviewing the manuscript.

Ralph R. Huffsey
Charlotte A. Huffsey

Marshall, Missouri
January 1970

Contents

Descriptive Electronics

CHAPTER 1

Direct Current and Voltage

1.1 ATOMIC THEORY

Electronic devices are physical instruments which utilize much of what is known about electricity and magnetism. In our study of electronics we should first consider the conduction of electricity, that is, the motion of certain fundamental atomic particles called electrons. The conduction of electrons may be along a wire, through semiconductors such as transistors, in a vacuum as in a TV picture tube, or in a gas as in a neon sign. Electrons move through these devices in very, very large numbers; however, we will need to know something of the properties of a single electron.

Without trying to be too definitive, let us take a brief look at atomic theory. We first observe matter on a large scale as the "stuff" we find around us: air, water, wood, metal, and so on. These materials are either chemical or physical combinations of a relatively few basic substances called elements. Elements are so basic that they cannot be further broken down and still maintain their physical and chemical properties. Man recognizes about 106 elements; iron, copper, and oxygen are examples. Each element is made up of one or more small particles called atoms. And it is the atom, the building block of the elements and all matter, with which we want to become familiar in order to better understand the conduction of electricity.

Atomic theory tells us that we can begin to understand the atom, this extremely small particle of matter, by considering something very large, the sun and the planets that revolve about it. Like our solar system, the atom has a dense central core, called the nucleus, which is made up of even smaller particles called protons and neutrons, while moving about the nucleus are electrons. The electrons travel in rather well-defined orbits, or energy levels, and the maximum number of electrons at each energy level is limited. The first orbit may have two electrons, the second eight, and the third eighteen. The electrons farthest away from the nucleus are less tightly bound to orbit about the nucleus.

An electron has a negative electrical charge and may be referred to as a negative ion. If an atom loses one of its outermost electrons, it is said to be positively ionized. If an extra electron becomes attached to an outer orbit of an atom, the atom is said to be negatively ionized. An atom which has not gained or lost any electrons is electrically neutral; it does not have any electrical charge. Now if each electron orbiting about the nucleus is negatively charged, there must exist a positive charge in the nucleus which counteracts the negative electrons and makes the atom electrically neutral. The protrons in the nucleus are positively charged and, in fact, the magnitude of the positively charged protons is equal to the magnitude of the negatively charged electrons. This implies that there are as many protons within the nucleus of an electrically neutral atom as electrons about the nucleus. The number of protons in the nucleus of an atom is called its atomic number. This number is characteristic of a particular element and identifies that element. For instance, the oxygen atom has 8 protons, the copper atom has 29.

Since the electrons move in orbits about the nucleus, rather than flying away from it, there must be a force of attraction between the positively charged nucleus and the negative electrons. In other words, unlike charges attract each other. The electrons themselves are all negatively charged and repel each other, that is, like charges repel each other.

The outermost electrons of an atom are referred to as valence electrons. How complete the outermost orbit is will, to some extent, determine whether the element will serve as an insulator or a conductor of electricity. Most good conductors, such as copper, have one or two valence electrons. Semiconductors, materials which are poor conductors, have four or five valence electrons. Insulators have their outer orbit, or shell, nearly full.

To review, we have seen that electrons are atomic-sized particles with a negative charge which orbit about the nucleus of an atom. This nucleus contains protons, which are positively charged, and neutrons, which have no charge. We have also mentioned the law of electrostatics: Like charges repel and unlike charges attract each other.

1.2 ELECTRON SOURCES

Earlier we mentioned that electronic devices utilize the conduction of electrons. Thus, it is important that we be able to obtain a copious supply of the negatively charged electrons. Batteries supply large numbers of electrons obtained from certain chemical reactions. Mechanical methods employing magnetic fields can also be a source of electrons. And light striking certain elements or compounds will liberate electrons.

All these devices have two sides. One side is negative, the other positive. From our previous discussion, we know that the negative side of the device has an excess of electrons and the positive side a deficiency of electrons. This is a similar situation, on a very large scale, to the negatively ionized atom with its excess of electrons and the positively ionized atom with its deficiency of electrons.

1.3 CURRENT

Since unlike charges attract each other, we now know that the negative charges, electrons, are attracted to the positive side, the electron deficient side, of the electron suppliers. However, there is opposition to the recombining of negative and positive charges within the battery, generator, or photoelectric cell. But if we connect a good conductor, such as a copper wire, to the external terminals of an electron supplier, there will be little opposition to the flow of electrons from negative to positive terminal along the wire. How do we know that electrons are flowing along our conductor? We cannot actually see them, but there are several indications of their motion along the wire. If you touch the wire, you will find it warm and may even be burned. A more useful method of indirectly observing the electron flow is to place an electric light bulb along the conductor and see it glow.

With this series circuit of battery, copper wire, and bulb all connected, we can make some other observations. If the bulb glows very brightly, we know the battery is supplying a large number of electrons. On the other hand, if the bulb has a feeble glow, then the battery is supplying just a few electrons. Another very important observation is that no electrons flow until the circuit is complete, that is, until there is an external path from negative to positive terminal of the battery.

In this experiment, if we are concerned only with having the bulb glow brightly, it is necessary to choose the right size battery and bulb. First let us consider the size of the battery in terms of the number of electrons which travel from negative to positive terminal through the light bulb. Electrons are so small, atomic size, that they cannot actually be seen to count. But suppose we could station ourselves somewhere along the circuit and see them pass. We could stay there as long as the circuit was complete and count all the electrons moving past. But the number would be so large that the job would still be impossible. An easier way would be to count just the number which go by our station in a second, then note for how many seconds the circuit is closed, or complete. Then to find the total number of electrons that passed, we could simply multiply the number which passed in one second by the total number of seconds. There is only one problem: Will the same number of electrons pass our station each second? If the light from the bulb increases or decreases in brightness, we know that the number of electrons traveling through it is correspondingly increasing or decreasing.

It now seems almost impossible to use even our second method to keep track of the negative electron flow. But it will be so important to measure the flow of electrons in the circuit that we introduce the concept of the ampere, a practical unit of electrical current. The ampere is equivalent to 6,250,000,000,000,000,000 (in scientific notation 6.25×10^{18}) electrons passing a point in the circuit each second. An instrument called an ammeter, which we will elaborate on later, is used to measure how much current, how many electrons each second, is in our circuit. Thus, by using the ampere, we observe the rate of flow of electrons each second, rather than the total number of electrons coming from the source. If the rate of flow of electrons is large—high amperage—then the bulb glows brightly.

1.4 VOLTAGE

The battery in the circuit has the potential ability to supply electrons to the bulb. If the potential of the battery is great, then the rate of flow of electrons will be large, or the current in amperes will be large. The potential ability of the battery to deliver fast moving electrons is called voltage. Observe at this point that if the potential (voltage) of the battery is large, then the amperage (rate of flow of electrons) is also large. This is referred to as a direct relation or direct proportion. That is, when voltage increases, current in amperes increases; when voltage decreases, current decreases.

1.5 RESISTANCE

Now, using the same circuit, let us assume that the battery has a large enough potential, or voltage, to supply a sufficiently large current to keep the bulb glowing brightly. We experiment by exchanging the copper wire for a much longer piece of the same wire. If the replacement is very long, the bulb will glow a little less brightly. Another experiment we might try is to exchange the original copper wire for a copper wire that is much smaller in diameter. Again the bulb is not as bright. We know the battery has the ability to deliver fast moving electrons, which caused the bulb to glow brightly. But something has happened: The longer wire or the smaller diameter wire has affected the circuit. An explanation of what has occurred may be that the copper wire, which we considered a good conductor of electrons, when changed, slowed the flow of electrons or, in other words, decreased the current. Although copper is a good conductor of electrons, if we change its dimensions drastically, as in the two experiments above, even the good conductor will oppose the flow of electrons. We say that it offers resistance to the flow. Obviously, if we exchanged the copper wire for some poorer conductor, the poorer conductor would offer more resistance to the flow of electrons, and again the bulb would dim.

Before we lose track of our experimental observations, remember that when we increased the opposition, or resistance, to electron flow, the bulb gave off less light. In other words, increasing the resistance of the circuit caused a decrease in the current. This is an inverse relation or inverse proportion.

1.6 OHM'S LAW

We now have two proportions involving current, or the rate of flow of electrons. The first is the direct proportion between voltage and current. When voltage increases, current increases; when voltage decreases, current decreases. The second is the inverse proportion between resistance and current. When resistance increases, current decreases; when resistance decreases, current increases.

Since both of these proportions concern the current in the circuit, they may be combined to form a single statement known as Ohm's Law. Ohm's Law states: The current in a circuit is directly proportional to the voltage and inversely proportional to the resistance. Since this is such an important law in electronics and because we will use it over and over, we prefer a shorter way to state it. We use symbols to abbreviate the quantities that appear in our statement. The letter E is used for voltage, R for resistance, and I for current. Ohm's Law then is stated as: I is directly proportional to E and inversely proportional to R, or, in equation form,

$$I = \frac{E}{R}$$

With the aid of arrows to indicate the direction of change of E, I, and R (an arrow pointed up$\uparrow$ will signify an increase, an arrow pointed down↓ a decrease), we can clearly see that the above equation has the same meaning as the statement of Ohm's Law in words. If the resistance R is not changed, then the current I increases with increasing voltage E. $I\uparrow = E\uparrow/R$. This is our direct proportion. Now if we let the voltage stay the same and increase the resistance R, the current I will decrease. $I\downarrow = E/R\uparrow$. Here we are dividing the constant voltage E by an increasing resistance $R\uparrow$. For example, let us suppose we have a 12-volt battery and 3 units of resistance, called ohms in honor of George Ohm who first observed the relation we now call Ohm's Law. If we divide 12 volts by 3 ohms, we get a current I of 4 amperes. Now if we increase the resistance ($R\uparrow$) to 6 ohms and divide this increased resistance into the constant 12 volts, we obtain 2 amperes of current. Note that the current decreased ($I\downarrow$) from 4 to 2 amperes with an increase in resistance ($R\uparrow$) from 3 to 6 ohms. This is an inverse proportion.

Notice that by solving the equation $I = E/R$ for E and for R, we can obtain two other forms of Ohm's Law.

$$E = IR$$

and

$$R = \frac{E}{I}$$

1.7 SERIES CIRCUITS

Rather than try to draw a battery and wire with resistance to electron flow, we use schematic symbols, shorthand representations of these components. 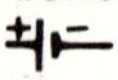 represents a battery. Notice that the shorter bar is the negative side of the battery, the side with an excess of electrons. A resistor has the schematic symbol  . Now combining these, we have a schematic diagram of an electrical circuit. (See Figure 1.1.)

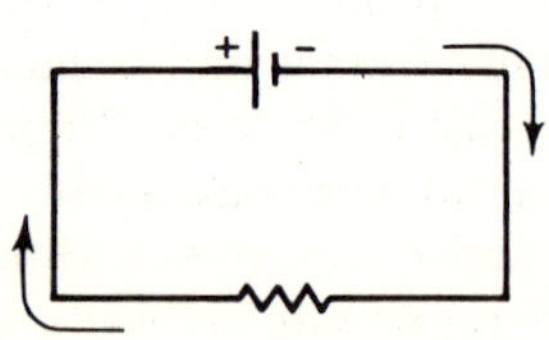

FIGURE 1.1 *Direction of electron flow from negative to positive.*

Consider a slightly more complex circuit which has two resistors but still has only one path for the flow of current. Since there is only one path for the current in this circuit, the same current passes through each resistor. This is called a series circuit. (See Figure 1.2.) For example, suppose the voltage E is 10 volts, resistance R_1 is 2 ohms, and resistance R_2 is 3 ohms. If we shorten the line in our diagram connecting R_1 and R_2, they may be taken together as one large resistance. Thus, the total resistance is the sum of the two resistances. $R_T = R_1 + R_2 = 2$ ohms $+ 3$ ohms $= 5$ ohms. Returning to Ohm's Law, we divide to find the total current. $I = E/R_T = 10$ volts/5 ohms $= 2$ A (A is a convenient abbreviation for amperes). From the definition of a series circuit we know that the current in each resistor is 2 A. Returning to the second form of Ohm's Law, we can determine the voltage drops, V_1 and V_2, across R_1 and R_2, respectively. $V_1 = IR_1 = 2$ A $\times$ 2 ohms $= 4$ volts,

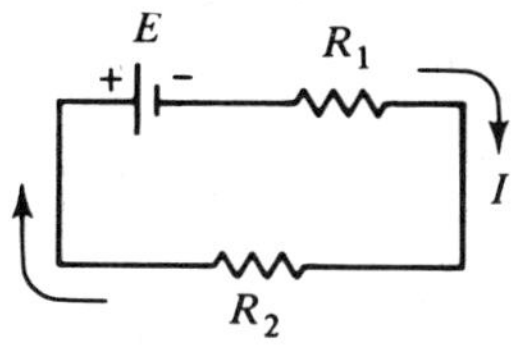

FIGURE 1.2

and $V_2 = IR_2 = 2$ A $\times$ 3 ohms $= 6$ volts. Notice that adding $V_1 = 4$ volts and $V_2 = 6$ volts, we have 10 volts which was the total voltage supplied by the battery:

$$E = V_1 + V_2$$

Summarizing, for series circuits, circuits with only one path for the flow of current, we have the following three relations:

$$R_T = R_1 + R_2 + R_3 + R_4 + \cdots + R_{\text{Last series resistor}}$$

$$I = \frac{E}{R_T}$$

$$E = V_1 + V_2 + V_3 + V_4 + \cdots + V_{\text{Last series resistor}}$$

We have written these relations a little differently than in the example. The reason for this is that a series circuit may include more than two resistors. The equations above can be used for a series circuit with any number of resistors.

We would like to make one additional comment about the third relation $E = V_1 + V_2 + \cdots + V_{\text{Last series resistor}}$. The values V_1, V_2, $\ldots$, $V_{\text{Last series resistor}}$ are sometimes referred to as the voltage drops around the circuit. Since voltage is the potential ability to supply electrons, the term voltage drop implies that each resistor in the series circuit uses some of this potential and thus causes a decrease, or drop, in voltage.

1.8 PARALLEL CIRCUITS

We have just looked at the properties of a series circuit, one which has only one path for the flow of current. Now we want to examine another type of circuit, which has more than one path for the flow of current, called a parallel circuit.

Consider the following parallel circuit made up of two resistors and a battery. (See Figure 1.3.) When the current leaving the battery reaches

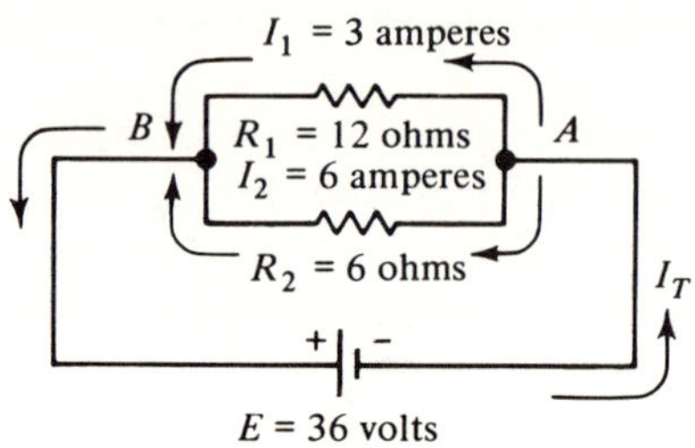

FIGURE 1.3

point A, it divides, not necessarily in half, but some electrons travel through R_1 and some through R_2. At point B the currents through R_1 and R_2 recombine and return to the positive terminal of the battery. The parallel circuit differs from the series circuit in this respect: The current in each component does not equal the total current; instead, as we see from the diagram, the total current entering at junction point A is equal to the sum of the currents in R_1 and R_2. $I_T = I_1 + I_2$. Using the values $I_1 = 3$ A and $I_2 = 6$ A, we can find the total current entering point A or leaving point B. $I_T = I_1 + I_2 = 3\text{ A} + 6\text{ A} = 9$ A. Applying Ohm's Law ($E = IR$), we find the voltage drops for R_1 and R_2. $V_1 = I_1R_1 = 3\text{ A} \times 12$ ohms $= 36$ volts, and $V_2 = I_2R_2 = 6\text{ A} \times 6$ ohms $=$ 36 volts. Thus, the voltage drop across R_1 is the same as the voltage drop across R_2, and this is the same as the voltage of the battery, that is, the same voltage appears across all branches of a parallel circuit.

We now know the total current $I_T = 9$ A, and the total voltage $E = 36$ volts. Using the form of Ohm's Law that is solved for resistance, we can find the total resistance of the circuit. $R_T = E/I_T = 36$ volts/9 A $= 4$ ohms. This result may seem a little strange when you recall that $R_1 = 12$ ohms and $R_2 = 6$ ohms are both larger than the total resistance. Can we empirically find an equation for the total resistance? Experimenting, we notice that the product $R_1 \times R_2 = 12 \times 6 = 72$ is much too large; the sum $R_1 + R_2 = 12 + 6 = 18$ is also larger than R_T. But if we divide the sum into the product, we get $72/18 = 4$ ohms, which is the value of R_T. Writing this procedure as an equation, we have

$$R_T = \frac{R_1 \times R_2}{R_1 + R_2}$$

These relations for parallel circuits, circuits with more than one path for the flow of current, may be summarized in a form similar to those for series circuits. Again, these relations are written in a generalized form, which provides for parallel circuits with more than two resistors.

$$I_T = I_1 + I_2 + I_3 + I_4 + \cdots + I_{\text{Last parallel resistor}}$$
$$E = I_T R_T$$

The relation for the total resistance of a parallel circuit in terms of its component resistors $R_1, R_2, \ldots, R_{\text{Last parallel resistor}}$ is more complicated. A practical procedure is simply to repeat the application of the equation $R_T = (R_1 \times R_2)/(R_1 + R_2)$ for each pair of resistors. Another method of obtaining total resistance of a parallel circuit is to use the equation

$$\frac{1}{R_T} = \frac{1}{R_1} + \frac{1}{R_2} + \frac{1}{R_3} + \frac{1}{R_4} + \cdots + \frac{1}{R_{\text{Last parallel resistor}}}$$

For circuits with two resistors in parallel, this equation becomes $1/R_T = 1/R_1 + 1/R_2$, which is easily shown algebraically to be equivalent to the equation $R_T = (R_1 \times R_2)/(R_1 + R_2)$.

Consider the following circuit with three resistors in parallel. (See Figure 1.4.) We wish to find the total resistance, the total current, and

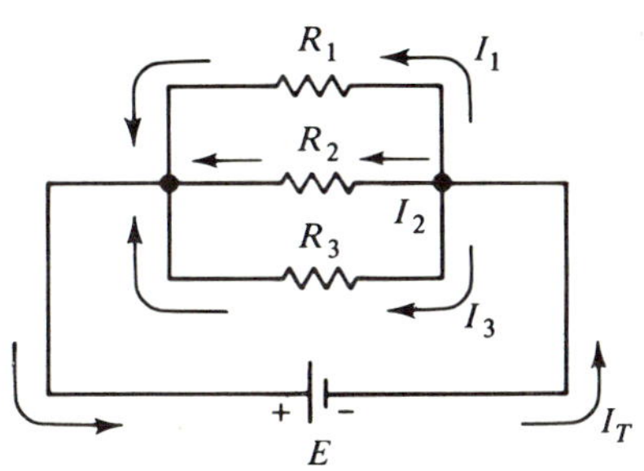

FIGURE 1.4 $E = 18$ *volts;* $R_1 = 9$ *ohms;* $R_2 = 18$ *ohms;* $R_3 = 3$ *ohms.*

the current through each resistor. First use the "product divided by the sum" technique to find the total resistance. The first pair of parallel resistors has combined resistance $R_{1,2} = (R_1 \times R_2)/(R_1 + R_2) = (9 \times 18)/(9 + 18) = 162/27 = 6$ ohms. Combining the 6 ohms of $R_{1,2}$ with R_3 to determine R_T, $R_T = (R_{1,2} \times R_3)/(R_{1,2} + R_3) = (6 \times 3)/$

$(6 + 3) = 18/9 = 2$ ohms. Knowing E and R_T, apply Ohm's Law to find $I_T = E/R = 18$ volts/2 ohms $= 9$ A. Have you noticed that Ohm's Law may be applied to either parallel or series circuits whether you are working with current, voltage, and resistance of a single component or total current, total voltage, and total resistance of the circuit?

To finish the problem and determine the current through each resistor, we apply Ohm's Law to each resistor in the circuit to find I_1, I_2, and I_3.

$$I_1 = \frac{E}{R_1} = \frac{18 \text{ volts}}{9 \text{ ohms}} = 2 \text{ A}$$

$$I_2 = \frac{E}{R_2} = \frac{18 \text{ volts}}{18 \text{ ohms}} = 1 \text{ A}$$

$$I_3 = \frac{E}{R_3} = \frac{18 \text{ volts}}{3 \text{ ohms}} = 6 \text{ A}$$

Making a final check, the total current should equal the sum of the currents in each branch of the parallel circuit. $I_T = I_1 + I_2 + I_3 = 2 \text{ A} + 1 \text{ A} + 6 \text{ A} = 9 \text{ A}$. Before leaving this problem, you should note that in parallel circuits the smallest resistor, $R_3 = 3$ ohms, has the largest current, $I_3 = 6$ A, while the largest resistor, $R_2 = 18$ ohms, has the smallest current, $I_2 = 1$ A. This type of relation, you remember, is an inverse proportion.

1.9 SERIES–PARALLEL CIRCUITS

Now let us look at one more example, which combines both series and parallel components. in one circuit. We do not need any new equations to analyze these compound circuits, but it is necessary to be very attentive in order to recognize which components are in series and which are in parallel. In this analysis, we reduce the complex circuit to an equivalent circuit with a single resistor connected to the battery.

Initially, we recognize that R_1 and R_2 are in parallel. (See Figure 1.5.) This fact is clear if we recall that a parallel branch presents two or

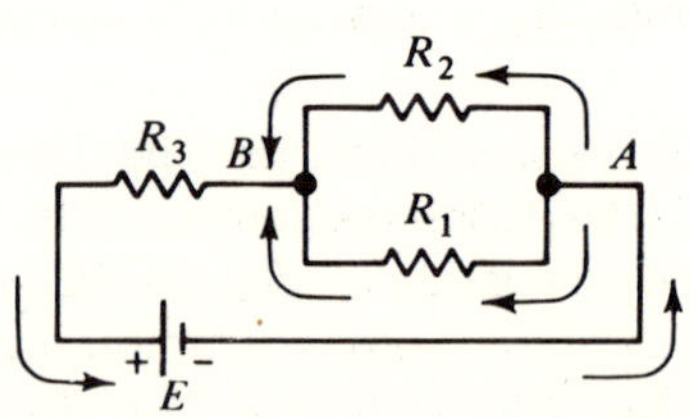

FIGURE 1.5 $E = 60$ *volts;* $R_1 = 40$ *ohms;* $R_2 = 40$ *ohms;* $R_3 = 10$ *ohms.*

more paths for the flow of current and notice that the current divides at junction A in our schematic diagram and recombines at B. R_1 and R_2 can be reduced to one equivalent resistor using the "product divided by the sum" technique. $R_{1,2} = (R_1 \times R_2)/(R_1 + R_2) = (40 \times 40)/(40 + 40) = 1600/80 = 20$ ohms. Schematically, this replaces the parallel branch of R_1 and R_2 with the single resistor $R_{1,2}$.

Note that when resistors of equal resistance are in parallel, the equivalent resistance is one half of the resistance of one of the component resistors.

Now consider the equivalent circuit, with R_3 and $R_{1,2}$ in series. (See Figure 1.6.) The series relation for resistance is $R_T = R_1 + R_2 +$

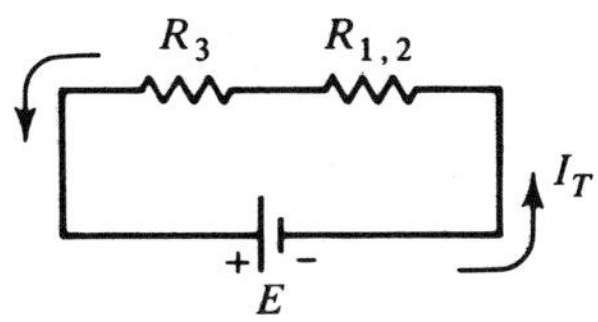

FIGURE 1.6

$R_3 + R_4 + \cdots + R_{\text{Last series resistor}}$. Therefore, the total resistance of the original series–parallel circuit is $R_T = R_{1,2} + R_3 = 20$ ohms $+ 10$ ohms $= 30$ ohms. We have now reduced the complex circuit to the circuit shown in Figure 1.7. To find the total current, we apply Ohm's Law to

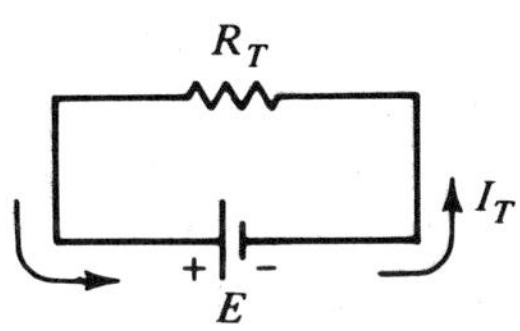

FIGURE 1.7

this equivalent circuit.

$$I_T = \frac{E}{R_T} = \frac{60 \text{ volts}}{30 \text{ ohms}}$$
$$= 2 \text{ A}$$

Working backward through the successive simplifications to the original circuit, we can find the current and voltage for each component. From Figure 1.6 we determine current and voltage for R_3 and for the parallel equivalent $R_{1,2}$. For this series circuit the current is the same: 2 A in each component. Now use Ohm's Law to find voltage drops for $R_{1,2}$ and R_3.

$$V_{1,2} = IR_{1,2} = 2 \text{ A} \times 20 \text{ ohms} = 40 \text{ volts}$$
$$V_3 = IR_3 = 2 \text{ A} \times 10 \text{ ohms} = 20 \text{ volts}$$

Making an intermediate check of our work, we note that in this series circuit, the sum of the voltage drops around the circuit equals the total voltage: $E = V_1 + V_2 + V_3 + V_4 + \cdots + V_{\text{Last series resistor}}$. In this case, $E = V_{1,2} + V_3 = 40 \text{ volts} + 20 \text{ volts} = 60 \text{ volts}$.

We now know the values of the current and voltage for R_3. Returning to the original circuit, we know from above that the voltage for the parallel branch $V_{1,2} = 40$ volts. For this parallel circuit, the voltage drop across each branch is 40 volts. Use Ohm's Law to find the current in each resistor.

$$I_1 = \frac{V_1}{R_1} = \frac{40 \text{ volts}}{40 \text{ ohms}} = 1 \text{ A}$$

$$I_2 = \frac{V_2}{R_2} = \frac{40 \text{ volts}}{40 \text{ ohms}} = 1 \text{ A}$$

These values of I_1 and I_2 are equal, just as we should have expected, since neither branch offers a preferential path, or an easier path, for the flow of current.

1.10 BATTERIES

At this point in our study of electronics we are depending on batteries for our supply of voltage. Earlier we indicated several methods for obtaining a copious number of electrons; however, even this list was not all inclusive. Electrons are also produced: (1) when certain crystals are squeezed (this phenomenon is known as piezoelectricity), and (2) thermoelectrically, when the junction of two unlike metals is heated. Each of these sources, in addition to the more common chemical battery, produces very small voltages. Therefore to obtain larger voltages or currents, combinations of batteries, individually called cells, are used.

Connecting cells in series, positive terminal to negative terminal, increases the voltage, as shown by the diagram, schematic, and equation. To be more precise, a battery is made up of two or more cells connected in series. (See Figure 1.8.)

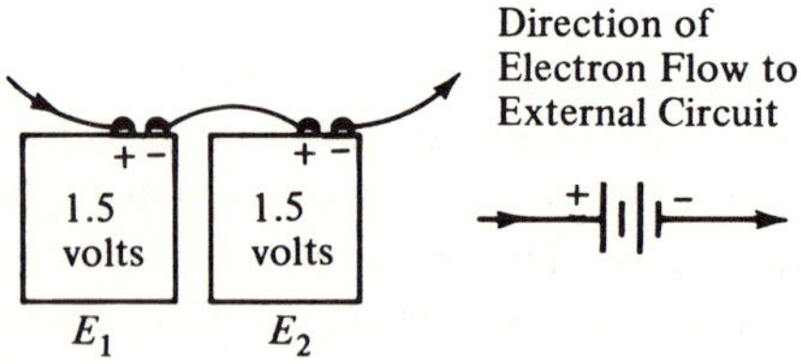

FIGURE 1.8 $E_{\text{Terminal}} = E_1 + E_2 = 1.5$ *volts* $+ 1.5$ *volts* $= 3.0$ *volts*.

To increase the current of the electron source, the cells should be connected in parallel, positive terminal to positive terminal and negative terminal to negative terminal. Again this is shown by diagram, equation, and schematic. (See Figure 1.9.)

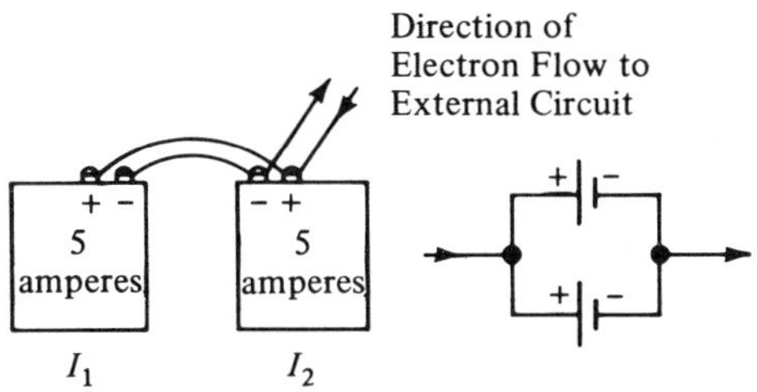

FIGURE 1.9 $I_{\text{Terminal}} = I_1 + I_2 = 5 \text{ amperes} + 5 \text{ amperes} = 10 \text{ amperes}.$

1.11 POWER

In addition to supplying a large number of electrons, we want our cell or battery to offer a constant potential, or a steady voltage, usually called dc voltage. This constant potential is desirable in that we want our electronic devices to work for us at a steady rate. A battery-powered transistor radio would not be pleasant to listen to if it were too loud one instant and barely audible the next.

You have probably noticed that many electric light bulbs have printed on them not only the manufacturer but also the voltage and the number of watts. The rating of the bulb in watts is, as you know, an indication of how brightly the bulb will glow; but, electrically, the wattage tells us the best steady rate at which the bulb will work. The work done, in this case, is the light given off. The watt then is a measure of the rate at which work is done each second. This time rate of doing work is called power.

The means of determining the power of a battery, or the number of watts it will deliver, is quite simple. The number of watts of power is equal to the voltage of the battery times the current it supplies. To write this in equation form, we let the letter P represent power.

$$P = EI$$

Now let us determine, for the following schematic, the power supplied by the battery and the power used by each resistor. (See Figure 1.10.) The total resistance of the series circuit is $R_T = R_1 + R_2 = 1.25$ ohms $+ 1$ ohm $= 2.25$ ohms. Knowing E and R_T, we apply Ohm's Law to find $I = E/R_T = 4.5$ volts/2.25 ohms $= 2$ A. Now we can calculate the power supplied by this 4.5-volt battery, which you should recognize

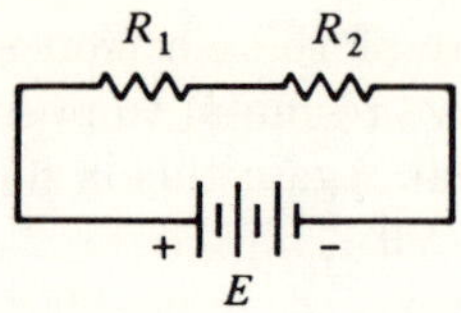

FIGURE 1.10 $E = 4.5$ *volts;* $R_1 = 1.25$ *ohms;* $R_2 = 1.0$ *ohm.*

is made up of three 1.5-volt cells connected in series. $P_T = EI = 4.5$ volts $\times$ 2 A = 9 watts.

To find the power used by each resistor, using the equation $P = EI$, we will have to know the voltage drop and current for each resistor. Since this is a series circuit, the current is 2 A. Using Ohm's Law, we calculate the voltage drop for each resistor. $V_1 = IR_1 = 2$ A $\times$ 1.25 ohms = 2.5 volts; $V_2 = IR_2 = 2$ A $\times$ 1 ohm = 2 volts. Now we can find the wattage rating for each resistor. $P_1 = V_1I = 2.5$ volts $\times$ 2 A = 5 watts; $P_2 = V_2I = 2$ volts $\times$ 2 A = 4 watts. The total power supplied by the battery is equal to the sum of the power used by each resistor in series. $P_T = P_1 + P_2 = 5$ watts + 4 watts = 9 watts.

In finding power, we used the equation $P = EI$. But since Ohm's Law states that $E = IR$, we can obtain another expression for power by substituting this value for E in our original equation. Hence, $P = EI = (IR)I = I^2R$. Thus, we could have found P_1 by writing $P_1 = I^2R_1 = (2\text{ A})^2 \times (1.25\text{ ohms}) = 5$ watts.

We have noted that electric light bulbs are rated in terms of power in watts and also in terms of voltage. Most electric appliances, such as toasters and irons, are rated in the same way. Resistors, such as you may find in your radio, television, or record player, are rated in terms of their resistance in ohms and their power in watts. The value, or size, of the resistance is indicated by brightly colored bands painted around the resistor. In Appendix II you will learn to determine resistance using the international color code for resistors. The power rating of a resistor is indicated by its physical size, as shown in Figure 1.11.

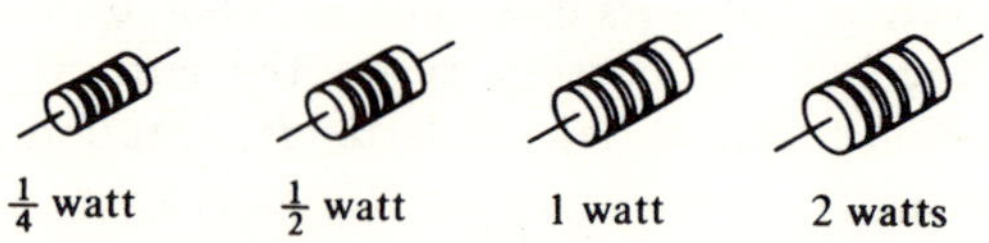

FIGURE 1.11

1.12 REVIEW

We have worked a number of problems in these last pages, and some of them may have seemed a little difficult. But all the different problems had one principal objective: to familiarize you with Ohm's Law, the relation involving dc voltage, current, and resistance. In working with series, parallel, and series–parallel circuits, we emphasized both the direct proportion between current and voltage and the inverse proportion between current and resistance. In the future, we will encounter such devices as voltage dividers, ammeters, voltmeters, and bridge circuits which will directly apply our knowledge of Ohm's Law. And in all electronic circuits, a fundamental understanding of the two proportions just mentioned is essential. Of course, we will continue to emphasize these relations as they are at the foundation of all electronics.

1.13 KIRCHHOFF'S LAWS

Nearly all resistive circuits can be solved by the series and parallel relationships we have developed. In fact, if you wish, you may omit the following work and go right into the study of ac voltage. For the small percentage of circuits which defy classification as either series or parallel, we can reach a solution by applying two relationships mentioned earlier. The first idea is taken from series circuits: Remember we said $E = V_1 + V_2 + V_3 + V_4 + \cdots + V_{\text{Last series resistor}}$. This is restated as follows: The total voltage, or supply voltage, is equal to the sum of the voltage drops around the series circuit. The second idea comes from our study of parallel circuits; we observed that the current entering a junction divided into the various branches and wrote $I_T = I_1 + I_2 + I_3 + I_4 + \cdots + I_{\text{Last parallel resistor}}$. Restating this, we say that the current entering a junction equals the current leaving the junction. These two relationships are called Kirchhoff's Laws: (1) The supply voltage equals the sum of the voltage drops around a series circuit, and (2) the current entering a junction equals the current leaving the junction.

Now we apply Kirchhoff's Laws to a circuit. We consider two possible closed paths for the flow of current in the circuit, one for I_1, the other for I_2, as shown by the arrows in Figure 1.12. Notice that in resistor R_2 there are two currents, I_1 and I_2, which flow in opposite direction. There

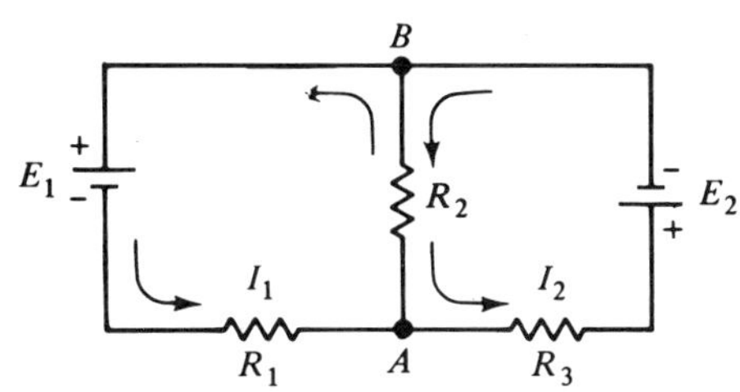

FIGURE 1.12 $E_1 = 2$ *volts;* $E_2 = 4$ *volts;* $R_1 = 10$ *ohms;* $R_2 = 10$ *ohms;* $R_3 = 5$ *ohms.*

are two junction points, A and B, in the circuit, and we check to see that the current entering each junction also leaves it. Hence the circuit satisfies Kirchhoff's second law. Next we apply the first law, the voltage relationship, to each voltage supply.

First, the voltage E_1 is equal to the sum of the voltages dropped by R_2 and R_1 as current I_1 moves around the first path. The voltage drops at R_2 and R_1 will be expressed as current times resistance using Ohm's Law ($E = IR$). However, as we saw earlier, there are two currents in R_2 moving in opposite directions, and the expression for the voltage drop at R_2 must take into account both of these currents.

$$E_1 = I_1R_2 - I_2R_2 + I_1R_1$$

The first two terms ($I_1R_2 - I_2R_2$) give the voltage drop for R_2; the negative sign is used because the flow of current from E_2 is in the opposite direction to that from E_1. We know the values for E_1, R_1, and R_2 and can substitute them in this equation

$$2 \text{ volts} = 10 \text{ ohms} \times I_1 - 10 \text{ ohms} \times I_2 + 10 \text{ ohms} \times I_1$$

We simplify this equation by adding the like terms (the two terms involving I_1):

$$2 \text{ volts} = 20 \text{ ohms} \times I_1 - 10 \text{ ohms} \times I_2 \qquad (1)$$

Next we apply Kirchhoff's second law to the voltage E_2. E_2 should equal the sum of the voltage drops at R_3 and R_2, again taking into account both currents in R_2.

$$E_2 = I_2R_3 + I_2R_2 - I_1R_2$$

Notice that in this loop, since we are following current I_2, the negative sign goes with current I_1, because it is in the opposite direction to I_2. Substituting,

$$4 \text{ volts} = 5 \text{ ohms} \times I_2 + 10 \text{ ohms} \times I_2 - 10 \text{ ohms} \times I_1$$

Simplifying, by adding like terms involving I_2,

$$4 \text{ volts} = 15 \text{ ohms} \times I_2 - 10 \text{ ohms} \times I_1 \qquad (2)$$

Now rewrite Equations (1) and (2).

$$2 \text{ volts} = 20 \text{ ohms} \times I_1 - 10 \text{ ohms} \times I_2$$
$$4 \text{ volts} = -10 \text{ ohms} \times I_1 + 15 \text{ ohms} \times I_2$$

Each equation has two values which we do not know, I_1 and I_2. But we can combine the two equations in such a way as to eliminate one of the unknowns. Multiply eacn term in Equation (2) by 2.

$$8 \text{ volts} = -20 \text{ ohms} \times I_1 + 30 \text{ ohms} \times I_2 \qquad (2a)$$

We add like terms on each side of Equations (1) and (2a).

$$\begin{array}{rcl} 2 \text{ volts} &=& 20 \text{ ohms} \times I_1 - 10 \text{ ohms} \times I_2 \quad (1) \\ 8 \text{ volts} &=& -20 \text{ ohms} \times I_1 + 30 \text{ ohms} \times I_2 \quad (2a) \\ \hline 10 \text{ volts} &=& 0 \text{ ohm} \times I_1 + 20 \text{ ohms} \times I_2 \\ 10 \text{ volts} &=& 20 \text{ ohms} \times I_2 \end{array}$$

Now we have an equation with just one unknown. Using Ohm's Law for current ($I = E/R$), we can find the value of I_2. $I_2 = 10$ volts/20 ohms $= 0.5$ A. Knowing the value of I_2, we can return to either Equation (1) or (2) to find I_1. In Equation (1), we write 0.5 A in place of I_2 to obtain

$$2 \text{ volts} = 20 \text{ ohms} \times I_1 - 10 \text{ ohms} \times 0.5 \text{ A}$$
$$2 \text{ volts} = 20 \text{ ohms} \times I_1 - 5 \text{ volts}$$

Now add 5 volts to both sides of this equation.

$$2 \text{ volts} + 5 \text{ volts} = 20 \text{ ohms} \times I_1 - 5 \text{ volts} + 5 \text{ volts}$$
$$7 \text{ volts} = 20 \text{ ohms} \times I_1$$
$$I_1 = \frac{7 \text{ volts}}{20 \text{ ohms}} = 0.35 \text{ A}$$

We now know the current in R_1 ($I_1 = 0.35$ A) and the current in R_3 ($I_2 = 0.5$ A). To find the current in R_2 we recall that I_1 and I_2 both travel through R_2 in opposite directions, so we simply subtract the smaller current from the larger. Current in $R_2 = I_2 - I_1 = 0.5 \text{ A} - 0.35 \text{ A} = 0.15$ A. To find the voltage drop across each resistor, use Ohm's Law $E = IR$. For resistor R_1, $V_1 = I_1R_1 = 0.35 \text{ A} \times 10 \text{ ohms} = 3.5$ volts. For resistor R_2, $V_2 = IR_2 = 0.15 \text{ A} \times 10 \text{ ohms} = 1.5$ volts. For resistor R_3, $V_3 = I_2R_3 = 0.5 \text{ A} \times 5 \text{ ohms} = 2.5$ volts.

Questions

(1) Describe the following: electron, positive ion, valence electron.

(2) We emphasized the direct and inverse relations involved in Ohm's Law. Can you think of other examples of direct and inverse proportion?

(3) Expand, in terms of voltage, current, and resistance, the following two analogies for electrons flowing along a wire: people in a crowded hallway; water flowing in pipelines.

(4) You might be interested in reading encyclopedia articles about the lives and work of George Ohm, Alessandro Volta, and André Ampère.

(5) State Ohm's Law in your own words.

(6) In a series resistive circuit, what quantity is the same at all points in the circuit? What quantity equals the sum of the voltage drops around the circuit?

(7) In a parallel resistive circuit, what quantity is the same for all the branches? Will the total resistance be larger or smaller than the smallest resistor in the circuit?

(8) If a series–parallel circuit, consisting of two resistors in parallel connected to one series resistor, has an additional resistor added in parallel, how will the total resistance and total current be changed, assuming that the total voltage remains unchanged?

(9) In any circuit, what four values can be known for any resistor?

Problems

(1) Solve the following series circuit for the voltage drop across each resistor. (See Figure 1.13.)

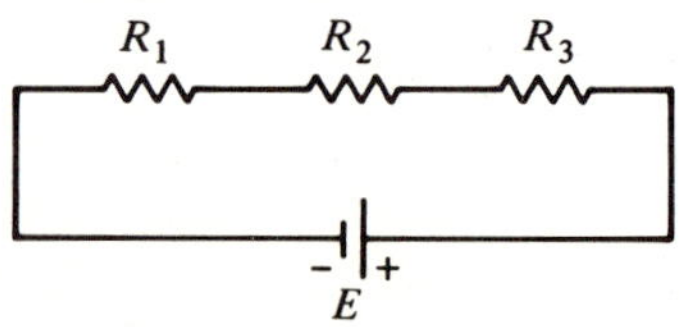

FIGURE 1.13 $R_1 = 10$ *ohms;* $R_2 = 20$ *ohms;* $R_3 = 30$ *ohms;* $E = 90$ *volts.*

(2) Solve the following parallel circuit for the current in each resistor and the total current. (See Figure 1.14.)

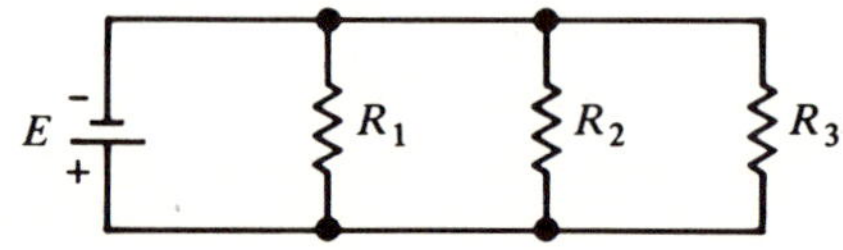

FIGURE 1.14 $R_1 = 2$ *ohms;* $R_2 = 4$ *ohms;* $R_3 = 3$ *ohms;* $E = 48$ *volts.*

(3) In the following series–parallel circuit, what is the voltage drop across R_2? (See Figure 1.15.)

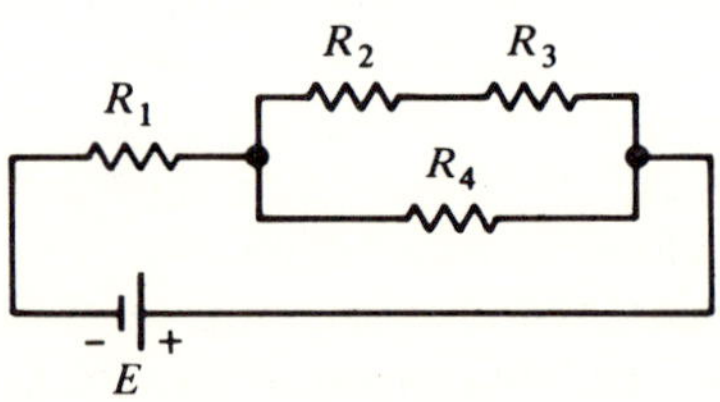

FIGURE 1.15 $R_1 = 30$ *ohms;* $R_2 = 15$ *ohms;* $R_3 = 5$ *ohms;* $R_4 = 20$ *ohms;* $E = 100$ *volts.*

(4) What is the voltage drop across each of the resistors in the following circuit? (See Figure 1.16.)

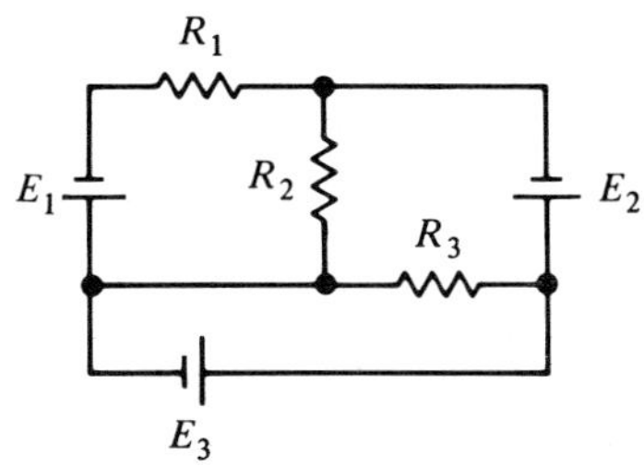

FIGURE 1.16 $R_1 = 5$ *ohms;* $R_2 = 10$ *ohms;* $R_3 = 20$ *ohms;* $E_1 = 10$ *volts;* $E_2 = 20$ *volts;* $E_3 = 30$ *volts.*

Suggested Experiments

(1) Resistance. In Appendix II you will find an explanation of how to read the resistance of a color coded resistor. Practice reading the resistance and tolerance of several resistors until you can do it easily. You might calculate the upper and lower values of the resistance from the tolerance. This should suggest to you that, in substituting one resistor for another in a circuit, it is not always necessary to use an exactly equivalent resistor.

(2) Resistors in Series. With a voltmeter and an ammeter, check the equations which state that in a series resistive circuit, the current is the same at all points, and the sum of the voltage drops about the circuit equals the supply voltage.

(3) Resistors in Parallel. As in the preceding experiment, check the Ohm's Law implications that in a parallel resistive circuit the voltage drops across each branch are equal and the sum of the currents in each branch equals the total current.

(4) Series–Parallel Resistive Circuit. As a continuation of Experiments (2) and (3), verify Ohm's Law for each component in a more complex series–parallel circuit, using a voltmeter and an ammeter. You may wish to extend these three experiments to include calculation of the power dissipated by each resistor. Is the size ($\frac{1}{4}$ watt, $\frac{1}{2}$ watt, and so on) of each resistor adequate?

(5) Resistance Using a Wheatstone Bridge. In the chapter on test instruments, you will find a description of a circuit called the Wheatstone bridge, which is used to find the value of an unknown resistance. Use the bridge to confirm the resistance of a known resistor or of several resistors in series or parallel.

CHAPTER 2

Alternating Current and Voltage

2.1 MAGNETISM

Have you ever used a magnetic compass? This device has been very important to travel around the earth. As you know, the magnetic compass consists of a small bar magnet suspended on a pin so that it can turn freely. When allowed to come to rest, the magnet points toward the North Pole. The earth's magnetism and the magnetism of the compass interact in such a way that the compass points in a direction determined by the earth magnet. If we place a small permanent magnet near the compass, we can make the compass needle point east or west or in any other direction. The small permanent magnet, when nearby, has a greater effect on the compass than the earth does.

Remember, in Chapter 1, we had a cell or battery with a copper wire connecting its positive and negative terminals. One indication that electrons were traveling along the wire from negative to positive terminal was that the wire became hot. The magnetic compass provides another indication of this current flow. If we place a compass under the wire, the needle will move. When the battery is disconnected, the needle returns to its north–south position. (See Figure 2.1.) What has happened? Electrons moving along a wire caused the compass needle to move. We have already observed that a magnet will affect another magnet, in other words, that a small permanent magnet causes a compass needle to move. These two observations lead us to a rather obvious conclusion: There must be something magnetic about electrons moving along a wire.

Let us try another experiment. You will need some insulated copper wire (what we call bell wire will be fine), an iron nail, and a battery.

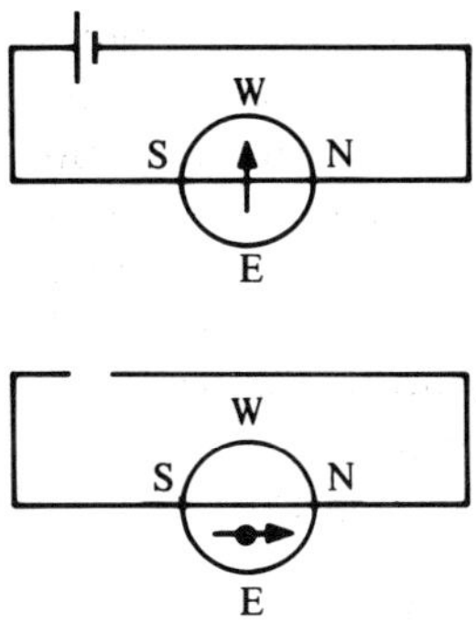

FIGURE 2.1

Leaving a little wire free at one end, wrap the wire around the nail, moving up and down the nail several times. Be careful always to turn the wire in the same direction. When you have finished, you should have some extra wire left at each end. Try to pick up some iron tacks or small nails. Unless the nail is already slightly magnetic, it should not pick up any of these iron objects. Push back the insulation on the ends of the wire and connect them to the terminals of the battery. Now you will be able to pick up the tacks; the nail acts like a magnet. In fact, it is called an electromagnet. The point of the experiment was to show, in another way, that electrons moving along a wire have magnetism.

These experiments will be helpful in understanding the ideas we discuss, and we hope that you have an opportunity to try some of the ones mentioned here or at the end of each chapter. Because of the unavailability of equipment, you may not be able to perform all the experiments we discuss in the text, but we hope you will not let that discourage you and will continue to do as many as are feasible. The following experiment is one that is rather sensitive in that at first it is difficult to be aware of anything happening.

2.2 MAGNETICALLY INDUCED ELECTRON FLOW

Take a piece of copper wire and connect its ends to form a closed loop. Move a horseshoe or bar magnet in and out of the loop of wire. (See Figure 2.2.) Does the wire get hot like it did when it was connected to a

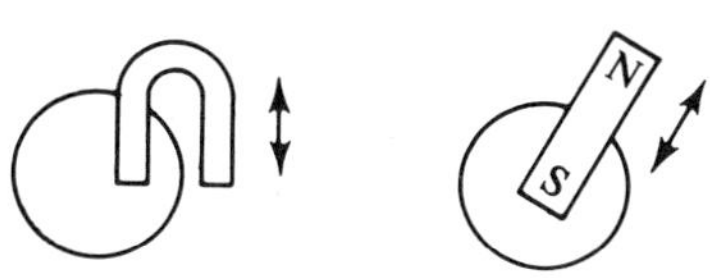

FIGURE 2.2

battery? No, of course not. Unless we can greatly extend our ability to recognize changes in temperature, we could move the magnet in and out of the loop a million times and still not notice any temperature increase. If anything is happening in the wire, we will need a more sensitive indicator to observe it. We will have to either extend our sense of temperature change or rely on one of our other senses.

If we connect a light bulb to the wire, forming a loop with the bulb in it, and again move the magnet in and out of the loop, we just might cause the bulb to glow. (See Figure 2.3.) We have a better chance to

FIGURE 2.3

detect a glow in the bulb if we perform the experiment in a dark room. Here we rely not only on our sense of sight but also on our ability to recognize the slightest contrast between light and dark.

Getting back to our experiment, another way to improve the chances of recognizing that something is happening along the wire is to increase the number of loops of wire through which the magnet passes. (See Figure 2.4.)

FIGURE 2.4

The galvanometer is a device which is very sensitive to changes along the wire loop. Later we will explain how the galvanometer works, but for now all we need to know is that it is an instrument which is sensitive to the flow of electrons in an electrical circuit. Let us replace the light bulb with a galvanometer and pass the permanent magnet in and out of the coil of wire. (See Figure 2.5.) The pointer on the galva-

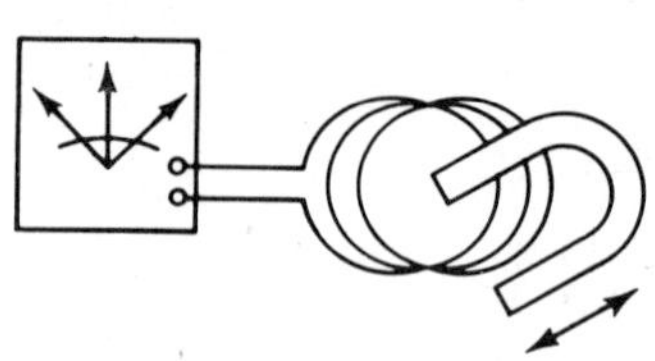

FIGURE 2.5

nometer deflects from one side to the other as the magnet moves in and out. Finally it seems that the experiment worked. Actually it worked all the time; at first we just did not have an indicator sensitive enough to show us that a magnet moving in and out of a loop of wire causes a flow of electrons along the wire. From our experiments with magnetism, we hope you have made the following observations: (1) A magnet is produced when current flows in a wire, and (2) a magnet moved near a wire produces an electrical current in the wire.

2.3 MAGNETIC FIELDS

We will use the above two observations to show how ac voltage is produced or generated. Unlike dc voltage, ac voltage is not a steady potential ability to deliver electrical current, but instead it increases and decreases in this ability. Admittedly, electronic devices operate on dc voltages, and we will eventually change the ac voltage back to dc with the help of certain electronic circuits called dc power supplies. The advantage of ac voltage is that the current it produces can be transmitted over long-distance wires more easily than the steady current produced by dc voltage. In fact, the voltage used in homes, schools, and businesses to operate lights and appliances is ac voltage and is usually generated miles away.

Now let us look at some more experiments. You may have done this first one before. Place a piece of paper over a permanent magnet and sprinkle iron filings on the paper. The filings align themselves to form interesting patterns. These patterns, made by the iron filings, represent something which we cannot see but which is there nevertheless. We call it the magnetic field around a permanent magnet. In Figure 2.6, notice that the iron filings are concentrated near the poles and are much farther apart at points farther away from the magnet. The magnetic field is not uniform; apparently, it is stronger where we find a greater concentration of iron filings.

2.4 CUTTING MAGNETIC FIELD LINES

Remember the difficult time we had in trying to show that an electrical current was produced by moving a permanent magnet in and out of a coil of wire? With our iron filing picture of a magnetic field, we can better visualize what was happening. It now becomes clear that we were "cut-

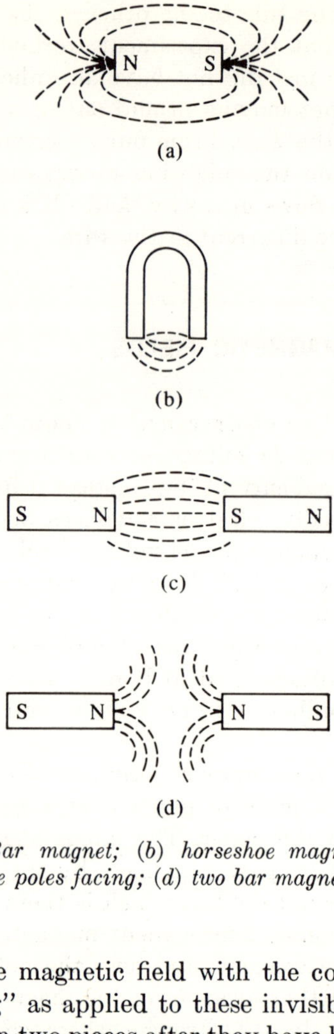

FIGURE 2.6 *(a) Bar magnet; (b) horseshoe magnet; (c) two bar magnets, unlike poles facing; (d) two bar magnets, like poles facing.*

ting" the lines of the magnetic field with the coil. Do not be confused by the term "cutting" as applied to these invisible magnetic field lines; the lines are not left in two pieces after they have been "cut." As we move the magnet in and out of the coil, it is noted that the galvanometer needle deflects farther when the coil passes through the stronger portion of the magnetic field where the lines are more concentrated.

Moving the magnet in and out of the coil caused the galvanometer needle to move back and forth on either side of its center rest position. The electrons in the wire were moving back and forth; we were generating ac (alternating) voltage, which produced this alternating current. It is easier to move the coil through the magnetic field lines, rather than to continue to move the heavy permanent magnet in and out of the coil. Either way we still cut the lines with the coil to generate ac voltage.

2.5 THE AC GENERATOR

You should study Figure 2.7 carefully. It is used to show two things: First, how ac voltage or ac current is generated; second, how we can represent the generated ac voltage or current with a picture or graph. In studying the diagrams, first consider the location of the coil in the magnetic field to determine whether many magnetic field lines are being cut. Then recall that the more lines cut, the farther the galvanometer needle is deflected, or the more voltage is generated. Next, looking at the graph of voltage generated, notice that indeed the more lines cut, the farther the voltage is from zero.

The figure shows coil $ABCD$ making one complete rotation inside a magnetic field and thus producing ac voltage. The arrows show direction of rotation of the coil and direction of current flow in the coil.

In Figure 2.7 (a), sides AB and CD are not cutting any magnetic field lines as they move parallel to the lines, and no voltage is generated. As the coil moves from its position in Figure 2.7 (a) to that in Figure 2.7 (b), sides AB and CD begin to cut across some of the lines, and voltage is generated. As sides AB and CD move perpendicular to the magnetic lines [Figure 2.7 (c)], they cut the greatest number of lines, and the voltage is at its largest value. In Figure 2.7 (d), the number of lines being cut is reduced; so is the ac voltage. In Figure 2.7 (e), as in Figure 2.7 (a), no lines are being cut, and no voltage is generated. Notice that AB is now in the position of CD when we started. As the coil moves from its position in Figure 2.7 (e) to Figure 2.7 (f), AB and CD begin to cut across some of the lines, and again voltage is generated. However, AB (and CD) cuts the lines in the opposite direction to that of Figure 2.7 (b), so electrons move in the opposite direction, and a negative voltage is generated. In Figure 2.7 (g), as in Figure 2.7 (c), sides AB and CD are cutting the largest number of lines, and the voltage is as negative as it will become. In Figure 2.7 (h) we are cutting fewer lines with our coil, and the voltage begins to move back towards zero. Finally in Figure 2.7 (i), the coil is back to its original position, and we see the completed graph of ac voltage generated in one rotation or cycle.

2.6 WAVEFORMS OF AC VOLTAGE AND CURRENT

The direction of the current produced by the coil moving in the magnetic field was indicated by arrowheads on the coil. Notice that in Figures 2.7 (a) through 2.7 (e), the arrows point from D around the coil toward A, but in Figures 2.7 (f) through 2.7 (i), they point in the opposite direction from A toward D. This should not be a surprise; we have already

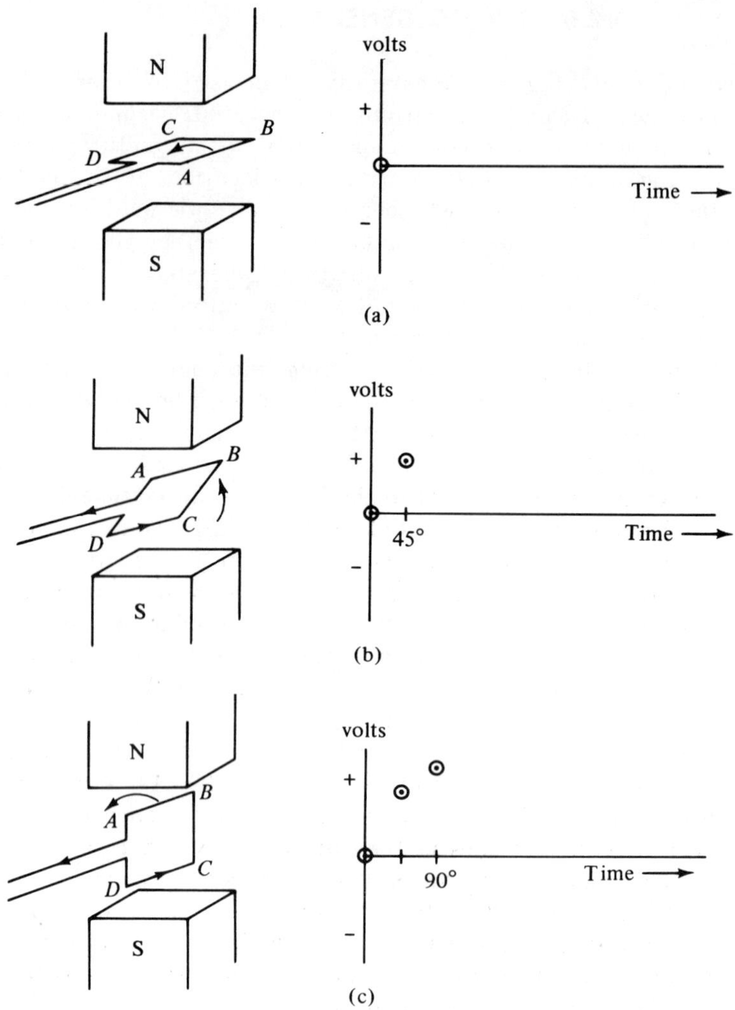

FIGURE 2.7 (*a*) *The coil in its original position—parallel to the lines;* (*b*) *the coil has been rotated through an angle of* 45°*;* (*c*) *the coil has rotated* 90° *from its original position in* (*a*)*;*

seen that the current produced is ac or alternating current. Of course, the voltage is also ac, and both current and voltage can be pictured in the form of a wave. The graphs accompanying the figures show this wave at various points in the rotation of the coil. Connecting the points, we have one wave, that is, a graph of the voltage through one complete rotation of the coil. The graph of ac current for one rotation looks exactly

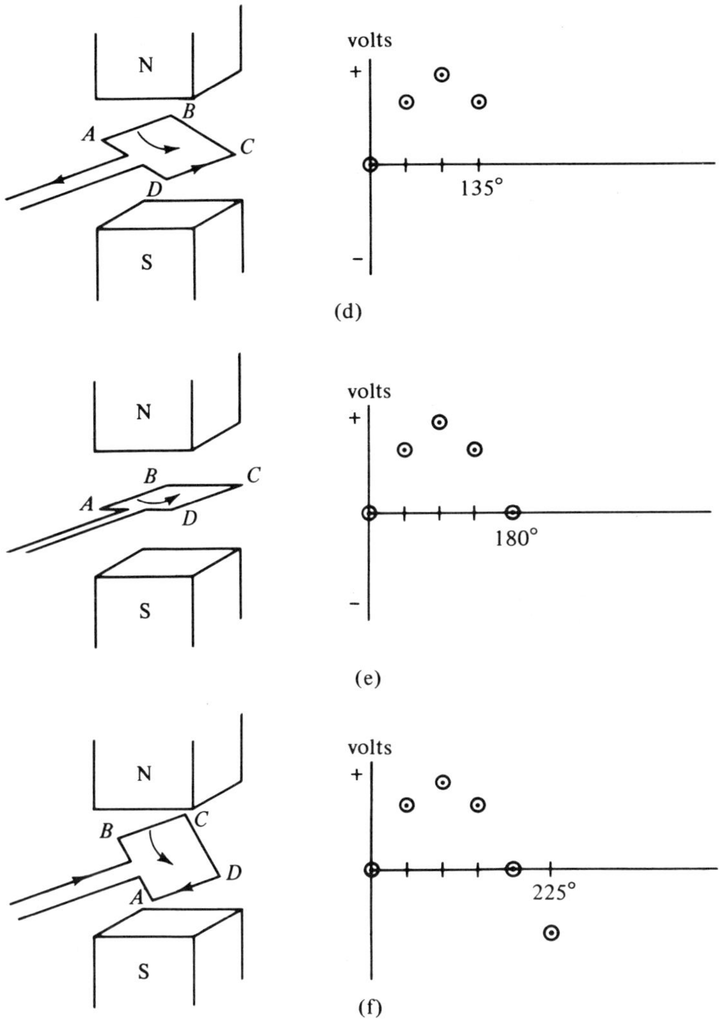

FIGURE 2.7 (*d*) *the coil has rotated* 135° *from original position;*
(*e*) *the coil has completed one half of a rotation or has rotated* 180° *from original position;*
(*f*) *rotation:* 225° *from position* (*a*);

the same as that for ac voltage. In Figure 2.7 (i), the coil is ready to repeat its rotation and thus produces another wave pattern like the first. One complete rotation of the coil, or one wave, is referred to as one cycle. The ac voltage produced in the United States generates 60 of these cycles each second. The number of cycles generated each second

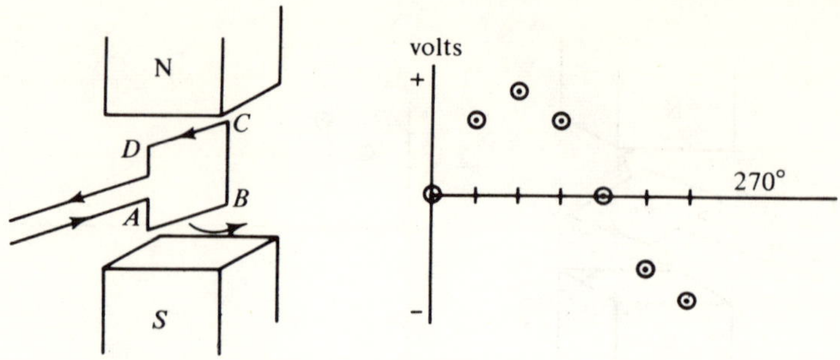

(g)

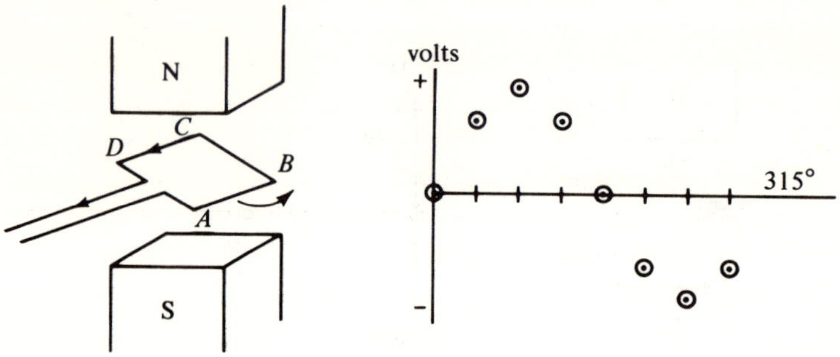

(h)

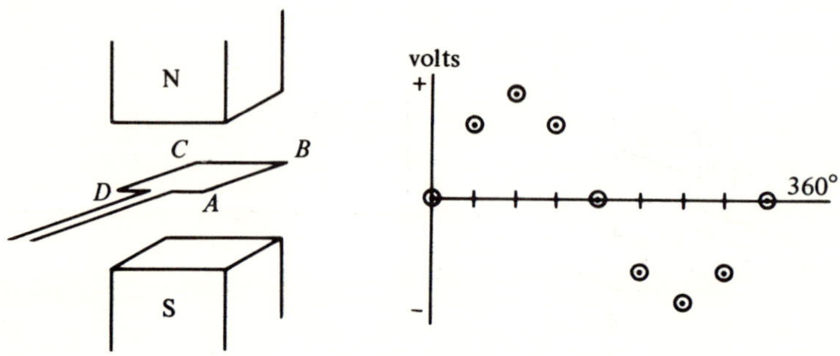

(i)

FIGURE 2.7 (*g*) *rotation:* 270° *from position* (*a*);
(*h*) *rotation:* 315° *from position* (*a*);
(*i*) *one complete rotation back to original position* (*a*): *The coil has moved through* 360°.

by an ac generator is called the frequency and is given the symbol f. Hence the current used in most homes has a frequency of 60 cycles each second or 60 Hz ($f = 60$ Hz). Radio and television call numbers or channels actually represent the frequency at which a particular station is allowed to broadcast.

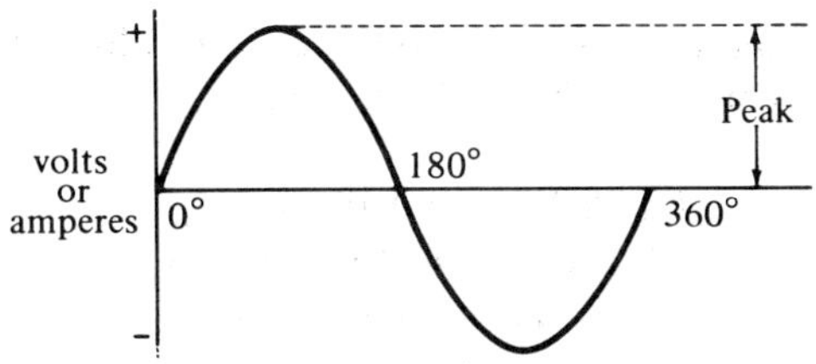

FIGURE 2.8

2.7 EFFECTIVE AC

Now we can generate ac voltage and current and picture them in the form of a wave. Since we know that dc voltage and current from a source such as a battery have a constant output, as shown in Figure 2.9, how can we compare ac and dc? Remember, in our first experiment with a copper wire connected to a battery, the wire became warm to the touch. If we were to connect a wire to an ac source, it would get warm too. We can make the comparison mentioned above by finding how much ac voltage or current will produce as much heat as a given amount of dc

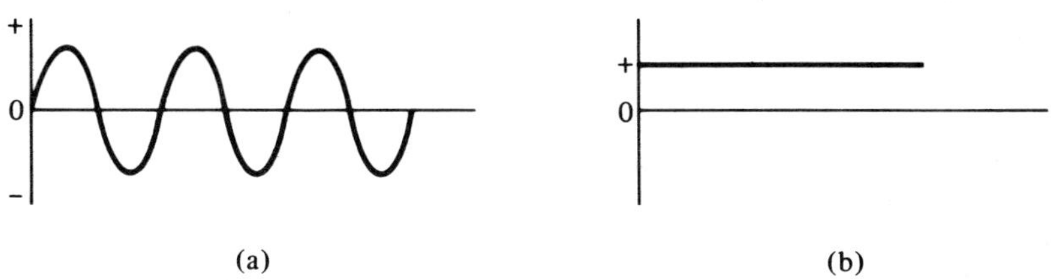

FIGURE 2.9 *(a) ac voltage or current; (b) dc voltage or current.*

voltage or current. Because ac oscillates from a maximum, or peak value, to smaller values, we might reasonably expect that we will need more ac than dc. Careful experiments comparing the heating effects of ac and dc have shown that 141.4 volts ac (at its maximum value) will produce as much heat as 100 volts dc, or we may say 1.414 volts ac = 1.00 volt dc. (See Figure 2.10.) If we reverse the experiment to find how much dc

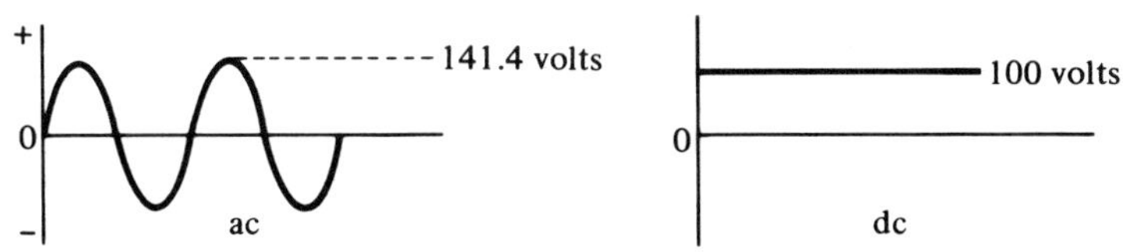

FIGURE 2.10

is equivalent to a peak value of 100 volts ac, the value will be 70.7 volts dc or 1.00 volt ac = 0.707 volt dc. The same relationships are true for current:

$$
\begin{aligned}
1.414 \text{ A ac} &= 1.00 \text{ A dc} \\
1.414 \text{ volts ac} &= 1.00 \text{ volt dc} \\
1.00 \text{ A ac} &= 0.707 \text{ A dc} \\
1.00 \text{ volt ac} &= 0.707 \text{ volt dc}
\end{aligned}
$$

We define the effective ac voltage or effective ac current to be equal to the dc equivalent of the ac peak value. Thus, when the effective ac value is 1.00 volt, the peak ac value is 1.414 volts (1.414 times the effective value), or $V_{\text{Peak ac}} = 1.414 \times V_{\text{Effective ac}}$. Furthermore, when the peak ac value is 1.00 volt, the effective ac value is 0.707 volt (0.707 times the peak value) or $V_{\text{Effective ac}} = 0.707 \times V_{\text{Peak ac}}$. For example, most small appliances operate on an effective ac voltage of 110 volts. The peak value of this voltage is

$$
\begin{aligned}
V_{\text{Peak ac}} &= 1.414 \times V_{\text{Effective ac}} \\
V_{\text{Peak ac}} &= 1.414 \times 110 \text{ volts} \\
V_{\text{Peak ac}} &= 155 \text{ volts}
\end{aligned}
$$

In the future when we want to compare power in an ac circuit, we will continue to use the two equations $P = IE$ and $P = I^2R$ which were developed for dc. However, we must use effective ac voltage and effective ac current in these equations. Fortunately, the instruments used to measure ac voltage and ac current (the ac voltmeter and ac ammeter, respectively) measure their effective values. It is important to remember that even though these instruments read a single value (the effective ac value), the voltage or current they measure is alternating.

Questions

(1) Describe the mechanical method of generating ac.

(2) Which is greater: 100 volts peak ac or 100 volts effective ac?

(3) What similarities are there between the modern compass and "lodestone"?

(4) You might enjoy reading encyclopedia articles about Hans Christian Oersted and Michael Faraday.

CHAPTER 3

Capacitors

3.1 CAPACITANCE

If we separate two conductors of electrons by a poor conductor or insulating material, such as air, paper, glass or oil, we have an electronic component called a capacitor. The schematic for a capacitor is similar to that for a cell. (See Figure 3.1.) The capacitor, as its name implies, has

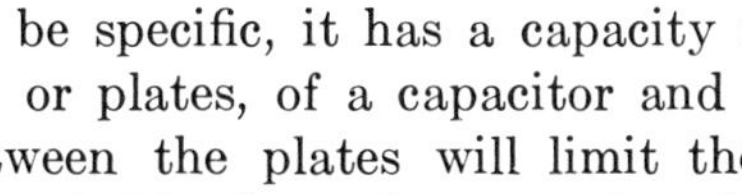

FIGURE 3.1 *(a) Cell; (b) fixed capacitor; (c) variable capacitor.*

some capacity; to be specific, it has a capacity for electrons. The size of the conductors, or plates, of a capacitor and the type of insulating material used between the plates will limit the number of electrons the capacitor can hold; in other words, these factors limit its capacitance.

Recalling the law of electrostatics will help us to understand why the capacitance depends on the size of the plates and the effectiveness of the insulating material. When electrons are placed on one plate, they will get as far apart as possible following the principle that like charges repel. Thus, larger plates allow for more electrons and a greater capacitance. The tuning knob on a radio is connected to a variable capacitor whose capacitance is changed by adjusting the area of the plates facing each other.

When the first plate is negatively charged, the second plate will become positively charged, again following the principle that like charges repel. In this case, free electrons (those not tightly bound to the nucleus of the atom) on the second plate will be repelled by the negative charge on the first plate and will move away. The second plate is left positively charged due to this deficiency of electrons. Now since unlike charges attract, the negatively charged electrons on the first plate are attracted to the positively charged second plate. Only the insulating material keeps the electrons from jumping, or arcing, to the positive plate. A capacitor with a better insulator can hold more electrons and thus has a greater capacitance.

We see then that the above-mentioned physical properties of a capacitor affect the number of electrons it can hold, or its capacitance. Capacitance is directly proportional to the number of electrons, or the charge Q.

Now suppose we keep the same number of electrons on the plate of a capacitor, but we apply an increasing voltage to the capacitor. This voltage has the potential ability to overcome somewhat the electrostatic repulsion between the electrons on the plate. Overcoming this repulsion means that the increased voltage will push the electrons closer together; they will use a smaller area of the plate. In effect, increasing the voltage decreases the size of the plate used by the electrons. But we have seen that plate size affects capacitance, and, in fact, decreasing the plate size will decrease the capacitance. From these two observations, we see that increasing the voltage decreases the capacitance; capacitance is inversely proportional to voltage V.

If we let capacitance be represented by the letter C (capacitance is measured in units called farads), then we can restate the two proportions described above as: C is directly proportional to Q and inversely proportional to V. In equation form, this becomes

$$C = \frac{Q}{V}$$

The Q in this equation represents the charge on the capacitor measured in coulombs. One coulomb is just a name for 6.25×10^{18} electrons, so charge in coulombs still represents the number of electrons on the plate. You probably remember seeing this number of electrons before. One ampere was defined as 6.25×10^{18} electrons passing a point on a conductor each second. Now that we have a shorter name for this number of electrons, we can define one ampere as one coulomb each second. In equation form, current in amperes is $I = Q/t$ where Q is number of coulombs and t is time measured in seconds.

3.2 CAPACITORS IN SERIES

We can place capacitors in series or parallel, just as we did with resistors. A circuit made up of capacitors in series has the same voltage relationship as series resistors; namely, the total voltage equals the sum of the voltage drops across each capacitor. $E = V_1 + V_2 + V_3$. In more general form, $E = V_1 + V_2 + V_3 + V_4 + \cdots V_{\text{Last series capacitor}}$. (See Figure 3.2.)

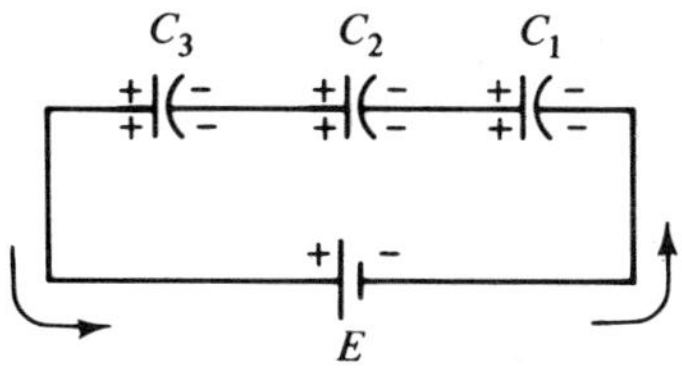

FIGURE 3.2

Now when the plate on the right side of C_1 becomes negatively charged, the negative electrons on the left plate of C_1 are repelled to the right plate of C_2. This leaves a positive charge on the left plate of C_1 equal to the original negative charge on the right plate of C_1. With the right plate of C_2 now charged negatively, its left plate will become positive by the same procedure just described for C_1. This process continues along the series circuit until the last negatively charged electrons are repelled to the positive terminal of the battery. The number of electrons, or the charge on each capacitor, is the same. In equation form, $Q_T = Q_1 = Q_2 = Q_3$ or $Q_T = Q_1 = Q_2 = Q_3 = Q_4 = \cdots = Q_{\text{Last series capacitor}}$.

The relation for total capacitance in a series circuit is similar to that for total resistance of resistors in parallel. You recall that the "product divided by the sum" for each pair of capacitors is $C_{1,2} = (C_1 \times C_2)/(C_1 + C_2)$.*

We have three relationships for capacitors in series:

$$E = V_1 + V_2 + V_3 + V_4 + \cdots + V_{\text{Last series capacitor}}$$
$$Q_T = Q_1 = Q_2 = Q_3 = Q_4 = \cdots = Q_{\text{Last series capacitor}}$$

"Product divided by sum" for each pair of capacitors, or

$$\frac{1}{C_T} = \frac{1}{C_1} + \frac{1}{C_2} + \frac{1}{C_3} + \frac{1}{C_4} + \cdots + \frac{1}{C_{\text{Last series capacitor}}}$$

* The total capacitance may also be found by using the equation:

$$\frac{1}{C_T} = \frac{1}{C_1} + \frac{1}{C_2} + \frac{1}{C_3} + \frac{1}{C_4} + \cdots + \frac{1}{C_{\text{Last series capacitor}}}$$

3.3 CAPACITORS IN PARALLEL

Turning to parallel capacitors, we see that the voltage relation is the same as that for parallel resistors. The voltage supplied to each capacitor is the same as the total voltage. $E = V_1 = V_2 = V_3$. In general, $E = V_1 = V_2 = V_3 = V_4 = \cdots = V_{\text{Last parallel capacitor}}$.

The negative electrons coming from the battery divide at junction A; the division depends on the capacitance of each parallel capacitor. Thus, the sum of the charges on the individual capacitors equals the total

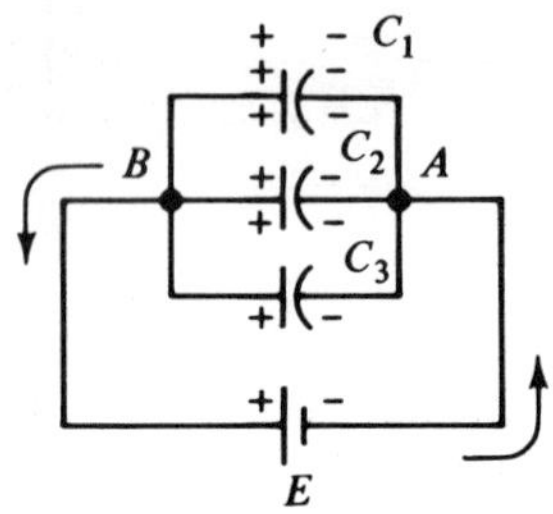

FIGURE 3.3

charge. $Q_T = Q_1 + Q_2 + Q_3$ or $Q_T = Q_1 + Q_2 + Q_3 + Q_4 + \cdots + Q_{\text{Last parallel capacitor}}$.

On closer observation of capacitors in parallel, we see that all the negative plates are connected; they appear essentially as one negative plate. The same is true of all the positive plates. Expressing this in equation form, $C_T = C_1 + C_2 + C_3$. For any number of capacitors in parallel, $C_T = C_1 + C_2 + C_3 + C_4 + \cdots + C_{\text{Last parallel capacitor}}$. Summarizing the relationships for capacitors in parallel:

$$E = V_1 = V_2 = V_3 = V_4 = \cdots = V_{\text{Last parallel capacitor}}$$
$$Q_T = Q_1 + Q_2 + Q_3 + Q_4 + \cdots + Q_{\text{Last parallel capacitor}}$$
$$C_T = C_1 + C_2 + C_3 + C_4 + \cdots + C_{\text{Last parallel capacitor}}$$

We have not worked any problems to help justify the relationships for series and parallel capacitors, hoping that the explanations have, to some extent, replaced the problem situation. We do want to emphasize the following ideas from our equations: For capacitors in series, the total capacitance is less than the smallest individual capacitor; for capacitors in parallel, the total capacitance is equal to the sum of all the individual capacitances and hence is greater than the largest individual capacitor.

3.4 VOLTAGE AND CURRENT IN THE CAPACITOR

Looking at Figure 3.4, we can better understand how a capacitor behaves in a circuit. Electrons leaving the battery accumulate on one plate of the capacitor. They do not pass on to the second plate because there is an insulator, sometimes referred to as a dielectric, between the plates. As we have mentioned before, the negatively charged bottom plate repels electrons on the upper plate of the capacitor. These electrons move around to the positive side of the battery; thus we have a flow of current. Since the capacitance of a capacitor is limited, current will flow in this dc circuit only until the bottom plate is full of electrons or, as we will now say, until the capacitor is completely charged.

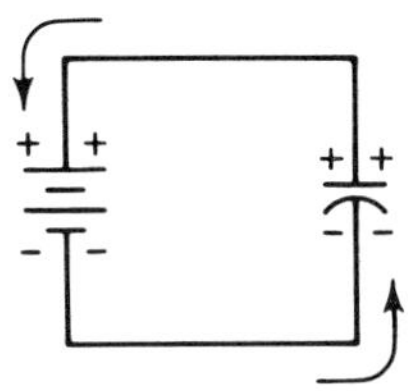

FIGURE 3.4

The instant the capacitor is connected to the battery, current begins to flow, charging the capacitor. At the very first instant, however, there are just a few electrons on the bottom plate of the capacitor, and the voltage across the capacitor is zero. As time passes, the current decreases, but with more electrons building up on the bottom plate, the voltage increases. Finally, when the capacitor is completely charged, no current flows (the current is zero), but the voltage is at its maximum since the capacitor is holding the maximum number of electrons possible. There is maximum current flow with zero voltage the instant the circuit is complete, and maximum voltage across the capacitor the instant the current stops. The current reaches maximum value before the voltage. An easy way to remember this important idea is to say that through a capacitor the current leads the voltage. We can use the word *ICE* to remind us of the current–voltage relationship in a capacitor. The letter I, current, is in front of, or leads, voltage, E, in a capacitor, C. Remember, in addition, that when current is at a maximum, voltage is zero, and vice versa.

When the capacitor is charged, the flow of current about the circuit stops. We recall that anything which reduces the flow of current in a circuit is referred to as resistance. The resistance of a capacitor, called capacitive reactance and given the symbol X_C, is quite different from the resistance of a resistor. In a resistor, the energy of the electrons moving in the circuit is changed to heat; with less energy, the electrons slow down, or the flow of current decreases. In a capacitor, the energy is stored by charging the capacitor. When a capacitor is connected to a battery or is placed in any dc circuit, it is charged very quickly. In fact, it charges so quickly and current flows for such a short time that, in effect, the capacitor blocks the flow of dc current.

On the other hand, suppose we consider a capacitor in an ac circuit. (See Figure 3.5.) Recall that for ac the direction of current flow changes

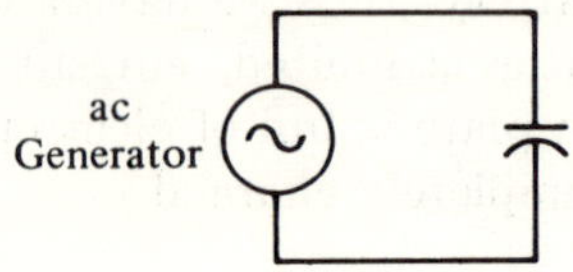

FIGURE 3.5

each half cycle. Therefore, in the first half cycle the bottom plate becomes negatively charged, the top plate positively charged. Then, as the wave moves from positive to negative, the capacitor discharges. In the second half cycle, the direction of current flow has changed, thus the top plate is negatively charged and the bottom positively charged. We see then that a capacitor does not block ac current. The principal function of the capacitor in electronic circuits is to block the flow of dc current and to pass the flow of ac current.

3.5 CAPACITIVE REACTANCE

Now let us return to the idea of the opposition to current flow of a capacitor and determine the factors that affect this capacitive reactance. (See Figure 3.6.) If we decrease the frequency (smaller number of cycles

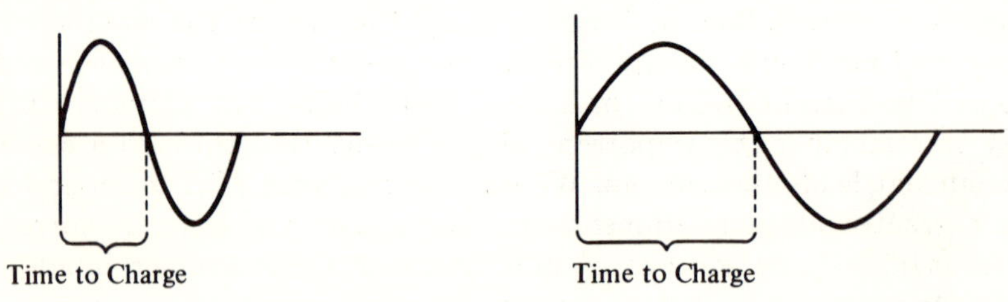

FIGURE 3.6

each second or hertz), the capacitor has more time to charge in one half cycle. (Notice the smaller the frequency, the fewer of these long cycles are generated each second.) But the more the charge builds up, the greater the resistance to current flow, so the capacitive reactance increases. For example, the frequency of dc is zero, and the capacitive reactance is infinitely large. In addition, decreasing the capacitance, or the number of charges the capacitor will hold, also increases the resistance to further build up the charge, or the capacitive reactance. Thus, X_C is inversely proportional to frequency f and capacitance C. Written as an equation, this becomes $X_C = 1/2\pi fC$, which is read capacitive reactance is equal to 1 over the quantity 2 pi f times C. The 2π is a proportionality constant, approximately equal to 6.28, which changes frequency in hertz (cycles each second) to radians (an angle measurement) each second.*

3.6 PHASE RELATION OF VOLTAGE AND CURRENT IN A CAPACITOR

We wish to consider the wave patterns for current and voltage in a capacitor in an ac circuit. Recalling the relationship (*ICE*) between current and voltage in a capacitor, we check the diagram of the waveforms to see that at the instant the circuit is closed, current is at a maximum, voltage is zero; as current falls to zero, voltage rises to its maximum. (See Figure 3.7.) The current and voltage are said to be 90° out of phase,

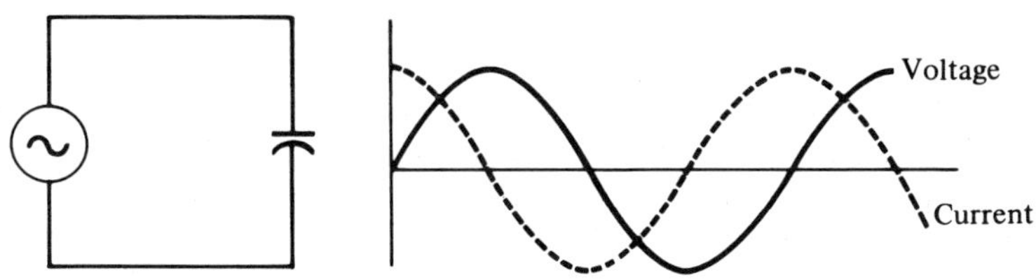

FIGURE 3.7 (*Note: Current begins to flow when the circuit is first closed.*)

that is, the phase angle is 90°. These are not temperature degrees but instead are directional degrees as on a compass or an ac generator (see Figure 2.7). The current in the capacitor leads the voltage by 90°.

Now replace the capacitor with a resistor. We know that the current neither leads nor lags the voltage in a resistor, and this is shown by the wave patterns for this current and voltage. (See Figure 3.8.) We say there

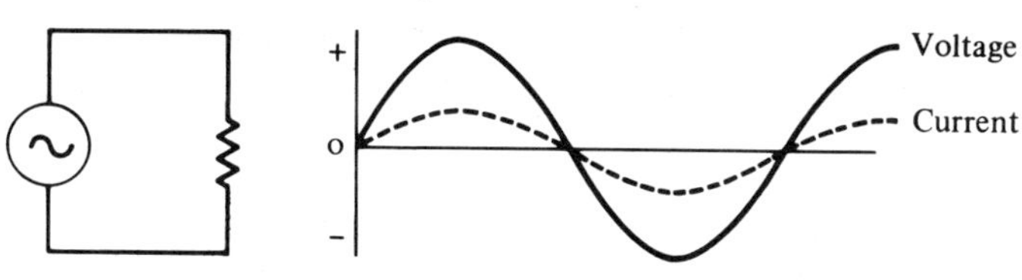

FIGURE 3.8

* There are 2π radians in one cycle.

is no phase shift between voltage and current in a resistor, or the phase angle is 0°. A resistor in an ac circuit serves as a reference for the phase angle between current and voltage in various other components.

3.7 SERIES RC CIRCUITS

Because of this phase shift, in our next problem we find when adding 50 volts to 50 volts, we get 70.7 volts. Consider the ac circuit with a resistor and capacitor in series. (See Figure 3.9.) In a series circuit the voltage

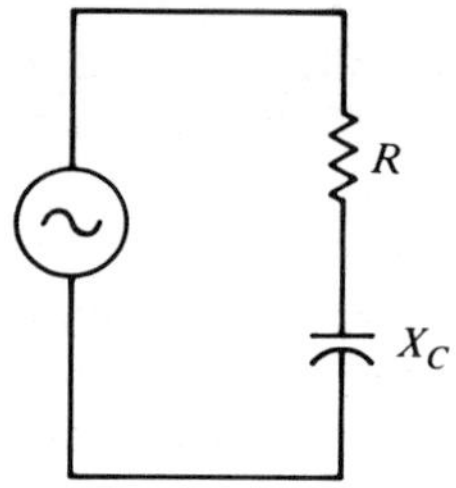

FIGURE 3.9 $R = 10$ *ohms;* $X_C = 10$ *ohms;* $I_{\text{Peak ac}} = 5$ *amperes;* $E_{\text{Peak ac}} = 70.7$ *volts.*

supplied, or total voltage, equals the sum of the voltage drops about the circuit. The voltage drops can be found using Ohm's Law. Across R the voltage drop is $V_R = IR = 5\text{ A} \times 10$ ohms. $V_R = 50$ volts. The voltage drop across the capacitor, which has a capacitive reactance or opposition to current flow of 10 ohms, is $V_C = IX_C = 5\text{ A} \times 10$ ohms. $V_C = 50$ volts. We would then expect the total voltage to be the sum of these two voltage drops, or 100 volts, rather than 70.7 volts.

To explain this dilemma, we must turn to the graphs of current and voltage in this circuit. In the series circuit, the current in the resistor, I_R, and the current in the capacitor, I_C, are in phase. Also, we have seen that the voltage in a resistor, V_R, and the current in the resistor, I_R, are in phase and serve as a reference for the phase angle between voltage and current in other components. Finally, combining the above with our relation *ICE* (current leads voltage by 90° in a capacitor), we reach the following conclusion: I_C, I_R, and V_R are all at a maximum when V_C is zero; when V_C is at a maximum, I_C, I_R, and V_R are zero. If we add the values of V_C and V_R at various points along the graph and plot this sum (which is, of course, E), we generate another wave. The maximum value of this new wave occurs when $V_R = 35.35$ volts and $V_C = 35.35$ volts—at which point $E = 70.7$ volts. This agrees with our original value for $E_{\text{Peak ac}}$. (See Figure 3.10.)

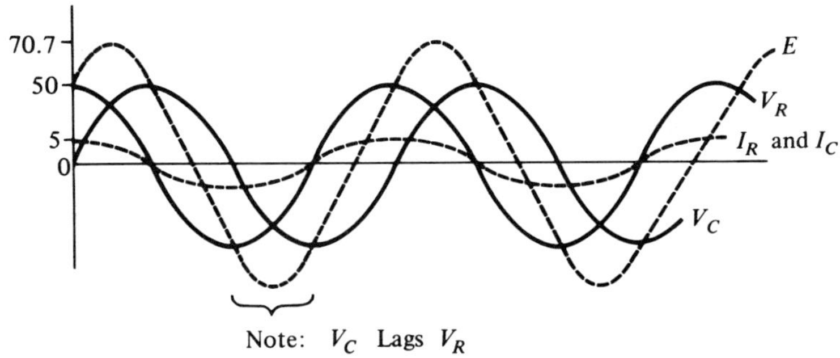

FIGURE 3.10 (*Note: V_C reaches its maximum after V_R.*)

You notice that to determine total voltage, we had to add the voltages V_R and V_C by using their graphs, rather than taking the simple arithmetic sum. There is another method of finding V_T besides the graphical sum that we used. Voltages have another characteristic other than their value, or magnitude; they also have a phase angle, which we have already described, associated with them. For example, V_C and V_R both had a magnitude of 50 volts. But the peak value for V_C occurred 90° after the peak value for V_R. If we use the phase angle of V_R as a reference (we say it is 0°), then V_C lags V_R by 90°. Looking at our graphs, we notice that the peak value of V_R occurs at 0°, and the peak value of V_C occurs 90° later. V_C always remains 90° behind V_R.

3.8 VECTORS

We will use arrows, or vectors, as a means of showing both magnitude and phase difference on a single diagram. The length of the arrow will indicate magnitude. This is accomplished by establishing a scale, for example, 1 centimeter = 25 volts. The direction the arrow points indicates phase angle: to the right is 0°, up is 90°, left 180°, and down 270°. (See Figure 3.11.) These arrows, whose magnitude and direction both have meaning, are called vectors.

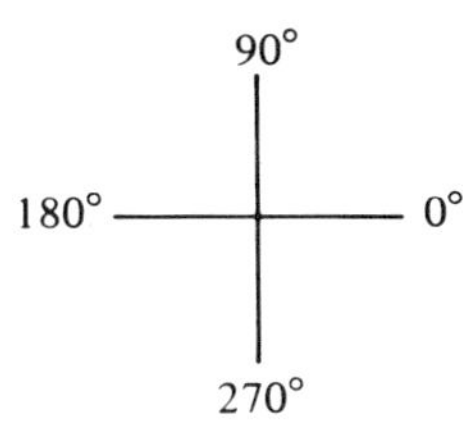

FIGURE 3.11

To write V_R and V_C as vectors, we first establish our scale 1 cm = 25 volts. Now since V_R serves as our reference, we draw an arrow 2 cm long at 0°. (See Figure 3.12.) V_C reaches its maximum 90° later than V_R,

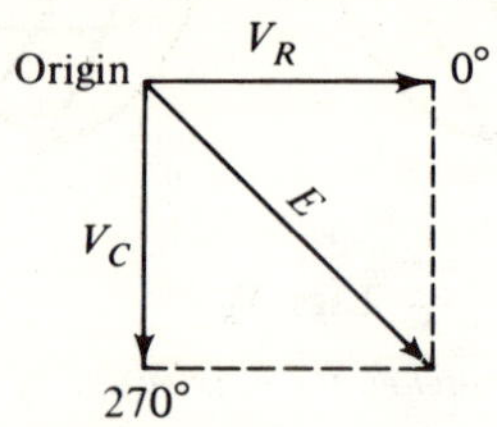

FIGURE 3.12

or V_C lags V_R by 90°, so we draw the arrow, or vector, representing V_C 90° below V_R. Thus, V_C is shown as a vector 2 cm long at 270°. Notice that both vectors begin at the same point, the origin. To add these two vectors, V_R and V_C, we draw one line from the tip of V_C parallel to V_R and a second line from the tip of V_R parallel to V_C. Next we draw a vector from the origin to the point at which these lines intersect. This new vector is called the resultant, E, and represents the sum of V_C and V_R. Measuring this vector, we find it to be 2.82 cm long. Since 1 cm = 25 volts, this represents a voltage of 70.7 volts or E = 70.7 volts, as found earlier.

Notice that the vector diagram shows E between V_C and V_R. Specifically, E is below V_R, or lags V_R, and it is above, or leads, V_C. Let us return to the graph of the waveforms to see that this is actually the case. E reaches its maximum after V_R but before V_C.

3.9 IMPEDANCE

Since we know E = 70.7 volts and I = 5 A in our example, we can apply Ohm's Law to find the total opposition to current flow. We will not give this "resistance" the symbol R, because only part of the opposition comes from the resistor, the rest is due to the capacitor (capacitive reactance). Instead, we will call this total opposition to current flow impedance and give it the symbol Z. Z = total opposition to current flow = E/I = 70.7 volts/5 A = 14.14 ohms.

However, we have said that for series resistive circuits, the total opposition is the sum of the component oppositions; so we might have suspected that total opposition = $R + X_C$ = 10 ohms + 10 ohms = 20 ohms. This is another apparent contradiction. You probably recognize that this difficulty is similar to the one just solved for voltages and can be overcome by considering the phase relation between V_C and V_R. Only when we have purely resistive circuits can we use the relationships for resistors in parallel and series. Thus, since our circuit includes a capacitor, we cannot say that the total opposition to current flow is 20 ohms, the arithmetic sum of the component oppositions. How then can we find impedance using the component oppositions? Just as with our voltages, we will add R and X_C vectorially.

In the series circuit, the current is the same in all the components. Thus, I_R and I_C are in phase (their phase difference is 0°), as is shown in Figure 3.10. The phase relation between V_R and V_C has already been established. Now since opposition to current flow is voltage divided by current, the phase relation between R ($= V_R/I_R$) and X_C ($= V_C/I_C$) is the same as that between V_R and V_C (the currents are in phase and thus do not affect the phase difference). As in the vector diagram for voltages, we establish a scale 1 cm = 5 ohms. R and X_C are drawn in the same direction as V_R and V_C, respectively. (See Figure 3.13.) The vector sum,

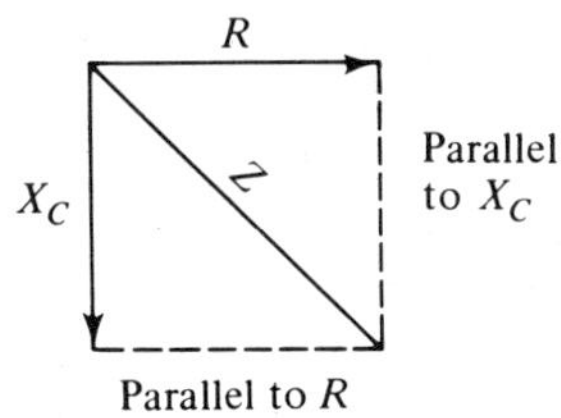

FIGURE 3.13

or resultant, of R and X_C, which is impedance Z, is found in the same way as E. Draw lines parallel to R from the tip of X_C and parallel to X_C from the tip of R. Then the vector from the origin to their point of intersection represents Z. Measuring its length and referring to the sale, we find $Z = 14.14$ ohms.

Observe that $R^2 = 10 \times 10 = 100$, $X_C{}^2 = 10 \times 10 = 100$, $Z^2 = 14.14 \times 14.14 = 200$ or $Z^2 = R^2 + X_C{}^2$.* Taking the square root of each side of this equation, $Z = \sqrt{R^2 + X_C{}^2}$. Thus, we can find impedance using this formula without the aid of a vector diagram. In our problem, $Z = \sqrt{100 + 100} = \sqrt{200} = 14.14$ ohms. We could have developed a similar formula for total voltage in series circuits $E = \sqrt{V_R{}^2 + V_C{}^2}$.

* Those who are familiar with the Pythagorean Theorem will recognize that our vector diagram contains two right triangles each with hypotenuse Z and sides X_C and R. Thus, the square of the hypotenuse equaling the sum of the squares of the lengths of the sides becomes, in our problem, $Z^2 = R^2 + X_C{}^2$.

In our example, $E = \sqrt{50^2 + 50^2} = \sqrt{2500 + 2500} = \sqrt{5000} = 70.7$ volts.

3.10 PARALLEL RC CIRCUITS

We now wish to consider another ac circuit, this one with a resistor and capacitor in parallel. (See Figure 3.14.) The same voltage is applied to

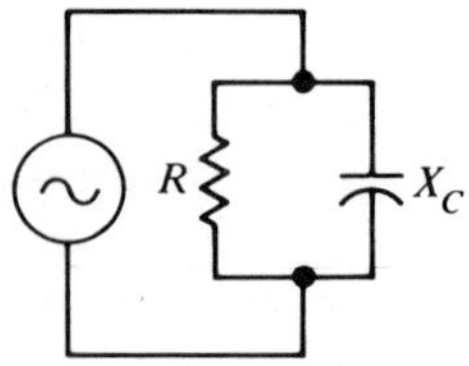

FIGURE 3.14 $R = 50$ *ohms;* $E = 100$ *volts;* $X_C = 50$ *ohms;* $I = 2.82$ *amperes.*

each parallel component. Thus, $V_C = V_R = E$, and there is no phase difference between V_C and V_R. We can find the current for each component by applying Ohm's Law: $I_R = V_R/R = 100$ volts/50 ohms $= 2$ A; $I_C = V_C/X_C = 100$ volts/50 ohms $= 2$ A. Again we must recognize that, since this circuit contains a capacitor, the total current will not be the arithmetic sum of the currents in each component. To find I_T, we must analyze the graph of the waveforms. I_R and V_R are in phase and will be used as a reference. (See Figure 3.15.) Notice that although the

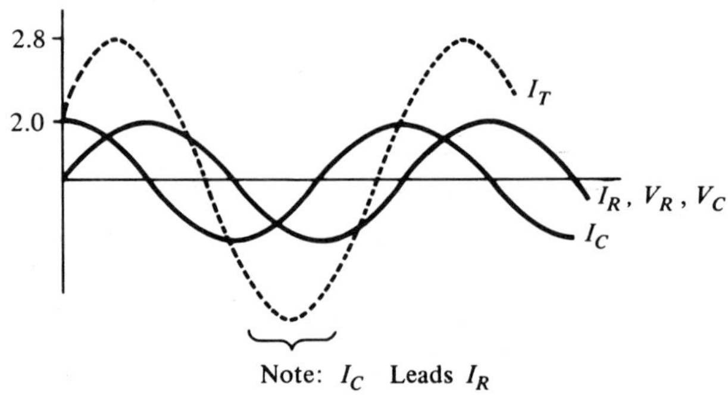

FIGURE 3.15 (*Note:* I_C *reaches its maximum before* I_R.)

peak value of I_R is 2 A and the peak value of V_R is 100 volts, we draw I_R and V_R with the same maximum (or amplitude); this reduces the number of waves on the diagram. Since $V_R = V_C$, the wave for I_R and V_R can also represent V_C. Using the relation *ICE*, we draw a wave representing I_C which leads voltage V_C by 90°. Adding I_R and I_C graphically as we have done before, we find a curve representing I_T whose peak value occurs when $I_R = I_C = 1.41$ A. Therefore, $I_T = 2.82$ A.

We can also find I_T using a vector diagram for our currents. (See Figure 3.16.) First establish a scale 1 cm = 1 A. I_R is the reference and is

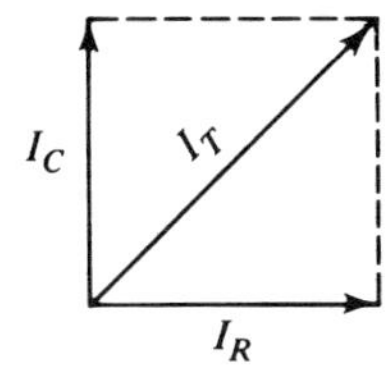

FIGURE 3.16

drawn 2 cm long at 0°. I_C leads I_R by 90° and so is drawn 90° above I_R. Drawing parallels to I_C and I_R to establish the resultant I_T, we measure it to find $I_T = 2.82$ A.

Note that $I_C^2 = 2 \times 2 = 4$, $I_R^2 = 2 \times 2 = 4$ and $I_T^2 = 2.82 \times 2.82 = 8$. Thus, we have $I_T^2 = I_C^2 + I_R^2$; $I_T = \sqrt{I_C^2 + I_R^2}$. For our problem, $I_T = \sqrt{2^2 + 2^2} = \sqrt{4 + 4} = \sqrt{8} = 2.82$ A.

Knowing $E = 100$ volts and $I_T = 2.82$ A, we apply Ohm's Law to find total opposition to current flow, or impedance: $Z = E/I_T = 100$ volts/2.82 A = 35 ohms. You should recall that the equation $Z = \sqrt{R^2 + X_C^2}$ was developed for series circuits and will not give impedance in ths example of a parallel circuit.

Questions

(1) The capacitive reactance of a given capacitor is 50 ohms when the frequency is 1000 Hz. If the frequency is doubled, how is the capacitance affected?

(2) Which has the smaller capacitance: two identical capacitors in series or in parallel?

(3) In the series RC circuit, we have noted that current leads voltage by 90° in the capacitor and that current and voltage in the resistor are in phase. What is the phase relation between the applied voltage and the voltage across the capacitor?

(4) In what type of RC circuit does the current in the resistor lag the current in the capacitor by 90°?

Suggested Experiments

(1) Capacitance. The capacitance bridge, which is described in Chapter 28 on test instruments, can be employed to measure the capacitance of a single capacitor or of several capacitors in series or parallel.

(2) Series RC Circuit. With the aid of an oscilloscope we may observe the phase relations in a series RC circuit. (See Figure 3.17.)

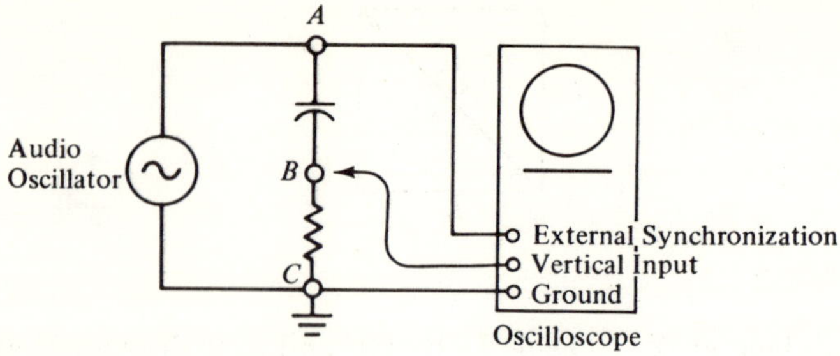

FIGURE 3.17

To observe the applied voltage waveform, attach the vertical input lead to A. Next move the vertical input lead to B. This shows the voltage waveform across B, that is, across the resistor. Interchange the resistor and capacitor and again attach the vertical input lead to B in order to observe the voltage waveform across the capacitor. With a careful comparison of the last two waveforms you will recognize the phase shift.

(3) Parallel RC Circuit. The phase relations in parallel RC circuits may be observed similarly. (See Figure 3.18.) In the parallel circuit, we

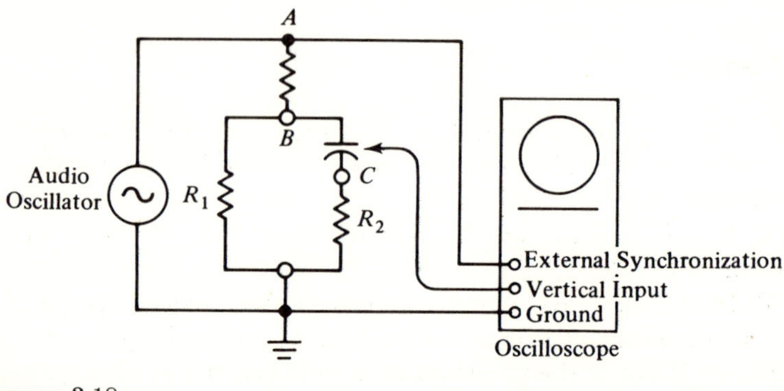

FIGURE 3.18

want to compare currents. To do this, we take advantage of the fact that the current and voltage in a resistor are in phase. If we attach the vertical input to point C, we observe voltage across R_2. But since this voltage is in phase with current in R_2 and, hence, with current in the capacitor, we are indirectly observing current in the capacitor. Moving the vertical input to point B, we observe current in R_1. If we exchange the resistor across AB with the parallel circuit across BD and place the vertical input at B, we measure the total current.

CHAPTER 4

Inductors

4.1 INDUCED CURRENT

In Chapter 2 we talked about producing a magnet by letting current flow in a wire; electrons flowing along a wire produce a magnetic field about the wire. In a later experiment we observed that a changing magnetic field, such as produced by moving a magnet, can cause current to flow in a nearby conductor; this is how we generated ac.

Now let us consider ac current flowing in a coil of wire. (See Figure 4.1.) As this current increases from zero to a maximum, the strength of

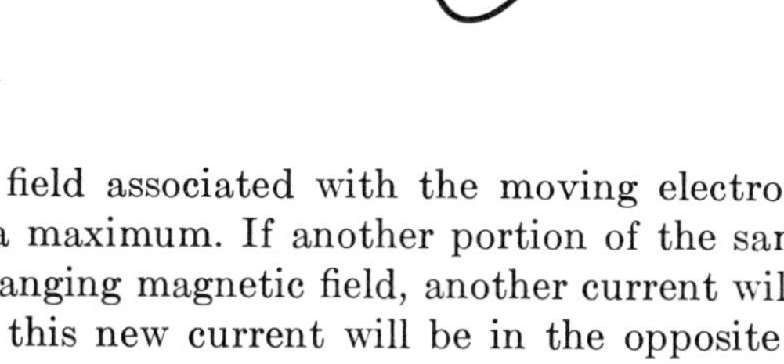

FIGURE 4.1

the magnetic field associated with the moving electrons also increases from zero to a maximum. If another portion of the same coil of wire is cut by this changing magnetic field, another current will be generated in the wire, but this new current will be in the opposite direction to the original ac current flow. This new current we call an induced current. (See Figure 4.2.)

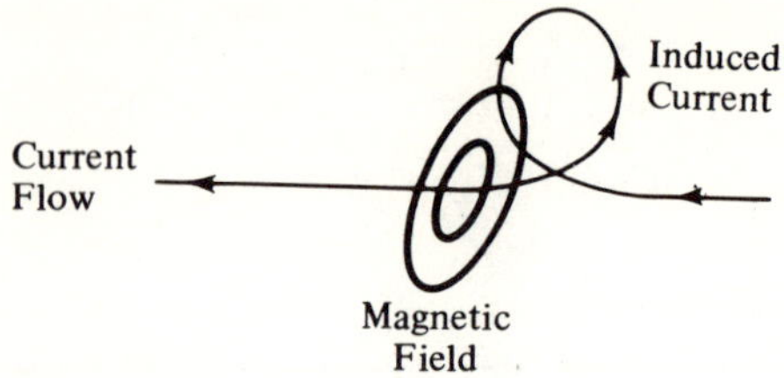

FIGURE 4.2

4.2 INDUCTIVE REACTANCE

Since the induced current is in the opposite direction to the original current, it opposes the flow of the original current. The generated ac current still flows in the coil, but it experiences some opposition due to the induced current; the situation is analogous to a few people trying to enter a theater when the crowd inside is leaving: These few people do not prevent anyone inside from leaving, but they do make leaving more difficult (they offer opposition). This form of opposition to current flow is called inductive reactance and given the symbol X_L.

Inductive reactance occurs in a coil of wire called an inductor. The schematics for two types of inductors, or coils, are shown in Figure 4.3.

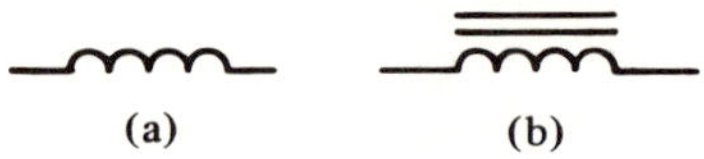

FIGURE 4.3 *(a) Fixed inductor; (b) iron core inductor (coil is wound around iron).*

A coil of wire may oppose the flow of electrons in two ways: first by its resistance R which changes some of the energy of the electrons to heat, and second by the inductive reactance X_L just discussed, which stores some of the energy of the electrons in the form of a magnetic field. If the resistance of the wire in the coil is zero and the opposition to current flow comes entirely from the inductive reactance, then we have what is referred to as a pure inductor. Most coils have such a small resistance that they may be treated as pure inductors. Coils which are not pure inductors are usually shown as a resistor and an inductor in series.

We have seen that the magnetic field around a coil carrying a current produces another current in the opposite direction. Since this induced current is in the same coil, we refer to the process as self-induction. From our study of ac generators, we recall that more current is generated when more magnetic field lines are cut or when they are cut more quickly.

The speed at which the magnetic field lines change is the same as the rate at which the ac current changes, that is, the frequency. Thus, the self-induced current, which depends on the speed with which the magnetic field lines cut the coil, increases as the frequency of the original ac current increases. Since this self-induced current opposes the original flow of current, the magnitude of this opposition (X_L or inductive reactance) is directly proportional to frequency f. The more turns in the coil, the more the magnetic field lines will be cut and thus the greater the self-induced current; the number of turns and other physical properties of the coil, such as whether it has an air core or an iron core, determine what we call its inductance L (measured in henrys). Increasing the inductance increases the self-induced current, and thus increases inductive reactance. X_L then is directly proportional to f and L, or $X_L = 2\pi fL$, where the 2π, as in the expression for capacitive reactance, is a proportionality constant which changes cycles per second (hertz) to radians per second.

4.3 PHASE RELATION OF VOLTAGE AND CURRENT IN AN INDUCTOR

Remember we have shown that in a capacitive circuit, the current leads the voltage. Now we will attempt to explain why, in an inductive circuit, the opposite is true, that is, the voltage leads the current. We have established that the changing magnetic field induces a current in the opposite direction to that supplied by the ac generator. (See Figure 4.4.)

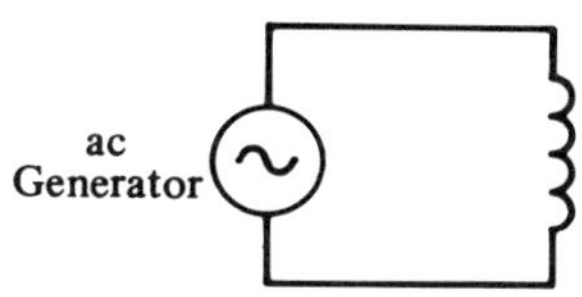

FIGURE 4.4

This induced current will have an induced voltage associated with it, because a flow of electrons indicates a potential difference between two points, one point negative, the other point positive or less negative. The induced voltage is, of course, in the opposite direction to the voltage drop, or applied voltage, across a pure inductor; it will have its peak values when the magnetic field lines are changing most rapidly, which occurs when the current is changing most rapidly. In considering the waveform of ac current, we see that the greatest change occurs when the current passes through zero (the current must change from positive to negative, or vice versa; this is the "steepest" section of the graph, implying that the current rises or falls most rapidly at this point.) Figure 4.5 shows induced voltage a minimum (and hence applied voltage, in the opposite direction, a maximum) at the point where current is increasing most rapidly (current goes from negative through zero to positive) and induced voltage a maximum (applied voltage a minimum) where current is decreasing most rapidly (from positive through zero to negative). Notice that the applied voltage reaches its maximum 90° before the current reaches its maximum.

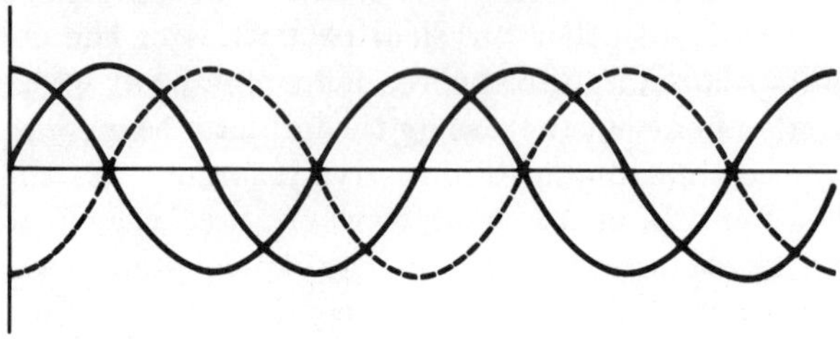

FIGURE 4.5 (*Note: Maximum voltage occurs* 90° *before maximum current.*)

In an inductor, the applied voltage leads the current by 90°. Thus, *ELI* reminds us that applied voltage E is ahead of, or leads, current I in an inductor L.

4.4 SERIES RL CIRCUITS

Now let us combine a coil and a resistor in a series circuit. (See Figure 4.6.) The frequency f and inductance L are such that inductive reactance

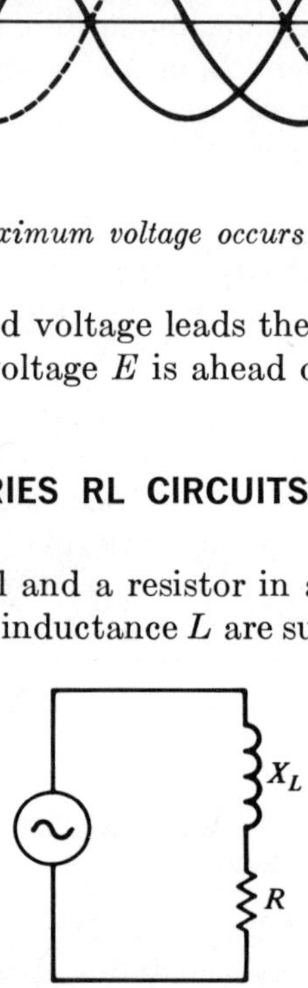

FIGURE 4.6 I_T = 5 *amperes (peak value); E* = 70.7 *volts (peak value); R* = 10 *ohms;* X_L = 10 *ohms.*

X_L = 10 ohms. We know the current in a series circuit is the same in all components, $I_T = I_R = I_L = 5$ A. Applying Ohm's Law, we can find the voltage drop across each component.

$$V_R = I_R R = 5 \text{ A} \times 10 \text{ ohms} = 50 \text{ volts}$$
$$V_L = I_L X_L = 5 \text{ A} \times 10 \text{ ohms} = 50 \text{ volts}$$

The sum of the voltage drops around a series resistive circuit is equal to the supply voltage; in this case, however, the phase relation between V_R and V_L is such that the two cannot be added arithmetically. Drawing the waveforms for the circuit, we first recall that current and voltage in the resistor (I_R and V_R) are in phase and are used as a reference. Since $I_R = I_L$, the same wave can represent both. Recalling that voltage leads current by 90° in an inductor (*ELI*), we plot V_L with its maximum occurring 90° before the maximum of I_L (and hence 90° before the maximum of V_R). If we add V_L and V_R at points along their graphs, we obtain E which has a maximum value of 70.7 volts, $E = 70.7$ volts. (See Figure 4.7.)

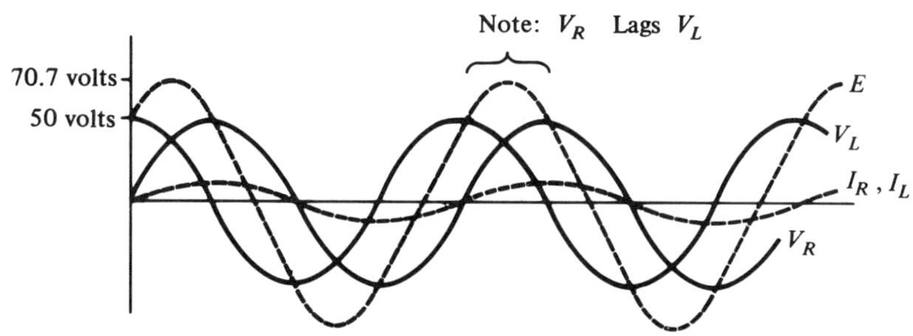

FIGURE 4.7 (*Note: Maximum* V_L *occurs* 90° *before maximum* V_R.)

This addition may be carried out vectorially as we did with capacitors. First we establish a scale, 1 cm = 25 volts. Next, we draw reference vector V_R at 0°. Since V_L leads V_R, it is drawn 90° above V_R. Constructing the parallel lines and measuring the resultant, we find $E = 70.7$ volts. (See Figure 4.8.)

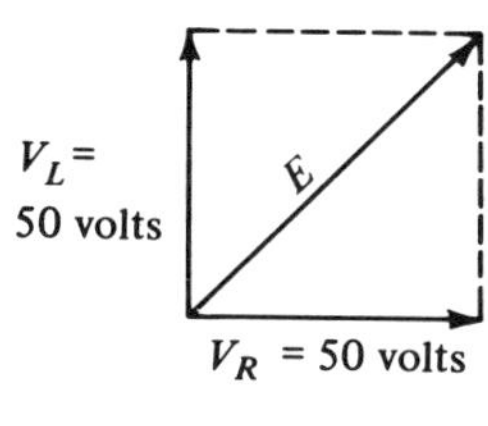

FIGURE 4.8

We might also observe that $E^2 = V_R^2 + V_L^2$ or $E = \sqrt{V_R^2 + V_L^2}$. In this problem then, $E = \sqrt{50^2 + 50^2} = \sqrt{2500 + 2500} = \sqrt{5000} =$ 70.7 volts.

Knowing total voltage and total current, we can find total opposition to current flow, or impedance, using Ohm's Law. $Z = E/I_T = 70.7$ volts/5 A = 14.14 ohms. Since $X_L = V_L/I_L$ and $R = V_R/I_R$ have the same phase relation as V_L and V_R, the impedance can also be found vectorially. (See Figure 4.9.) Again we could also have used the equation

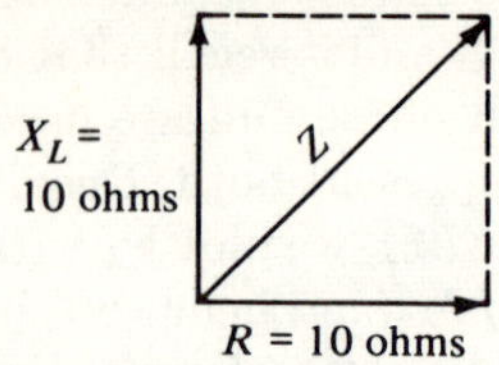

FIGURE 4.9 (*The resultant is 2.828 cm long. Thus the impedance Z is* 5(2.828) = 14.14 *ohms.*)

$Z^2 = R^2 + X_L^2$, or $Z = \sqrt{R^2 + X_L^2}$, since this is a series circuit. Substituting values from this problem, $Z = \sqrt{R^2 + X_L^2} = \sqrt{10^2 + 10^2} = \sqrt{100 + 100} = \sqrt{200} = 14.14$ ohms.

4.5 PARALLEL RL CIRCUITS

Now consider the following parallel combination of coil and resistor. (See Figure 4.10.) In the parallel circuit, the voltage applied to each component

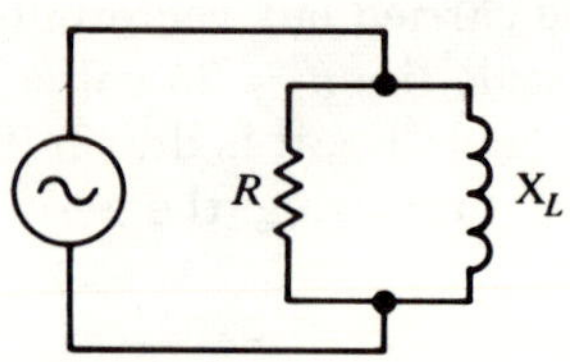

FIGURE 4.10 $E = 100$ *volts;* $R = 50$ *ohms;* $X_L = 25$ *ohms.*

is the same. $E = V_R = V_L = 100$ volts. Turning to Ohm's Law, we find the current in each component.

$$I_R = V_R/R = 100 \text{ volts}/50 \text{ ohms} = 2 \text{ A}$$
$$I_L = V_L/X_L = 100 \text{ volts}/25 \text{ ohms} = 4 \text{ A}$$

The wave pattern will show the voltages for the coil and resistor in phase ($V_L = V_R$), and, of course, I_R, the current in the resistor, is also in phase with V_R. Only the current in the coil, I_L, is out of phase. (See Figure 4.11.) V_L leads I_L by 90°. Thus, I_R also leads I_L by 90°. We can find I_T using the vector diagram in Figure 4.12. Or we find I_T using the equation $I_T = \sqrt{I_R^2 + I_L^2}$. $I_T = \sqrt{2^2 + 4^2} = \sqrt{4 + 16} = \sqrt{20} = 4.47$ A.

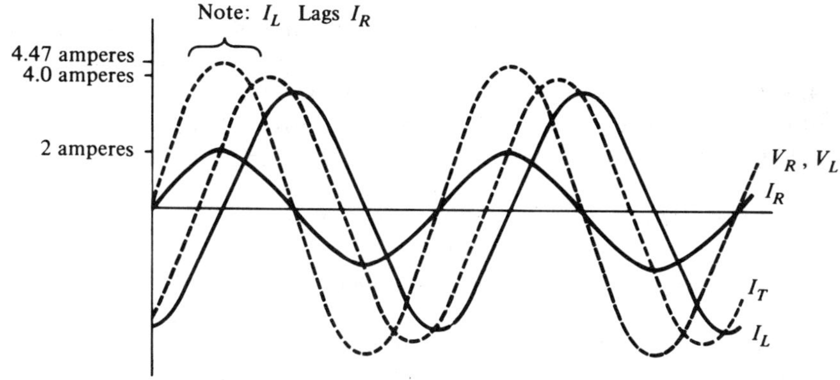

FIGURE 4.11 (*Note: Maximum I_R occurs 90° before maximum I_L.*)

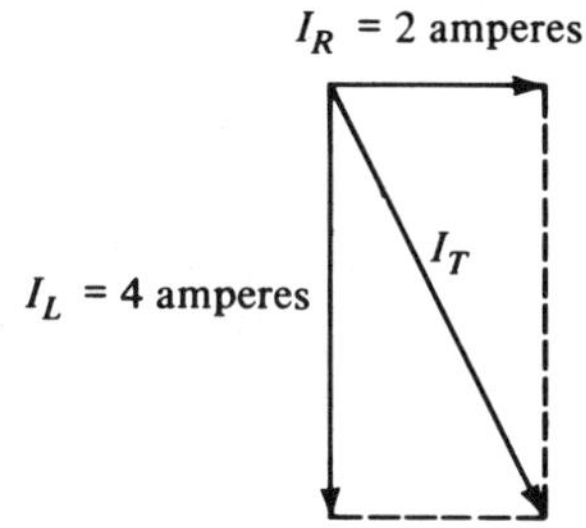

FIGURE 4.12 (*Resultant is 4.47 cm long; therefore, I_T = 4.47 amperes.*)

4.6 SERIES RCL CIRCUITS

In presenting these circuits with capacitance, inductance, and resistance, we want to emphasize the relations concerning phase and frequency. In series circuits we compare reactances, since the current is the same in all components. Consider the circuit in Figure 4.13 containing all three com-

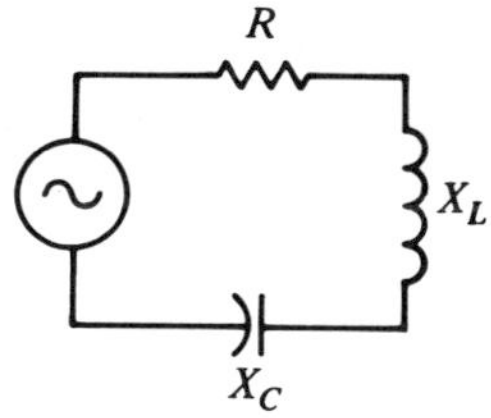

FIGURE 4.13 $E = 26$ *volts;* $I_T = 2$ *amperes;* $R = 5$ *ohms;* $X_L = 13$ *ohms;* $X_C = 25$ *ohms.*

ponents in series. The frequency of the ac generator is 100 Hz. An inductor or capacitor will have printed on it the value of its inductance L or capacitance C. It is necessary then to find X_L and X_C using this information and the equations $X_L = 2\pi fL$ or $X_C = 1/2\pi fC$. Suppose for the above circuit $L = 0.0207$ henry and $C = 0.0000637$ farad. This latter value would normally be given as 63.7 μF (microfarads), where 1 farad is equal to 1,000,000 microfarads. We can now find X_L and X_C for this circuit.

$$X_L = 2\pi fL = 6.28 \times 100 \times 0.0207 = 13 \text{ ohms}$$

$$X_C = \frac{1}{2\pi fC} = \frac{1}{6.28 \times 100 \times 0.0000637} = \frac{1}{0.04} = 25 \text{ ohms}$$

Again, we consider the waveforms for current and voltage in each component. As we have done in the past, we will draw I_R and V_R with the same amplitude to reduce the number of waves. First, we draw the reference wave representing current and voltage in the resistor (I_R and V_R). Since current is the same in all components, this wave also represents I_L and I_C (current in the inductor and capacitor). Recalling our two phase relations: (1) *ICE* (I_C leads V_C by 90°) and (2) *ELI* (V_L leads I_L by 90°), we graph V_C with its maximum occurring 90° after the maximum I_C and V_L with its maximum 90° before the maximum I_L. (See Figure 4.14.)

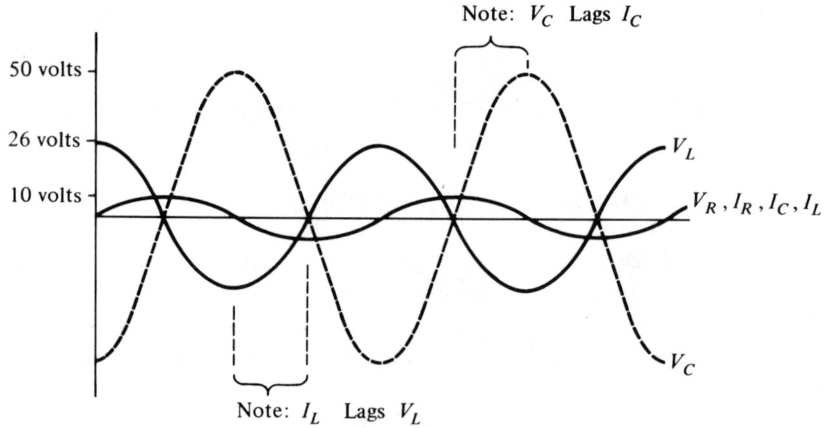

FIGURE 4.14 (*Note:* V_C *reaches maximum* 90° *after* I_C; V_L *reaches maximum* 90° *before* I_L.)

Let us study the graphs of V_L and V_C. When V_L has its peak positive value, V_C has its most negative value, and vice versa. Since the maximum value for V_L occurs at 0° and the maximum value for V_C occurs at 180°, we say that V_L and V_C are 180° out of phase. When one is on the positive half of its cycle, the other is on its negative half cycle, so V_L and V_C are in opposite directions. It appears then that the voltage of the inductor and the voltage of the capacitor tend to eliminate each other. We will pursue this idea with vectors, but first let us determine the magnitude of each voltage drop. Since this is a series circuit, $I_T = I_R = I_L = I_C = 2$ A. Applying Ohm's Law,

$$V_R = I_R R = 2 \text{ A} \times 5 \text{ ohms} = 10 \text{ volts}$$
$$V_L = I_L X_L = 2 \text{ A} \times 13 \text{ ohms} = 26 \text{ volts}$$
$$V_C = I_C X_C = 2 \text{ A} \times 25 \text{ ohms} = 50 \text{ volts}$$

Because of the phase relation, we would not expect the arithmetic sum of the voltage drops to equal E.

Establishing a scale 1 cm = 10 volts, we construct vectors for V_R, V_L, and V_C. (See Figure 4.15.) V_R, the reference vector, is drawn at 0°.

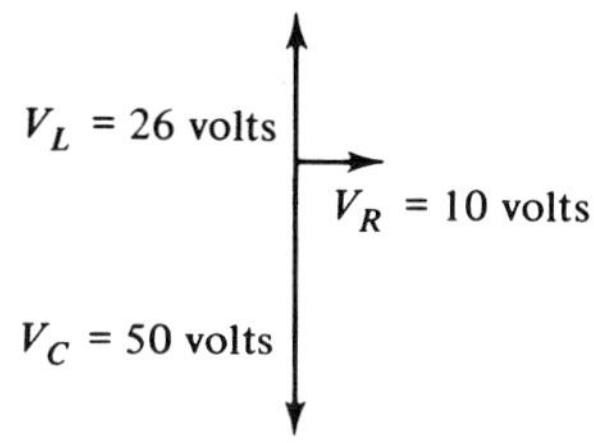

FIGURE 4.15

V_L leads V_R by 90°, so is drawn at 90°. V_C lags V_R by 90°, so is drawn at 270°, 90° below V_R. V_L and V_C are in opposite directions, just as we had observed from their waveforms. Since they are in opposing directions, we find the voltage across these two components by subtracting the smaller from the larger. The remainder in this problem is 50 volts − 26 volts = 24 volts in the direction of V_C (since V_C was larger). We now plot this 24 volts from the capacitor with V_R. Using these two voltages, we construct the resultant and measure to find E = 26 volts. (See Figure 4.16.)

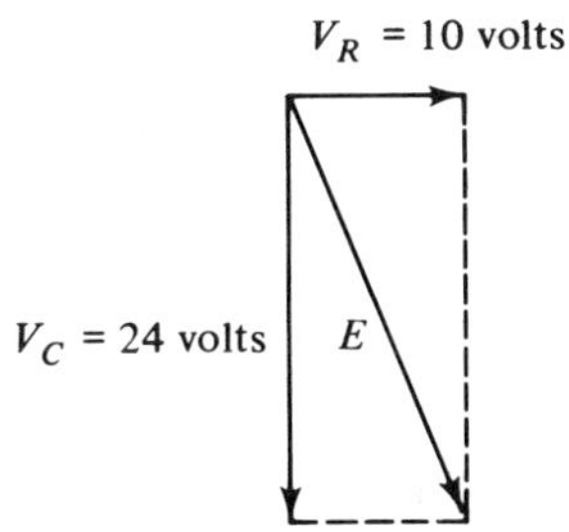

FIGURE 4.16

We can also develop an equation for E if we recognize that the total voltage squared is equal to V_R squared plus the difference of V_C and V_L squared. $E^2 = 26^2 = 676$; $V_R{}^2 = 10^2 = 100$; $(V_C - V_L)^2 = (50 - 26)^2 =$

$24^2 = 576$. Thus, $E^2 = V_R^2 + (V_C - V_L)^2$ or $E = \sqrt{V_R^2 + (V_C - V_L)^2}$. In this problem, $E = \sqrt{100 + 576} = \sqrt{676} = 26$ volts. We should also mention that if V_L were larger than V_C, the equation for total voltage could be written: $E = \sqrt{V_R^2 + (V_L - V_C)^2}$.

Previously, we were also interested in determining total opposition to current flow or impedance Z. Two methods suggest themselves. First, we can apply Ohm's Law to the total circuit: $Z = E/I = 26$ volts/2 A $=$ 13 ohms. A second method is to add the resistance and reactances vectorially. This is possible with a series circuit, because the phase relations for reactances are the same as those for voltage. Thus, the resistance is our reference at 0°; X_L, like V_L, leads the reference by 90°; X_C, like V_C, lags by 90°. X_L and X_C are 180° out of phase, and we subtract $X_C - X_L = 25 - 13 = 12$ ohms. (See Figure 4.17.) Finding the resultant, which represents impedance Z, we see $Z = 13$ ohms. (See Figure 4.18.)

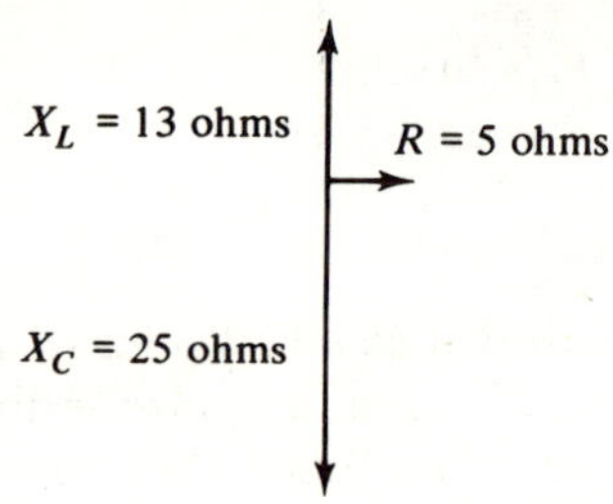

FIGURE 4.17

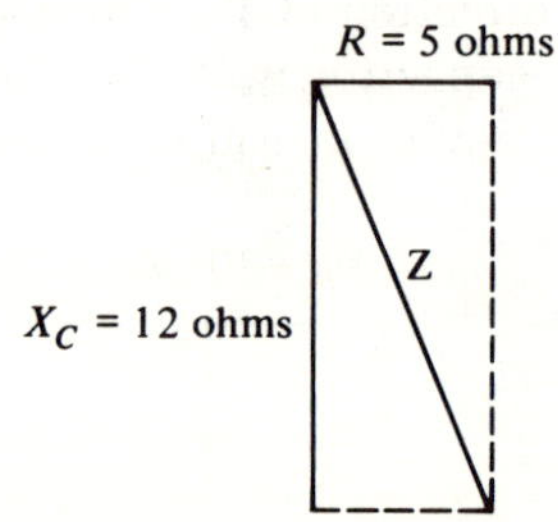

FIGURE 4.18

Before leaving this circuit, we want to indicate how it can be replaced with a simpler circuit that has the same characteristics. Two circuits with the same characteristics are said to be equivalent. The fact that X_C and X_L were 180° out of phase and could be subtracted leads us to consider the characteristic of the remainder, $X_C - X_L = 25 - 13 = 12$ ohms. The 12-ohm remainder is characteristic of the capacitor, since X_C was the larger value. Therefore, we may replace the coil of inductive reactance

13 ohms and the capacitor of capacitive reactance 25 ohms with a single capacitor of capacitive reactance 12 ohms to obtain a circuit equivalent to the one in our problem. (See Figure 4.19.)

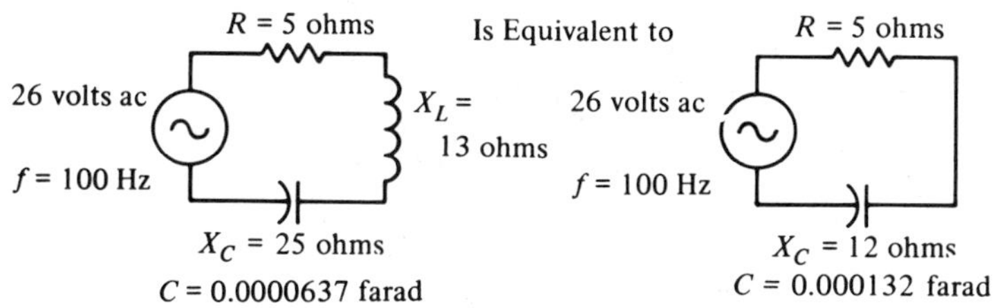

FIGURE 4.19

We can also find the size (capacitance) of the capacitor in the equivalent circuit by using the equation $X_C = 1/2\pi fC$. This equation may be solved algebraically for C (capacitance) to give $C = 1/2\pi fX_C$. Thus, for the capacitor in the equivalent circuit, whose capacitive reactance is $X_C = 12$ ohms, we have $C = 1/6.28 \times 100 \times 12 = 1/7536 = 0.000132$ farad. Note that there has been an increase in capacitance from the original circuit.

4.7 PARALLEL RCL CIRCUITS

Finally, we turn to a parallel circuit containing all three components. The ac generator produces a voltage of 48 volts at a frequency of 1000 Hz. Knowing the capacitance $C = 39.8$ μF and the inductance $L = 478$ μH (microhenrys), we can find capacitive reactance $X_C = 1/2\pi fC = 1/6.28 \times 1000 \times 0.0000398 = 4$ ohms, and inductive reactance $X_L = 2\pi fL = 6.28 \times 1000 \times 0.000478 = 3$ ohms. Notice that C must be changed to farads and L to henrys before using these equations. (See Figure 4.20.)

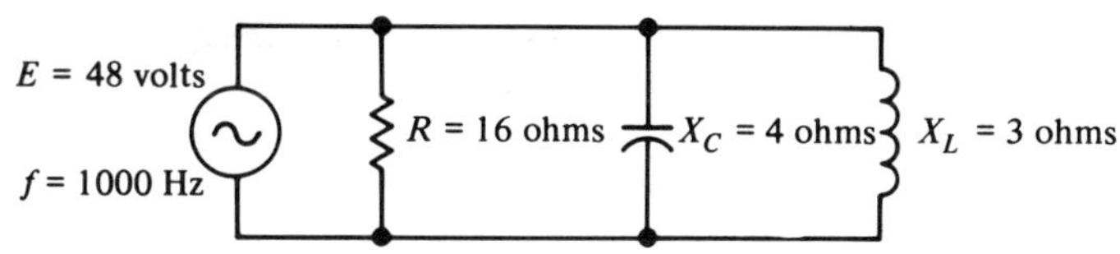

FIGURE 4.20

In a parallel circuit, the voltage across each component is the same. $E = V_R = V_C = V_L = 48$ volts. Therefore, we will compare current waveforms. V_R and I_R are in phase and furnish the reference wave; this wave also represents V_C and V_L. Using the phase relations *ICE* (I_C leads V_C) and *ELI* (I_L lags V_L), we draw the other waves. (See Figure 4.21.)

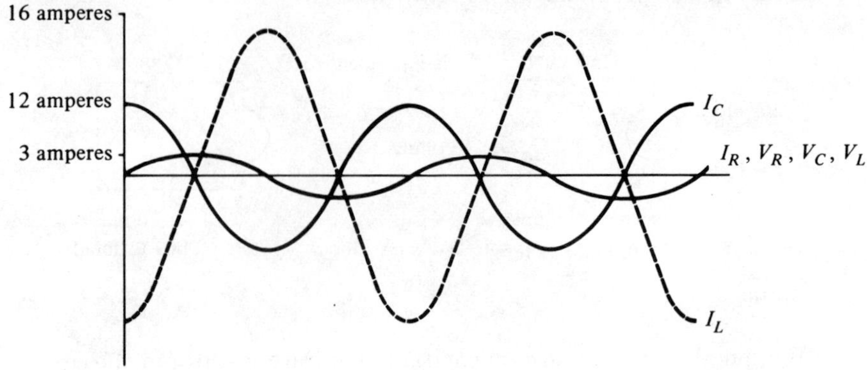

FIGURE 4.21

To determine the exact value of the current in each component we used Ohm's Law:

$$I_R = \frac{V_R}{R} = \frac{48 \text{ volts}}{16 \text{ ohms}} = 3 \text{ A}$$

$$I_C = \frac{V_C}{X_C} = \frac{48 \text{ volts}}{4 \text{ ohms}} = 12 \text{ A}$$

$$I_L = \frac{V_L}{X_L} = \frac{48 \text{ volts}}{3 \text{ ohms}} = 16 \text{ A}$$

Looking at the graph, we see that I_L and I_C are 180° out of phase and thus tend to cancel each other. Using the phase relations with I_R, the reference vector, at 0° to obtain a vector diagram, we make the same observation about I_C and I_L. (See Figure 4.22.) Subtracting I_C from I_L,

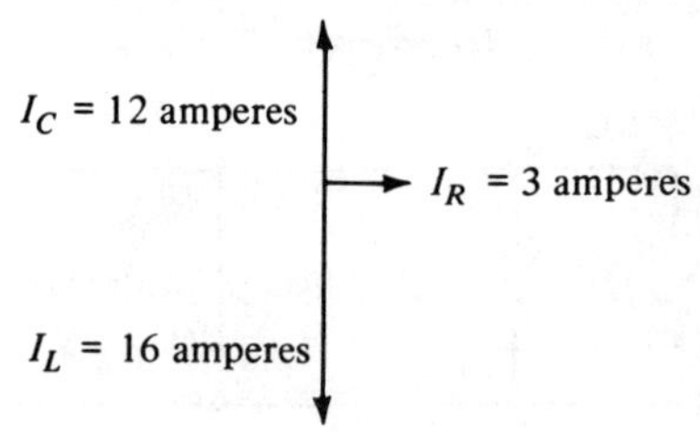

FIGURE 4.22

we obtain the reactive current 16 A − 12 A = 4 A in the direction of I_L. Next, we construct the resultant and measure to obtain I_T = 5 A. (See Figure 4.23.)

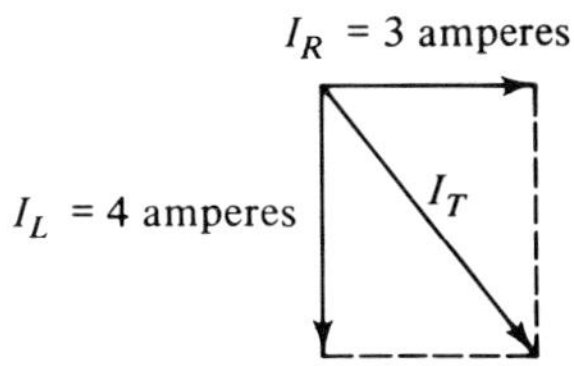

FIGURE 4.23

In equation form, we can write $I_T^2 = I_R^2 + (I_L - I_C)^2$, that is, the square of the total current equals the sum of the square of the current through the resistor and the square of the difference of the currents in the coil and capacitor. Taking the square root, $I_T = \sqrt{I_R^2 + (I_L - I_C)^2}$. For this problem, $I_T = \sqrt{3^2 + (16 - 12)^2} = \sqrt{3^2 + 4^2} = \sqrt{9 + 16} = \sqrt{25} = 5$ A.

Impedance can now be easily computed using Ohm's Law: $Z = E/I_T = 48$ volts/5 A $= 9.6$ ohms. Since this is a parallel circuit, we do not know the phase relations for the reactances and hence cannot find impedance vectorially as we did for series circuits.

The reactive current of 4 A was obtained by subtracting I_C from the larger value I_L, and hence this reactive current is characteristic of the inductor. If we replace the capacitor and inductor by an equivalent component, we recognize that the new component, in our particular problem, will be an inductor. In general, the equivalent component will have the characteristic of that parallel component which allows the largest flow of current. The value of X_L is found using Ohm's Law, knowing the voltage and the equivalent reactive current. (See Figure 4.24.)

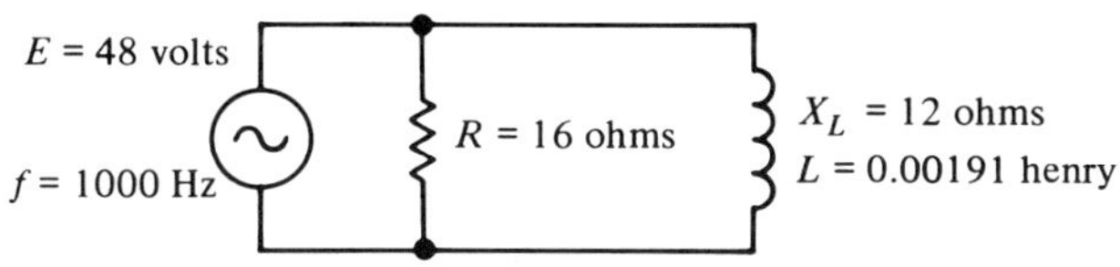

FIGURE 4.24 *Equivalent circuit.*

$X_L = V_L/I_{\text{Equivalent}} = 48$ volts/4 A $= 12$ ohms. We can also calculate the size of the coil using $X_L = 2\pi fL$ which can be written $L = X_L/2\pi f = 12/6.28 \times 1000 = 0.00191$ henry. We cannot reduce this circuit further to one component because of the 90° phase relation between currents I_R and I_L.

4.8 SUMMARY

Our aim has been to learn to recognize the equivalent nature of various electronic circuits. In a circuit containing both coils and capacitors, is the total effect inductive or capacitive? Will the opposition to current flow then be directly proportional to frequency, as is the case with inductive circuits, or will it be inversely proportional to frequency, as in capacitive circuits? We have looked at wave patterns depicting phase relations between voltage and current, at vector diagrams, and at mathematical solutions to RCL circuits (resistive, capacitive, inductive circuits) to justify the characteristic nature of the circuit. These ideas will be important when we talk about filter networks and high- and low-frequency amplifiers.

Now let us summarize some of our observations about series and parallel circuits. In an ac circuit containing capacitors and inductors in series, in order to determine the characteristic nature of the circuit, first find the reactance of each component at the given frequency. Then calculate the sum of the capacitive reactances and that of the inductive reactances. This gives an equivalent circuit with one capacitor and one inductor. The larger of these two sums determines the character of the circuit at the specified frequency. Consider the example in Figure 4.25. The sum

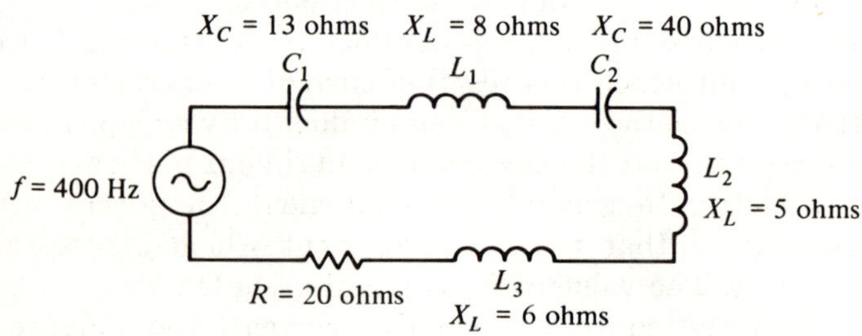

FIGURE 4.25

of the capacitive reactances is $13 + 4 = 17$ ohms. The sum of the inductive reactances is $8 + 5 + 6 = 19$ ohms. (See Figure 4.26.) Thus, the

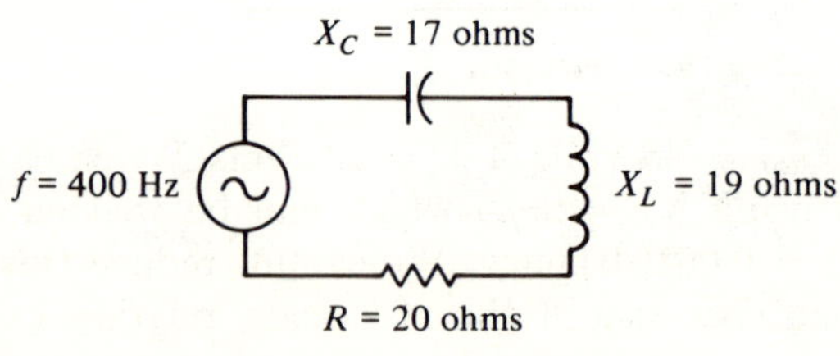

FIGURE 4.26

final equivalent circuit is inductive as shown in Figure 4.27. At a different frequency, this circuit could be capacitive since it would have different values for the reactances. We will discuss this in more detail when we work with resonant circuits.

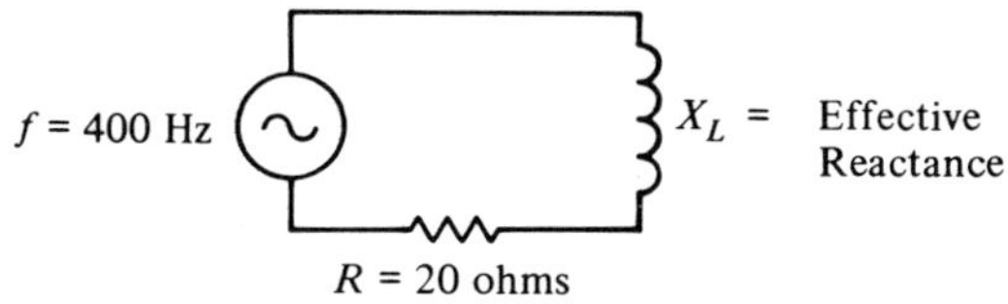

FIGURE 4.27

Determining the character of a parallel ac circuit also requires that we find the reactance of each component at the given frequency. Then to reduce the circuit to one inductor and one capacitor, we apply the same technique to parallel capacitive reactances and to parallel inductive reactances that we used on parallel resistances, that is, the product divided by the sum. Consider the circuit in Figure 4.28. First, the capacitive reactances $X_C = (X_{C1}X_{C2})/(X_{C1} + X_{C2}) = (18 \times 9)/(18 + 9) = 6$ ohms. Then the inductive reactances $X_L = (X_{L1}X_{L2})/(X_{L1} + X_{L2}) = (8 \times 8)/(8 + 8) = 4$ ohms. We thus have the equivalent circuit shown in Figure 4.29. This equivalent circuit has the characteristic of the smaller reactance (which allows more current flow), and thus the circuit is inductive at this frequency.

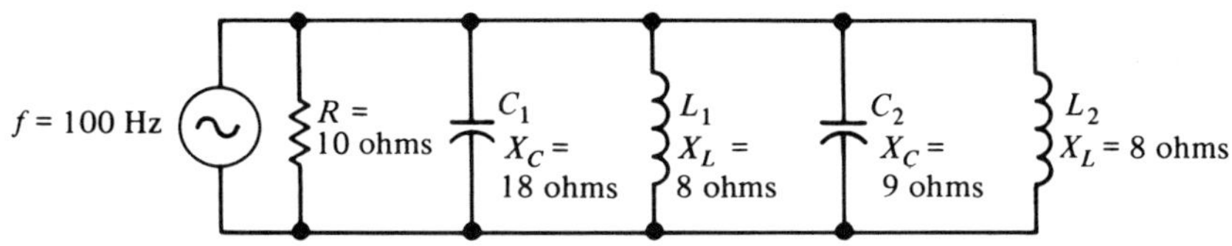

FIGURE 4.28

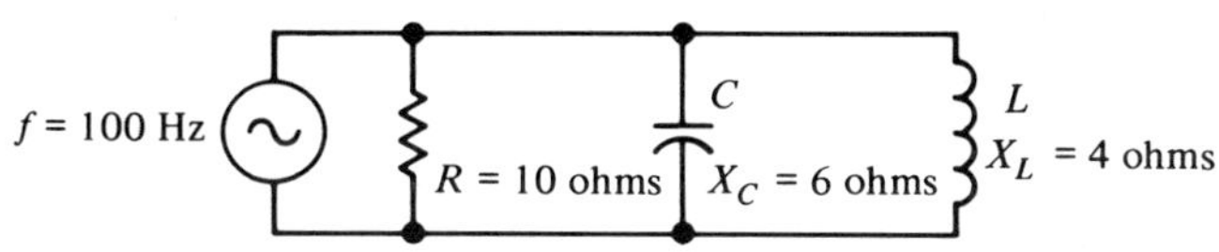

FIGURE 4.29

To determine the magnitude of the effective reactance, we need more information, such as total voltage. In this review we were primarily concerned with the overall character of the circuit and not the specific value of effective reactance. (See Figure 4.30.)

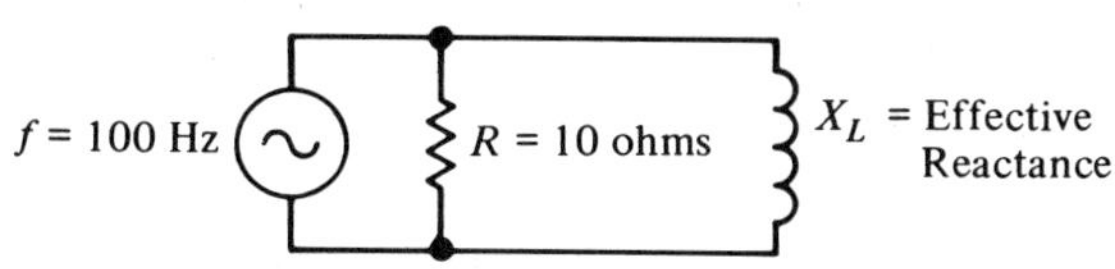

FIGURE 4.30

Questions

(1) How will the inductance of a coil be affected when the frequency of the incoming signal is tripled?

(2) In a series RL circuit, what waves are in phase with the current in the inductor?

(3) In a parallel RL circuit, does the total current lead or lag the current in the inductor? The current in the resistor?

(4) In the series RCL circuit, the total inductive reactance is the sum of the individual inductive reactances, and the total capacitive reactance is the sum of the individual capacitive reactances. How is the characteristic of this equivalent RCL circuit then determined?

(5) When a capacitive reactance and an inductive reactance are connected in parallel, how is the resultant reactance found?

Problems

(1) In a series RL circuit, $I_T = 10$ A, $R = 20$ ohms, and $X_L = 30$ ohms. What is the applied voltage? The impedance?

(2) In a parallel RL circuit, $E = 100$ volts, $R = 20$ ohms, and $X_L = 10$ ohms. What is the applied current?

(3) A series RCL circuit consists of $R = 5$ ohms, $X_L = 10$ ohms, $X_C = 15$ ohms. Is the circuit capacitive or inductive?

(4) A parallel RCL circuit consists of $R = 5$ ohms, $X_L = 10$ ohms, $X_C = 15$ ohms. Is the circuit capacitive or inductive?

Suggested Experiments

(1) Inductance. The inductance of a single inductor or of several inductors in series or parallel may be measured using the inductance bridge described in the chapter on test instruments (Chapter 28).

(2) Series RL Circuit. The phase relations in a series RL circuit can be observed in exactly the same way as was done with a series RC circuit in Experiment (2), Chapter 3. Simply replace the capacitor with a coil.

(3) Parallel RL Circuit. Replace the capacitor in Experiment (3), Chapter 3, with a coil to observe the phase relations in this circuit.

(4) Series RCL Circuit. We will observe the voltage phase relations for this type of circuit. Attach the vertical input to point *A* to obtain the applied voltage. The voltage across the capacitor is observed by attaching the vertical input to point *B*. Interchanging the coil and capacitor, we

can observe the voltage across the coil. Finally, exchange the coil and resistor to obtain the voltage across the resistor. (See Figure 4.31.)

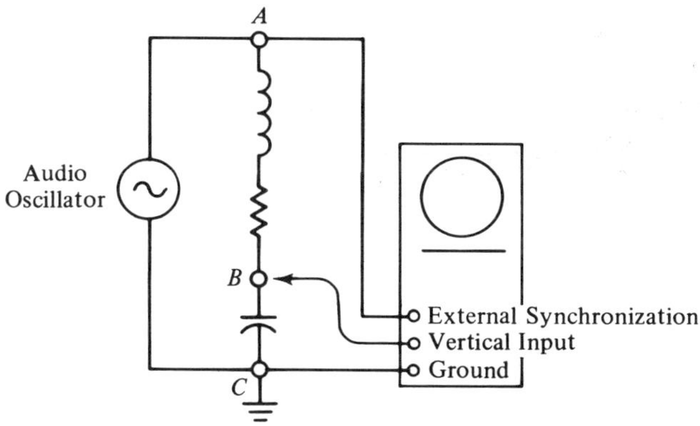

FIGURE 4.31

(5) Parallel RCL Circuit. As with parallel RC and RL circuits, we want to compare current waveforms. We do this indirectly by placing resistors in series with the capacitor and coil and observing the voltage waveforms, which are in phase with the currents, across these resistors. Current in the parallel resistor is observed when the vertical input lead is placed at A; current in the capacitor when the vertical input is at B; current in the coil when the vertical input is at C; total current when the vertical input is at D. (See Figure 4.32.)

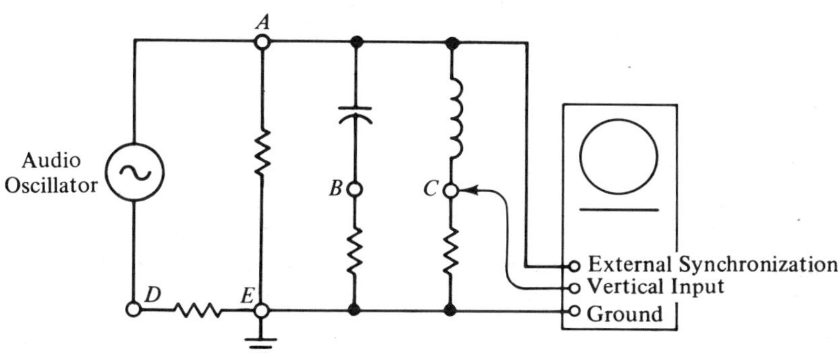

FIGURE 4.32

CHAPTER 5

Resonance

5.1 $X_L = X_C$

We have seen that the characteristic of a series RCL circuit is determined by the larger reactance and that of a parallel RCL circuit by the smaller reactance. You have probably been wondering what type of circuit we would have if there were not a larger reactance, in other words, if the inductive reactance and capacitive reactance were equal. This special case, when $X_C = X_L$ in either series or parallel circuits, is called resonance. Resonance will be employed for such diverse circuits as wave traps and oscillators.

The resonant point can be shown using several different graphs. First consider a graph showing the way reactance changes as frequency changes. The graph consists of two curves. (See Figure 5.1.) One, which is actually a straight line, comes from the equation $X_L = 2\pi fL$. For a particular coil, we know the value of L; we can then use the equation to find the value of X_L at different frequencies. Plotting these points and connecting them, we get the line. The other curve comes from plotting values of X_C for different frequencies, this time using the equation $X_C = 1/2\pi fC$ for a particular capacitor with capacitance C. Notice that on this curve, X_C is large when f is small and vice versa; this is the inverse relation already discussed.

The line representing X_L and the curve representing X_C have one point in common, the point at which they cross, or intersect, one another. At this point of intersection, the values of X_L and X_C are the same ($X_L = X_C$), and this is the resonant point. The frequency at which X_C

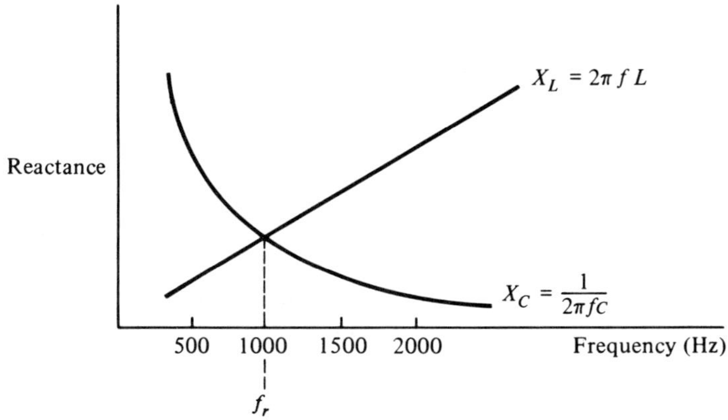

FIGURE 5.1

and X_L are equal, the resonant frequency f_r, is indicated below the point of intersection.

5.2 CIRCUIT CHARACTERISTICS

Earlier we indicated that an equivalent ac circuit may be either inductive or capacitive, depending on the frequency. Looking at Figure 5.1, we see that to the left of the resonant frequency, the capacitive reactance is larger than the inductive reactance. In a series circuit, the effective characteristic would thus be capacitive; but in a parallel circuit, where the effective characteristic is in the nature of the smaller reactance, the circuit would be inductive. If we increase the frequency beyond the resonant point, moving to the right of f_r on our graph, we note that now the inductive reactance is larger than the capacitive reactance. The same series circuit would now become inductive, and the parallel circuit would be capacitive.

5.3 SERIES RESONANCE

Now let us take a closer look at a series resonant circuit in terms of our vector diagram for reactances. We recall that the inductive reactance leads the resistance by 90° and the capacitive reactance lags the resistance by 90°. Thus, X_C and X_L are 180° out of phase. (See Figure 5.2.) Since $X_C = X_L$ in the resonant circuit, they cancel each other, leaving only resistance R. Thus, the total opposition to current flow, or impedance, of a series resonant circuit is the resistance, $Z = R$, which is the smallest possible impedance for this circuit. We can obtain the same

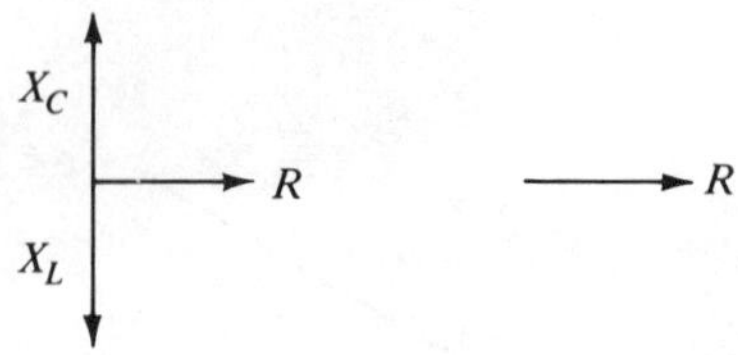

FIGURE 5.2

results, using the equation $Z = \sqrt{R^2 + (X_L - X_C)^2}$. Since $X_L = X_C$ in the resonant circuit, their difference $X_L - X_C = 0$. We have $Z = \sqrt{R^2 + 0^2} = \sqrt{R^2} = R$ for a series resonant circuit.

At the resonant point (when $X_L - X_C = 0$) the impedance has its smallest possible value for the series circuit. When impedance is a minimum, there will be a maximum flow of current. Figure 5.3 shows the

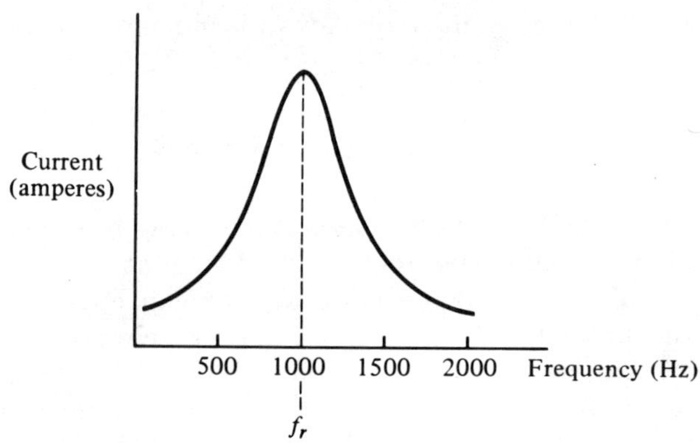

FIGURE 5.3

current flow for different frequencies in a series RCL circuit. Since current is at a maximum at the resonant point, we can find resonant frequency by looking directly below the highest point on the curve.

Consider the series RCL circuit shown in Figure 5.4. $X_C = 1/2\pi fC = 1/6.28 \times 10^3 \times 256 \times 10^{-7} = 6.28$ ohms. $X_L = 2\pi fL = 6.28 \times 10^3 \times$

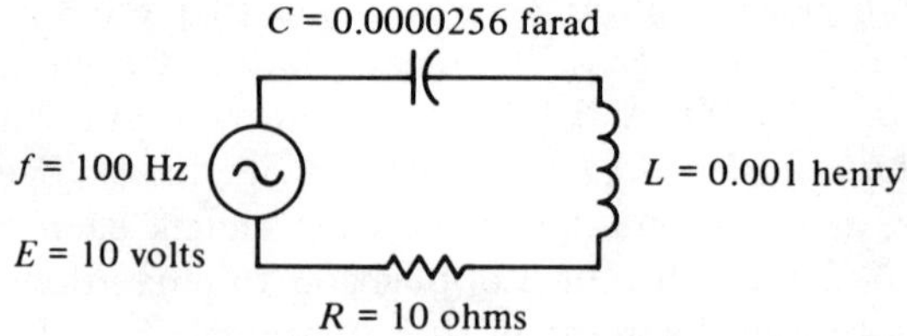

FIGURE 5.4

$10^{-3} = 6.28$ ohms. Since $X_L = X_C = 6.28$ ohms, this is a series resonant circuit. The resonant frequency is, of course, 1000 Hz. The current in all the components is the same as the total current in a series circuit. To find I_T we use Ohm's Law: $I_T = E/Z$. But we have shown that for a series resonant circuit, $Z = R$. Hence $I_T = E/R = 10$ volts/10 ohms $= 1$ A.

5.4 BANDPASS

You may have noticed with your radio that you can get your favorite station by turning the tuning dial (a variable capacitor) over a small range; the station can be received at more than one precise spot on the dial. If this range is large, the radio is not very selective, for if other stations broadcast at frequencies near the frequency of your station, there may be some interference between the two. The selectivity of the radio depends on the ability of a resonant circuit to accept or reject frequencies. Remember that the maximum current in a series RCL circuit occurs at the resonant frequency. The radio's resonant circuit will operate over a range of frequencies that produces a current of at least 70.7 percent of the maximum current. If the output of the resonant circuit is at least 70.7 percent of the maximum current, the frequency is said to be within bandpass or bandwidth and the radio will operate. This explains why your radio can get a station at frequencies near the broadcast frequency.

In the last example we had a maximum current of 1 A at the resonant frequency of 1000 Hz; 70.7 percent of this maximum current is 0.707 A. There are two points on the graph in Figure 5.5 where the current is 0.707 A. Looking directly below these two points, we find the minimum fre-

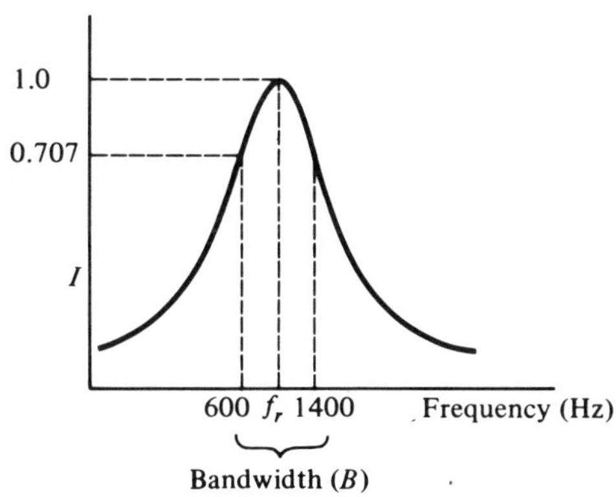

FIGURE 5.5

quency and the maximum frequency of the bandpass. On our graph we see that the bandwidth extends from 600 Hz to 1400 Hz. Subtracting, we can find the width of the bandpass (B). $B = 1400 - 600 = 800$ Hz.

If you move the dial outside the bandpass, you receive a different station. When you turn the tuning dial, you are actually changing the capacitance of a variable capacitor. The resonant frequency will, of course, be different for the changed RCL circuit, and thus the radio now receives a different station over a different bandpass. If the bandwidths of two stations, which are near each other on the dial, are too large, they might overlap, causing the interference mentioned earlier.

Obviously, a more selective radio has a smaller bandwidth associated with each resonant frequency. An indication of the selectivity of this and other resonant circuits, series or parallel, is the measure of relative merit or sharpness of resonance called the Q value. One method of determining the Q value of a resonant circuit is as follows. Q = Resonant frequency/Bandwidth or $Q = f_r/B$. If the value of Q is large (which will occur when B is small), the circuit has good selectivity and a narrow bandwidth. In the problem just considered, $Q = 1000 \text{ Hz}/800 \text{ Hz} = 1.25$, which is rather small, indicating poor selectivity.

We can change the Q value of a series resonant circuit, yet keep the same resonant frequency, by changing the value of the series resistor. Ohm's Law tells us that increasing the resistance decreases the current. Figure 5.6 shows the effect of resistance on bandwidth. Since increasing

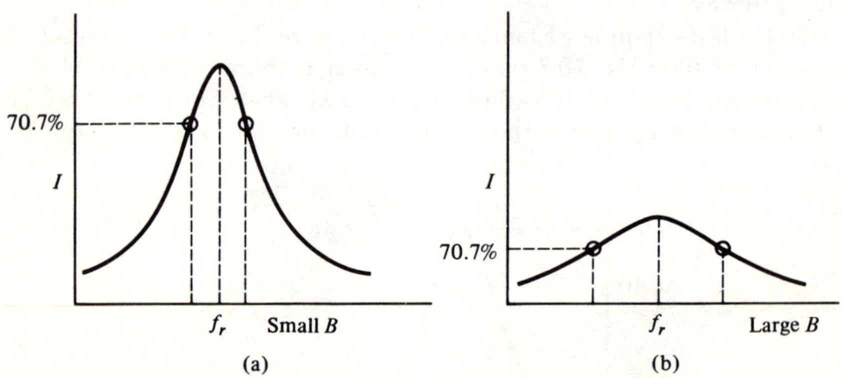

FIGURE 5.6 *(a) Small resistance, large I; (b) large resistance, small I.*

resistance in a series resonant circuit increases bandwidth, the Q value will decrease. R and Q are inversely proportional. A second equation for Q shows this inverse relation: $Q = X_L/R$. This equation gives the same Q value as $Q = f_r/B$.

5.5 PARALLEL RESONANCE

Figure 5.7 shows a parallel resonant circuit. In this circuit, the resistance R is quite small in comparison to the inductive reactance X_L and is usually taken to be the resistance of the coil (an impure inductor). At resonance, X_C and X_L are practically equal, and hence I_C and I_{Coil} are

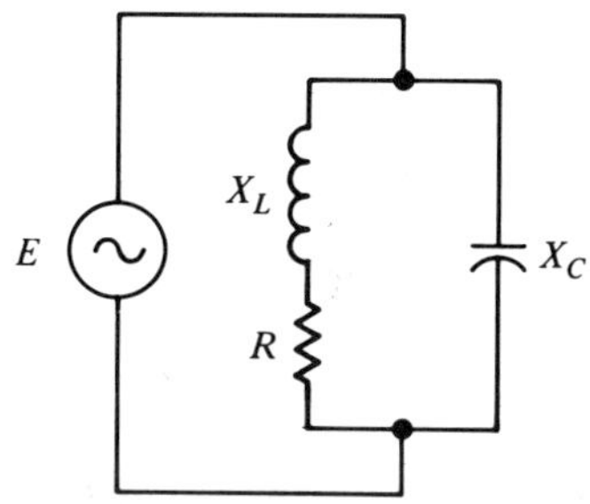

FIGURE 5.7

also approximately equal and nearly 180° out of phase. The small resistance in the inductive branch of the circuit accounts for our use of the words "approximately" and "nearly" instead of the exact relations we have seen in the series resonant circuit.

We recognize that the voltage drop across each branch of the parallel circuit is the same as the applied voltage E. Thus, the current of the capacitor is $I_C = E/X_C$ and leads E by 90° (ICE). Consider the impedance in the inductive branch of the circuit. Since X_L and R are in series, the same current flows in each one, but they are out of phase by 90°, with X_L leading as shown in Figure 5.8. Notice the impedance in the inductive

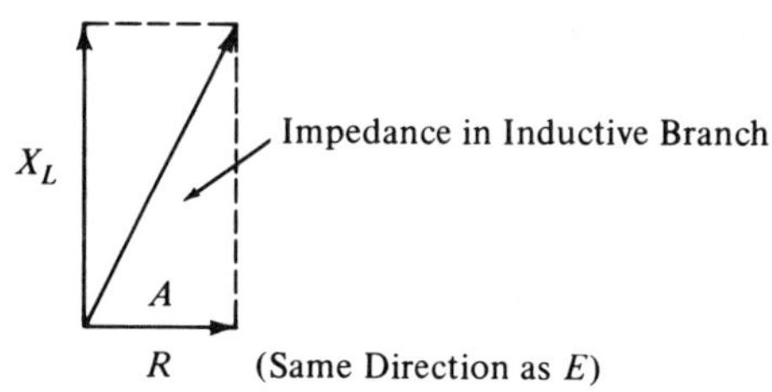

FIGURE 5.8

branch is slightly larger than X_L and leads R by angle A shown on the diagram. Since R is in the direction of E, the current in the inductive branch will lag E (ELI) by this same angle A (which is slightly less than

90°). The resultant I_T is found by drawing a line from the tip of I_C parallel to I_{Coil} and another from the tip of I_{Coil} parallel to I_C. The vector from the origin to the intersection of these two parallel lines represents I_T (Figure 5.9). Note that I_T is in the same direction as E.

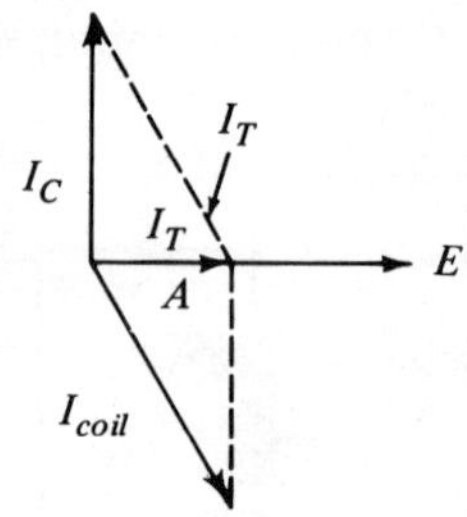

FIGURE 5.9

As we see in our diagram, the currents in the two branches at resonance are much larger than the current supplied by the source I_T. This implies that the circuit stores energy in the electric field of the capacitor, when charged, and in the magnetic field about the coil. In fact, the energy changes from one form to the other as the capacitor discharges and charges, and the current in the coil increases and decreases. The only energy lost, or dissipated, is in heating the resistor, and it is this energy that the generator must replace.

We see then that although neither X_C and X_L nor I_C and I_{Coil} are quite equal in a parallel resonant circuit, they can be taken as equal. For all practical purposes, $X_C = X_L$ and I_C and I_{Coil} are equal and 180° out of phase.

We have seen that I_T is quite small at resonance. If we graph I_T for different frequencies, the minimum value of the current occurs at the resonant frequency f_r (Figure 5.10). From Ohm's Law we know that a

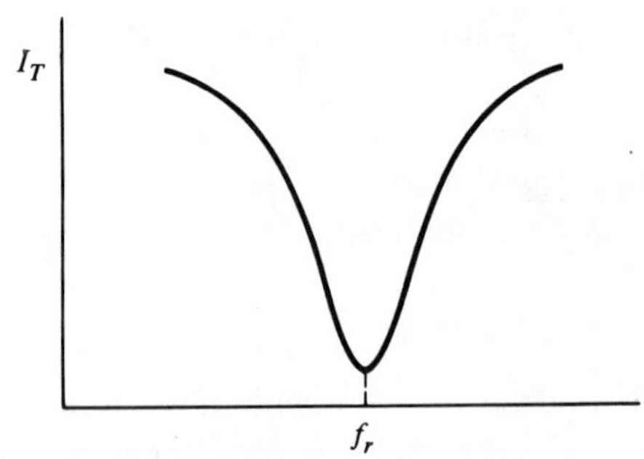

FIGURE 5.10

minimum current implies a maximum impedance. Thus, the impedance Z is a maximum at resonance in a parallel resonant circuit. Remember that the series resonant circuit offered a minimum opposition to the flow of current at resonance, whereas the parallel resonant circuit offers a maximum opposition to the flow of current at resonance. At frequencies on either side of resonance, both of these circuits act as variable resistors,

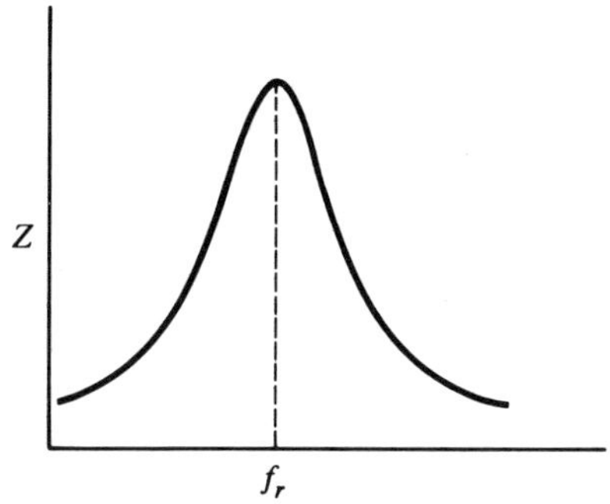

FIGURE 5.11 *Parallel resonance.*

that is, their impedance varies at different frequencies. The same relations for Q value, or selectivity, hold for parallel resonant circuits as for series resonant circuits. (See Figure 5.11.)

5.6 WAVE TRAPS

The problem of a nearby broadcaster's signal being so strong as to eliminate the reception of weaker signals at nearly the same frequency can be solved at the antenna of the receiver by devices called wave traps. There are two types of wave traps: series and parallel wave traps. Both are adjusted so their resonant frequency is the same as the strong undesirable signal. The series resonant circuit offers a minimum opposition to current flow at resonance. Thus, in the series wave trap, the resonant signal travels through the series resonant circuit, which offers minimum impedance, and bypasses the receiver, as shown in Figure 5.12(a). Any signal other than the resonant signal finds greater impedance through the series circuit and thus takes the other path. In the parallel wave trap, the resonant signal experiences maximum impedance and is dissipated before it reaches the receiver; any signal not at resonance finds less opposition and can get to the receiver. Later we will see how the parallel resonant circuit can act as an oscillator.

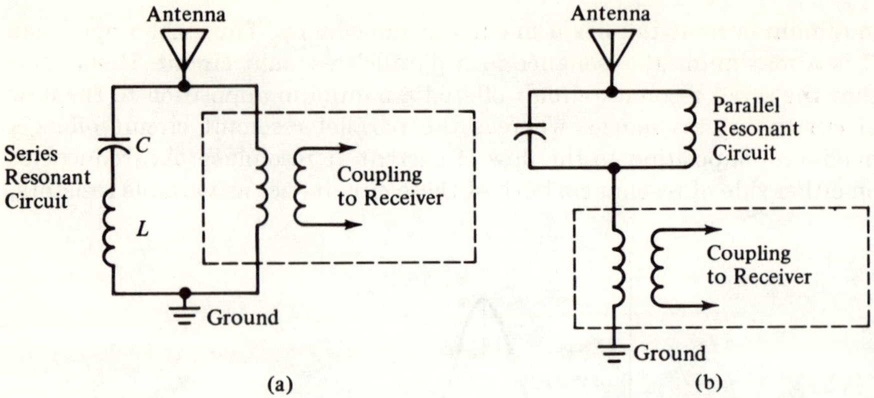

FIGURE 5.12 *(a) Series wave trap; (b) parallel wave trap.*

Questions

(1) What is the relation between inductive and capacitive reactance in a series resonant circuit? A parallel resonant circuit?

(2) At the resonant frequency, what are the relative magnitudes (large, small) of voltage, current, and impedance in a series LC circuit? A parallel LC circuit?

Suggested Experiments

(1) Series Resonant Circuits. In the series resonant circuit, the current is at a maximum and the impedance a minimum at resonance. In Figure 5.13, we will determine the resonant frequency by observing

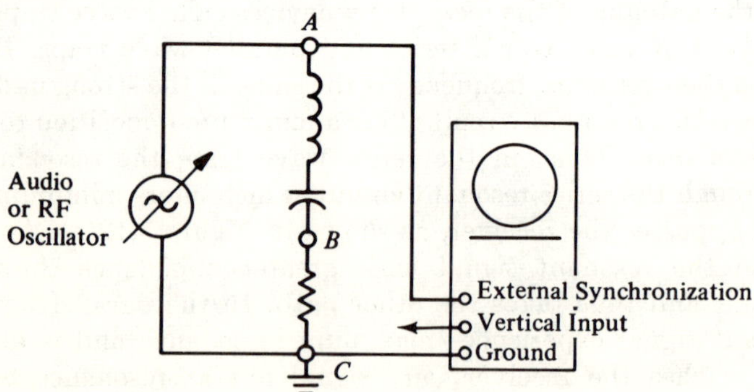

FIGURE 5.13

at what frequency maximum current flows through the circuit. We will do this indirectly by observing the maximum voltage across the resistor. The resonant frequency of your circuit may be out of the audio-frequency range and hence require a radio-frequency oscillator instead of an audio generator. It is of prime importance that the same applied voltage be maintained at all oscillator frequencies. To check this when a frequency change is made, connect the vertical input lead to point A and adjust the oscillator, not the oscilloscope, so that the same height, or amplitude, of the waves seen on the screen is maintained. Now connect the vertical input lead to point B and observe the amplitude of the current in the circuit. As resonant frequency is approached, the amplitude, or height, of the current across the resistor will increase to a maximum. When the height of the voltage across R is at a maximum (current at a maximum), the oscillator is adjusted to resonance for the series LC circuit. You may calibrate the oscilloscope for voltage readings and, knowing the resistance between B and C, calculate the current ($I = V/R_{BC}$) at resonance and at other frequencies. Calibration of an oscilloscope is discussed in Chapter 28, "Test Instruments." In addition, you may wish to plot the current versus the frequency, as we have done in the text. Calculation of Q may also be carried out using this graph.

(2) Parallel Resonant Circuits. In parallel resonance, the current is at a minimum and the impedance a maximum at resonance. Again, being sure to maintain the same voltage at each new frequency, we adjust the oscillator until we obtain a minimum output. When a minimum output is obtained, the oscillator is at the resonant frequency for the circuit. As in Experiment (1), the oscilloscope may be calibrated, the currents calculated, and a graph of current versus frequency drawn.

CHAPTER 6

Time Constants

6.1 CHARGING A CAPACITOR

Let us return briefly to a dc circuit containing a resistor and capacitor in series. Notice that the circuit contains a switch which may be closed to either position *A* or position *B* (Figure 6.1). With the switch in position *A* the flow of current is shown, and we see that the capacitor is charging. As we have already explained, it takes time to charge the capacitor, and,

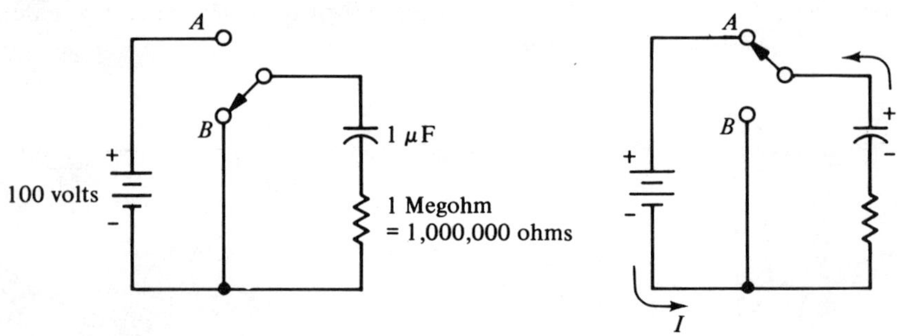

FIGURE 6.1

with the passage of time, the flow of current decreases while the voltage across the capacitor increases. (See Figure 6.2.) Since this is a series circuit, the sum of the voltage drops must equal the supply voltage. Thus, as the voltage across the capacitor increases, the voltage across the resistor must decrease so that at any instant the sum of V_R and V_C will equal the applied voltage. (See Figure 6.3.)

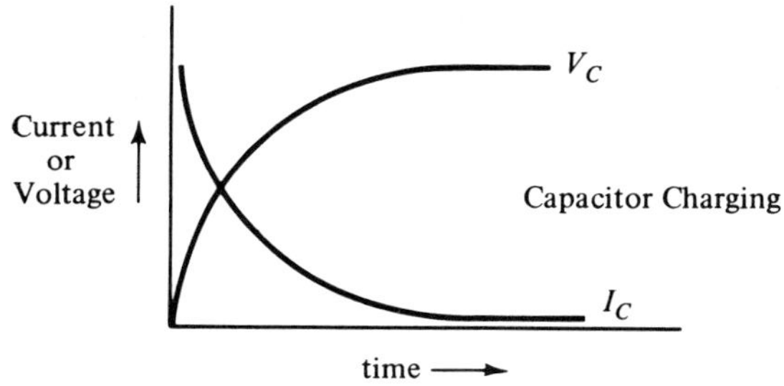

FIGURE 6.2 *Voltage across the capacitor V_C increasing and current in the capacitor I_C decreasing as capacitor charges.*

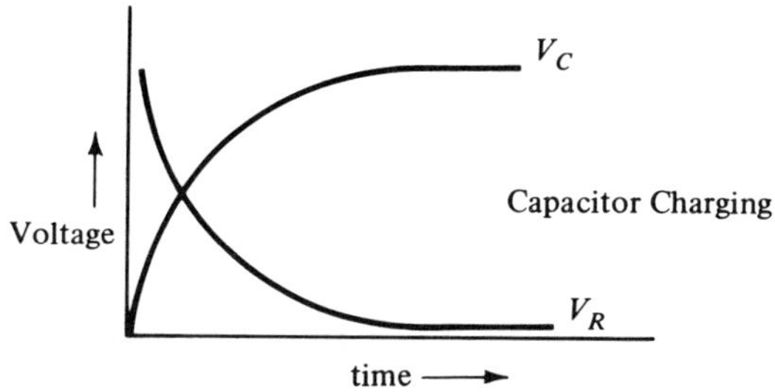

FIGURE 6.3 *Voltage across the capacitor V_C increasing and voltage across the resistor V_R decreasing as capacitor charges.*

If we recall three equations: (1) Ohm's Law, resistance is voltage divided by current, $R = E/I$, (2) capacitance is charge divided by voltage, $C = Q/V$ and (3) current is charge divided by time, $I = Q/t$, we can obtain an equation for time in terms of the resistance and capacitance of the series circuit. Note that $R \times C = E/I \times Q/V = Q/I =$ time. Q/I may not initially look like time, but, using the third equation, we see that $Q = It$. If we replace Q by It, we have $Q/I = It/I = t$, which is time. Thus, $R \times C = t$.

In our problem, $R = 10^6$ ohms and $C = 10^{-6}$ farad. $R \times C = 10^6$ ohms $\times\ 10^{-6}$ farad $= 1$ second. If we close the switch in our circuit to position A for 1 second, then open it, we will find that the capacitor has charged to 63 percent of its final value. Thus, after 1 second the voltage across the capacitor (V_C) is 63 percent of the applied voltage, and the voltage across the resistor is the remaining 37 percent of the total voltage, since the sum of V_C and V_R must equal the total applied voltage at any instant. This time ($R \times C$) required for the capacitor to charge 63 percent is called one time constant, abbreviated 1 TC. It takes five time constants (5 TC) to charge the capacitor to approximately 100 percent for all practical purposes. One time constant then is equal to the product of the series resistance and capacitance: 1 TC $= R \times C$. In our circuit

1 TC = 1 second, and the time for the capacitor to charge, or for the voltage across the capacitor (V_C) to equal the applied voltage, is 5 TC or 5 seconds. (See Figure 6.4.)

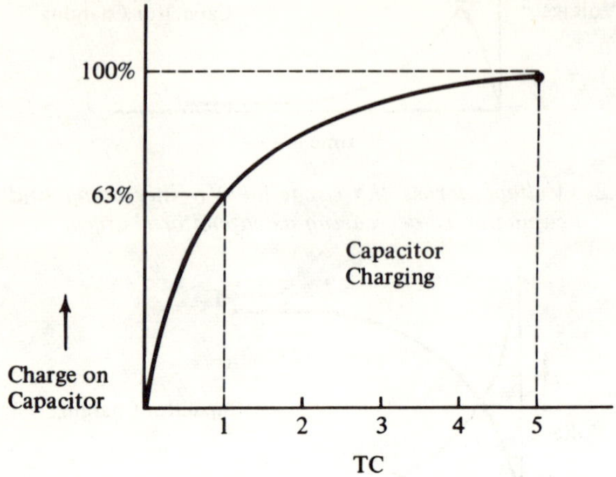

FIGURE 6.4 *Charge on capacitor with passage of time.*

Since the capacitor charges to 63 percent during one time constant (1 TC), it may seem to you that 5 TC is too long a time to charge the capacitor. Actually, during the second time constant, the capacitor will charge an additional 63 percent of the remaining charge (it charges an additional 63 percent of the remaining 37 percent; thus it charges an additional 0.63×37 percent = 23 percent). This brings the charge on the capacitor to just 86 percent (63 percent plus the additional 23 percent). During the third time constant, the capacitor again charges an additional 63 percent of the remaining charge, bringing the charge to 95 percent. After 4 TC the charge on the capacitor is 98 percent of the total charge, and after 5 TC the charge is approximately 100 percent. The time constant chart will be helpful in seeing these percentages, which hold for V_C as well as charge, and will be used in the near future when we discuss wave shaping circuits. (See Figure 6.5.)

6.2 DISCHARGING A CAPACITOR

Now let us assume the switch in our circuit has been in position *A* for 5 TC, charging the capacitor. We then move the switch to position *B*. (See Figure 6.6.) In this position, the supply voltage is disconnected from the circuit. The flow of electrons will be through the resistor to the positive side of the capacitor, thus discharging the capacitor. The time for the capacitor to discharge will also be 5 TC. (See Figure 6.7.) As the capacitor

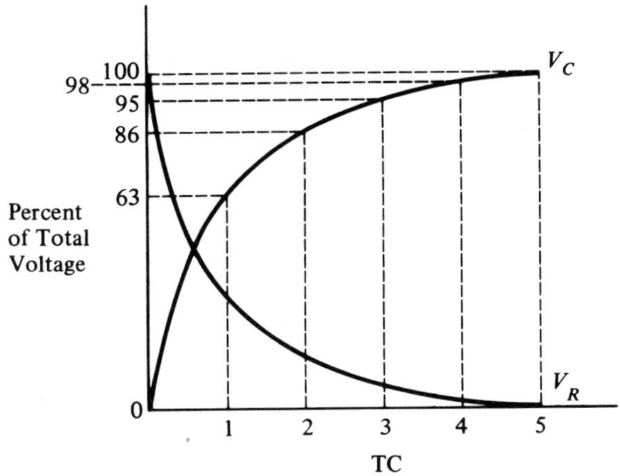

FIGURE 6.5 *V_C = voltage across capacitor when capacitor is charging; V_R = voltage across resistor when capacitor is charging.*

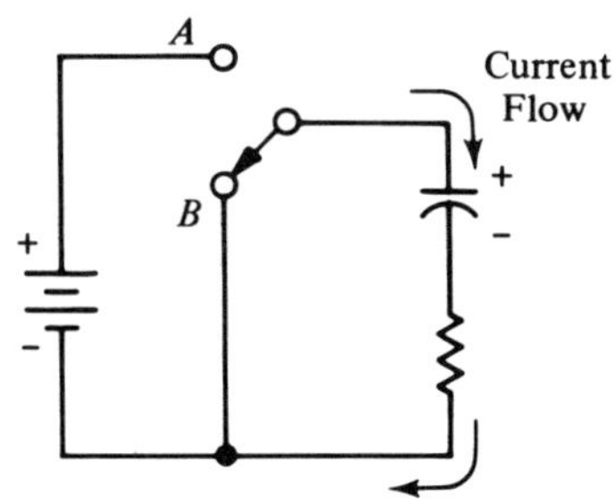

FIGURE 6.6

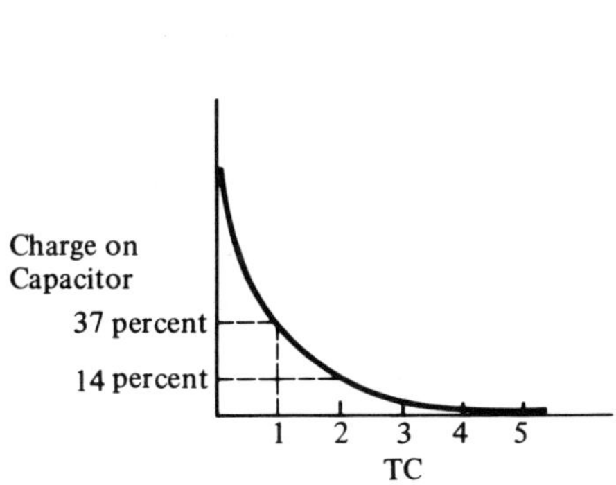

FIGURE 6.7 *Charge on capacitor with passage of time. (Note: Capacitor discharges 63 percent in 1 TC, 86 percent in 2 TC, and so on.)*

discharges, the potential, or voltage, across the capacitor decreases to zero. (See Figure 6.8.) The current, during the discharging of the capaci-

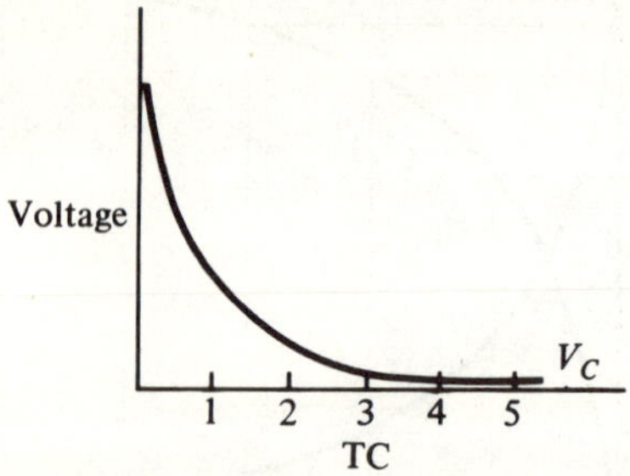

FIGURE 6.8 *Voltage across capacitor as capacitor discharges.*

tor, flows in the opposite direction to that during the charging of the capacitor and thus will be of opposite sign. The flow of current will gradually decrease and will be zero when the capacitor is discharged. Since this current is negative, due to its direction, the graph for current shows it going from its most negative value to zero during the discharge. (See Figure 6.9.) Since there is no applied voltage when the switch is in posi-

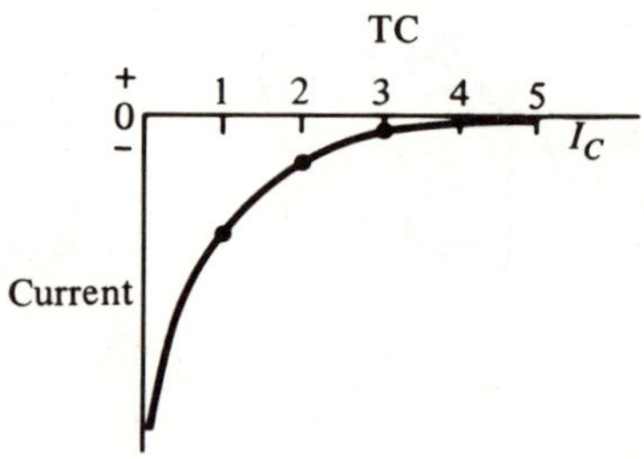

FIGURE 6.9 *Flow of current as capacitor discharges.*

tion B (the applied voltage is zero), the sum of V_R and V_C at any instant during discharge must equal zero. The current is in the opposite direction during discharge, so the voltage across the resistor is also in the opposite direction. V_R then begins at its most negative value and goes to zero. (See Figure 6.10.)

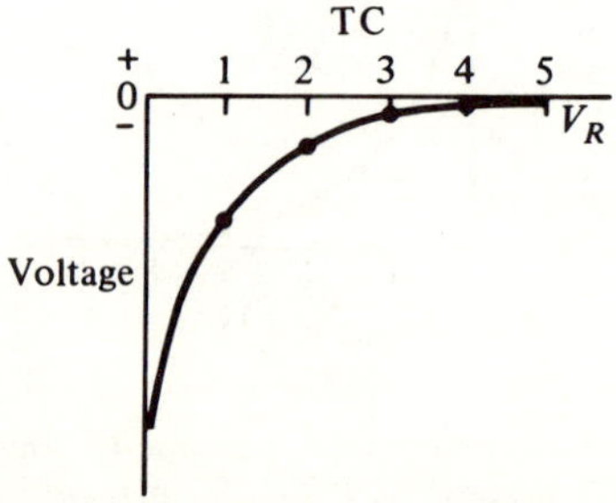

FIGURE 6.10 *Voltage across resistor as capacitor discharges.*

6.3 WAVEFORMS FOR A SERIES RC CIRCUIT

Now we want to combine the above graphs for voltage in order to draw wave patterns for the applied voltage, the voltage across the capacitor, and the voltage across the resistor during successive chargings and dischargings of the capacitor. In other words, we consider these voltages when the switch is moved to position *A* for 5 TC, then to position *B* for 5 TC, then back to position *A* for 5 TC, and so forth. First recall that the applied dc voltage is 100 volts at *A*, 0 volt at *B*, which gives the graph of applied voltage. (See Figure 6.11.) Next combine Figures 6.5 and 6.8 to obtain voltage across the capacitor (V_C). (See Figure 6.12.) Figures 6.5 and 6.10 give us the waveform for voltage across the resistor (V_R). (See Figure 6.13.)

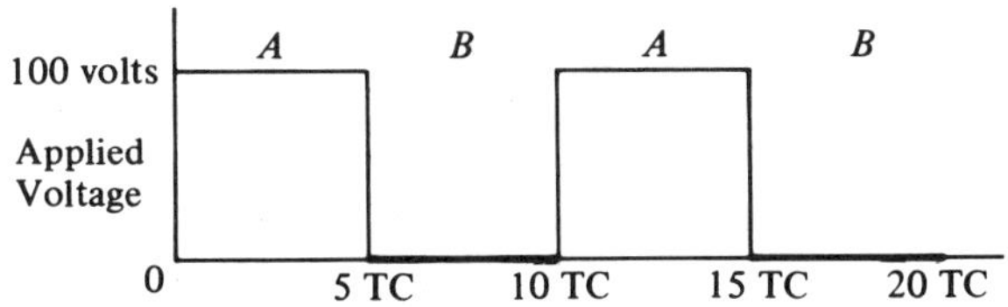

FIGURE 6.11

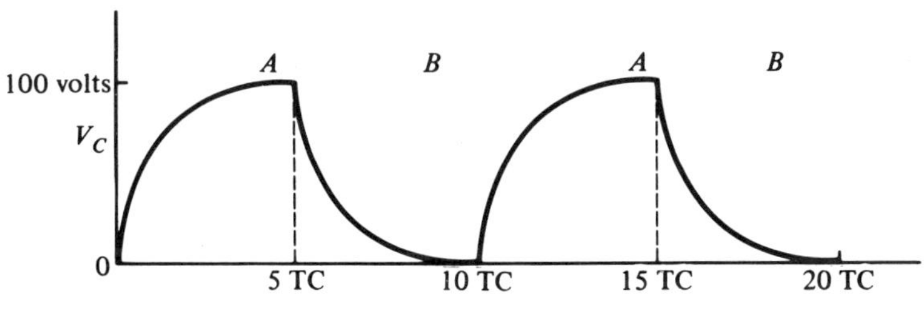

FIGURE 6.12

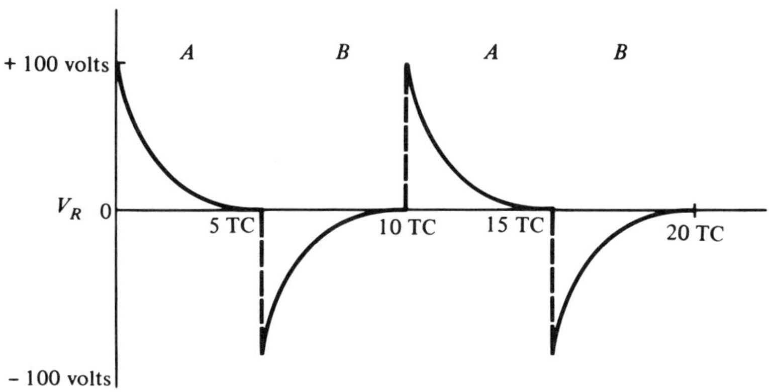

FIGURE 6.13

In sketching these curves, we allowed just enough time for the capacitor to charge and discharge. Now we want to consider the two alternatives: (1) that we do not allow enough time for the capacitor to charge or discharge, and (2) that we allow more time than necessary for the charging and discharging of the capacitor.

6.4 TC LARGE IN COMPARISON TO T

For the first alternative, when the time constant in seconds is large in comparison to the time we keep the switch in position *A*, the voltage across the resistor will not have time to move very far from its maximum value toward zero, that is, V_R will remain nearly the same as the applied voltage. Since the capacitor does not have much time to charge, there will not be much current flow through the resistor when the switch is moved to position *B*, and V_R will be close to zero. Again, V_R will be nearly equal to the applied voltage. An RC circuit of this type, which reproduces the input voltage across the resistor with minimum distortion, is called a coupling circuit and is often used in amplifiers to pass a signal from one stage of amplification to another with minimum distortion of the signal.

We will look at an example of this type of coupling circuit, but, to simplify the circuit, let us first introduce an automatic device to move the switch from *A* to *B* and back. This can be done mechanically, or, as we will see later, it can be done electronically by a device called a square-wave generator. For the time being we will just assume that we have a square-wave generator. Its output is shown in Figure 6.14. Notice that,

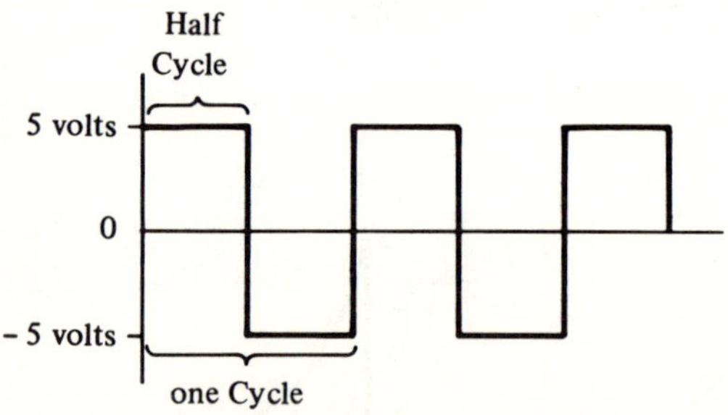

FIGURE 6.14 *Applied voltage from square-wave generator.*

whereas the switch changed the applied voltage from positive to zero, the square-wave generator changes applied voltage from positive to negative. To this source we attaoh a resistor and capacitor in series. (See Figure 6.15.)

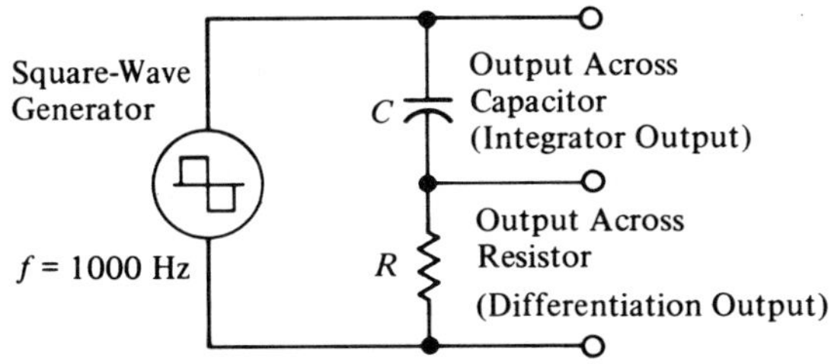

FIGURE 6.15 $E = 5$ *volts;* $C = 0.01\ \mu F = 0.01 \times 10^{-6}$ *farad;* $R = 10^6$ *ohms;* $f = 1000$ *Hz;* 1 $TC = R \times C = 10^6$ *ohms* $\times 0.01 \times 10^{-6}$ *farad;* 1 $TC = 0.01$ *second.*

Notice that the frequency of the square wave in our example is such that 1000 cycles are produced each second. We can find the time duration of one cycle by dividing the 1000 cycles into 1 second. Time for one cycle = 1 second/1000 cycles = 0.001 second for each cycle. The time for one complete cycle is usually referred to as the period, and, since it is a measure of time, it is given the symbol T. From this example, we can see that the time for one cycle (period T) is inversely proportional to the number of cycles each second (frequency f). In equation form, $T = 1/f$.

Earlier we said that a good coupling circuit is one in which the time constant is large in comparison to the time to switch from A to B and back to A. This switching is equivalent to one cycle of the square-wave generator, and the time for one cycle we now know is the period. One method of comparing the time constant (TC) and the period (T) is to divide TC by T. To couple signals with minimum distortion, ratios of at least 10 TC to 1 T are used. In the above circuit, the ratio is TC/T = 0.01 second/0.001 second = 10.

The time for one cycle (period T) is determined using the equation $T = 1/f$. The time for one-half cycle will be one half of the period. In our problem, time for one-half cycle $= \frac{1}{2}T = \frac{1}{2} \times 0.001$ second = 0.0005 second. Recalling that the time constant for this circuit is 1 TC $= R \times C = 0.01$ second, we see that the time required for one-half cycle is much less than 1 TC. If we can find what fraction of one time constant the one-half cycle represents, we can refer to the graph of V_R to determine by how much the voltage across the resistor decreases in one-half cycle. We want this decrease in V_R to be minimal so there will be little or no distortion for good coupling.

Looking at Figure 6.16, we see that after 0.0005 second (one-half cycle), the voltage across the resistor (V_R) has dropped from 5 volts to 4.8 volts, a decrease of 0.2 volt. The voltage across the capacitor (V_C) has, of course, reached 0.2 volt. Observe that V_R does not have time to

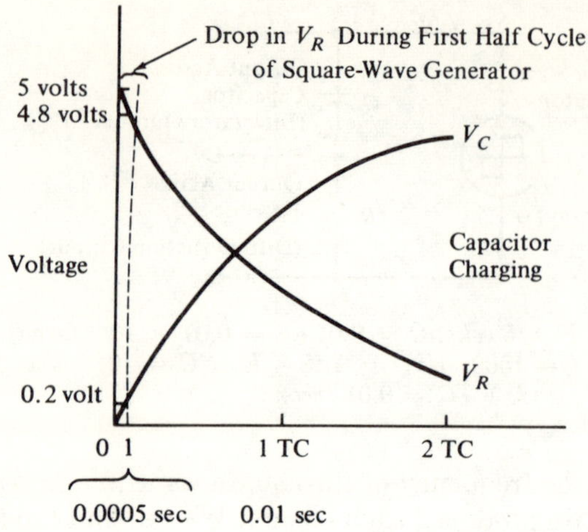

FIGURE 6.16

drop any further, because after 0.0005 second the generator goes into the second half of its cycle.

We can now consider the waveforms for the applied voltage, V_R, and V_C as shown in Figure 6.17. The following will help explain the values you see on the waveforms for V_R and V_C. The applied voltage drops from +5 volts to −5 volts, a drop of 10 volts, at the beginning of the second half cycle. This same drop is immediately seen across the resistor; therefore, at the beginning of the second half cycle, V_R drops 10 volts from +4.8 volts to −5.2 volts. We know that the sum of the voltage drops across the resistor and capacitor must equal the applied voltage at any instant. Thus, at the beginning of the second half cycle, when the capacitor is charged to 0.2 volt and $E = -5$ volts, we see that V_R must be −5.2 volts according to our equation. ($V_R + V_C = E$; −5.2 volts + 0.2 volt = −5 volts.) During the second half cycle, the capacitor discharges and charges again in the opposite direction; thus by the end of the second half cycle, $V_C = -0.2$ volt and V_R has gone to −4.8 volts. At the beginning of the third half cycle (or the beginning of the second full cycle), the applied voltage jumps to +5 volts, an increase of 10 volts. The voltage across the resistor (V_R) at the beginning of the third half cycle shows this same increase of 10 volts by jumping to 5.2 volts, 10 volts above its previous value of −4.8 volts. During the third half cycle, the capacitor discharges, then charges in the opposite direction to 0.2 volt while V_R drops to 4.8 volts by the end of the third half cycle. The fourth half cycle is a repeat of the second, the fifth like the third, and so on.

We have seen that during the first half cycle, voltage across the capacitor builds to 0.2 volt. During the second half cycle, the capacitor discharges, then begins charging in the opposite direction until V_C

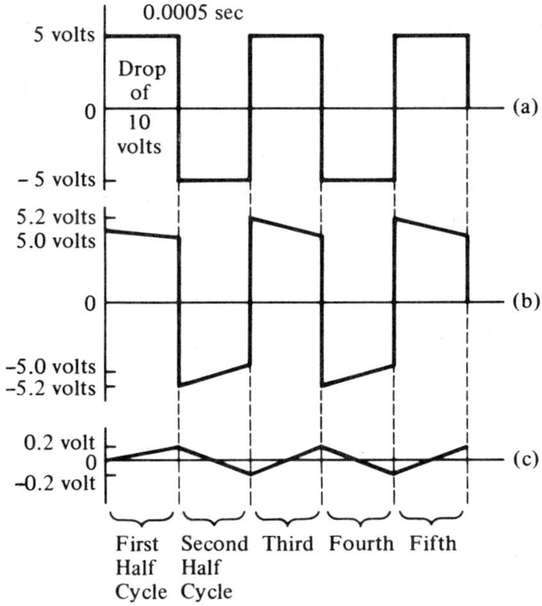

FIGURE 6.17 *(a) Applied voltage from square-wave generator; (b)* V_R*, voltage across resistor (differentiator output); (c)* V_C*, voltage across capacitor (integrator output).*

reaches -0.2 volt. You may have wondered how the time for the second half cycle, which is, of course, the same as the time for the first half cycle, could be adequate for this discharging and charging again. Remember that the time constant is used to measure the time necessary for all the applied voltage to be seen across the capacitor and is found by the equation 1 TC $= R \times C$. Thus, the value of the time constant depends on the values of resistance and capacitance. Resistance R is certainly the same in the first and second half cycles. To see that capacitance C is also unchanged in the second half cycle, we recall the relation $C = Q/V$. During the first half cycle, enough charge Q builds up on the capacitor to make V_C go from 0 volt to 0.2 volt. During the second half cycle, the voltage V_C changes by twice as much, from $+0.2$ volt to -0.2 volt, but the charge must also change by twice as much (the charge built up during the first half cycle must move to the opposite plate to discharge the capacitor, then an additional charge must build up on this opposite plate to charge the capacitor in the opposite direction). Thus, in the second half cycle, capacitance $C = 2Q/2V = Q/V$, which is the same as capacitance in the first half cycle. Doubling the charge Q and the voltage V leaves capacitance unchanged, and hence the time constant $(R \times C)$ does not change.

Let us compare the waveforms of the voltage from the square-wave generator and the voltage across the resistor. Notice that the input from the generator and V_R are nearly the same. (The graph of V_R was slightly exaggerated to show the small drop of 0.2 volt.) If the time constant had been even larger than ten times the period of the input wave, the voltage drop would have been less than 0.2 volt. For good coupling (minimum distortion) a time constant of at least ten times the period of the input signal is required.

6.5 TC SMALL IN COMPARISON TO T

Now let us consider the second alternative: The time constant is smaller than the period. (See Figure 6.18.) We see that in one ten-thousandth of a

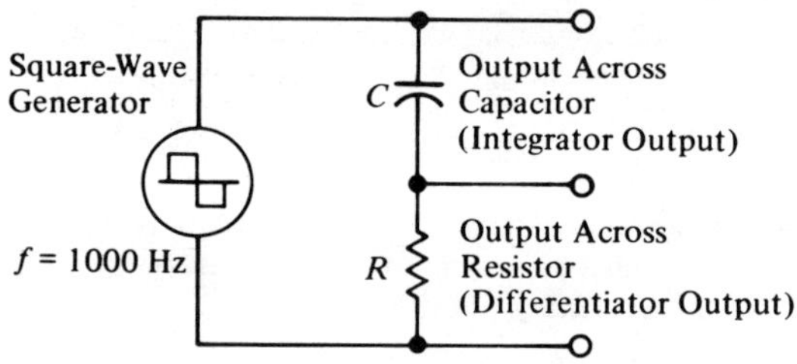

FIGURE 6.18 $E = 5$ *volts;* $C = 0.01\ \mu F = 0.01 \times 10^{-6}$ *farad;* $R = 10{,}000$ *ohms;* $f = 1000$ *Hz;* 1 $TC = R \times C = 10{,}000$ *ohms* $\times$ 0.01×10^{-6} *farad;* 1 $TC = 0.0001$ *second.*

second (1 TC = 0.0001 second) the voltage across the resistor will drop to 37 percent of its original value. This square-wave generator has the same frequency as the one in the preceding problem; we have already found its period T = $1/f$ = 1/1000 = 0.001 second.

The time constant is small in comparison to the period, thus giving the capacitor more than adequate time to charge and discharge. If there is plenty of time for this charging and discharging, the voltage across the resistor will fall or rise quickly, like a peak. Because of the waveform of the voltage across the resistor, this peaking, or differentiating, circuit is quite often used to start, or trigger, other circuits into operation; it is also used as a timer because it produces such sharp, well-defined pulses.

The ratio of time constant to period in this problem is TC/T = 0.0001/0.001 = 1/10; the time constant is one tenth as long as the period, a small enough ratio for a peaking circuit. The time for one-half cycle is $\frac{1}{2}$ T = $\frac{1}{2}$ (0.001 second) = 0.0005 second. Note that 0.0005 second is also equal to 5 TC (5 TC = 5 × 0.0001 second = 0.0005 second).

Now we will compare the waveforms for applied voltage, V_R, and V_C. Using the time constant graph of Figure 6.19, we see that V_R (voltage across the resistor) is at 5 volts just as the switch is closed and decreases to 0 volt after 5 TC or one-half cycle. At the same time, the capacitor is charging, and hence V_C increases from 0 volt to 5 volts. At the end of the first half cycle, the applied voltage drops 10 volts. (See

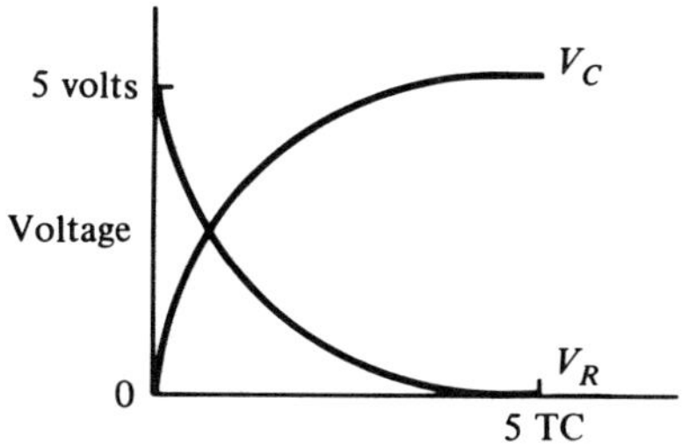

FIGURE 6.19

Figure 6.20.) As we have already seen, this drop is immediately observed across the resistor; therefore, at the beginning of the second half cycle, V_R drops from 0 volt to -10 volts. During the second half cycle, as we explained earlier, the capacitor discharges and charges again in the opposite direction; thus, by the end of the second half cycle, $V_C = -5$ volts and V_R has gone to 0 volt. At the beginning of the third half cycle, the applied voltage, and hence V_R, both increase 10 volts. Thus, V_R begins the third half cycle 10 volts above its previous zero value. During the third half cycle, the capacitor discharges, then charges in the opposite direction to 5 volts, while V_R drops to 0 volt. We see then that after the first half cycle, V_R changes in such a way that its peak value is double the supply voltage.

You have no doubt observed that in the last two problems, we have been studying the same circuit; we only changed the value of the time constant in the second problem. This circuit is called a differentiating circuit when its output is taken across the resistor. The name "differentiating" comes from the study of calculus and refers to a rate of change (V_R under certain conditions represents the rate of change of the input voltage).

The term integrating circuit (used when the output is taken across the capacitor) also comes from the calculus and refers to a sum of the input voltages. The integrator output (V_C) can be used to sort the input voltage when it is of varying frequency. If we examine the two waveforms for V_C, we see that when the input period is 10 TC or more, the output (V_C) has almost the same waveform as the input (Figure 6.20), but when

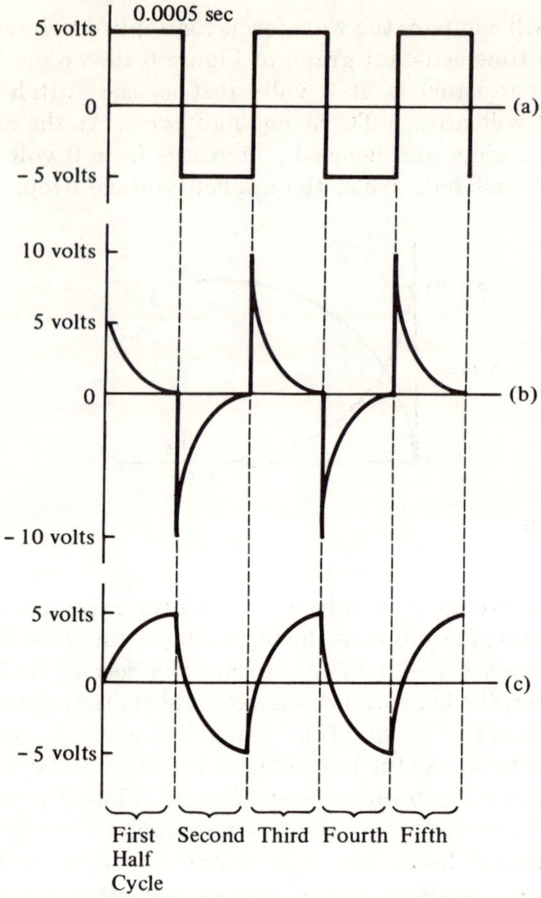

FIGURE 6.20 *(a) Applied voltage from square-wave generator; (b) V_R, voltage across resistor (differentiator output); (c) V_C, voltage across capacitor (integrator output).*

the input period is less than 1 TC, the output (V_C) is essentially zero (Figure 6.17), thus this input is eliminated.

6.6 WAVEFORMS FOR A SERIES RL CIRCUIT

Differentiating and integrating circuits can employ a coil and a resistor in series rather than the RC circuits we have just considered. Since coil and resistor circuits do not block the flow of dc, as do capacitors, they are not normally used for coupling. In the RL circuit in Figure 6.21, after the switch is moved to position *A*, time is required to build the magnetic

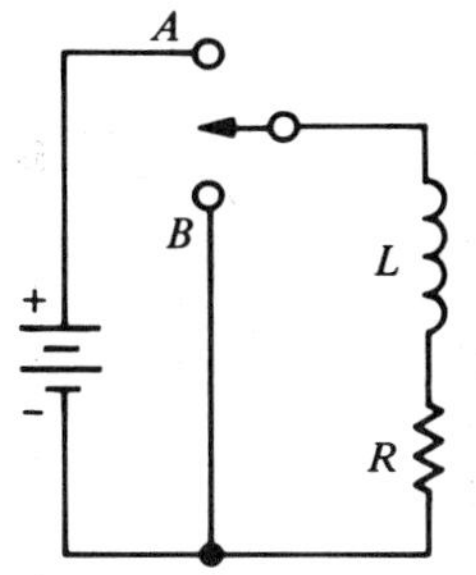

FIGURE 6.21

field about the coil. The time for the magnetic field to build to 63 percent of its maximum strength is also called one time constant (1 TC) and can be found by dividing the inductance of the coil by the resistance. 1 TC = L/R.* As we found with the RC circuit, the LC circuit requires five time constants (5 TC) to build, or to eliminate, a magnetic field about the coil. Current also builds to its maximum in the first 5 TC after the switch is closed.

Since $I = E/R$ and since current and voltage are in phase in a resistor, the instant switch A is closed, the voltage across the resistor begins building as more and more current passes through the resistor. Thus, voltage (V_R) has the same graph as current. (See Figure 6.22.) The sum of the voltage drops around a series circuit must equal the supply voltage. $E = V_R + V_L$. Since V_R is zero the instant the switch at A is closed, V_L must be a maximum. (See Figure 6.23.) As the voltage across the resistor (V_R) increases, the voltage across the coil (V_L) decreases. We can see this another way by recalling that as current increases, the magnetic field builds, "cutting" the coil and producing an induced voltage in the opposite direction to the applied voltage. As this induced voltage increases, the voltage drop across the coil, due to the supply voltage, decreases. (See Figure 6.24.)

Applying a square wave to the LC circuit, we consider the waveforms for voltage across the resistor (V_R) and voltage across the coil (V_L). (See Figure 6.25.) Notice that these waveforms are similar to those for the RC circuit, except the peaking wave is observed across the coil in the RL circuit rather than across the resistor as was the case in the RC circuit. Therefore, the differentiator output for the RL circuit will be across the coil. This circuit is quite often used as a timing circuit for cathode-ray tubes, such as picture tubes in a black-and-white TV, which employ electromagnetic deflection.

* It can be shown that inductance $L = Vt/I$. Dividing this by $R = V/I$, we have $L/R = (Vt/I)/(V/I) = t$ which is time.

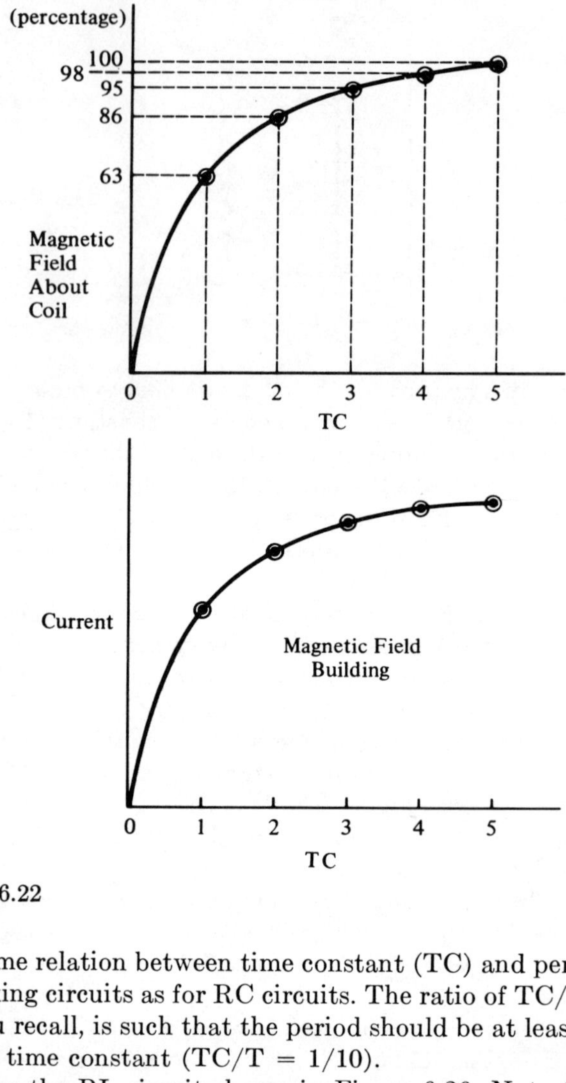

FIGURE 6.22

The same relation between time constant (TC) and period (T) exists for RL peaking circuits as for RC circuits. The ratio of TC/T for peaking circuits, you recall, is such that the period should be at least ten times as large as the time constant (TC/T = 1/10).

Consider the RL circuit shown in Figure 6.26. Note that TC/T = 0.001/0.02 = 1/20. The period is twenty times as large as the time constant; therefore, there will be more than enough time for the magnetic field to build or collapse. Remember in the earlier examples of peaking circuits, it was necessary to allow enough time for the capacitor to completely charge or completely discharge, whereas to obtain a coupling circuit or integrating circuit not enough time is allowed for the capacitor to charge and hence for the voltage across the resistor to decrease significantly.

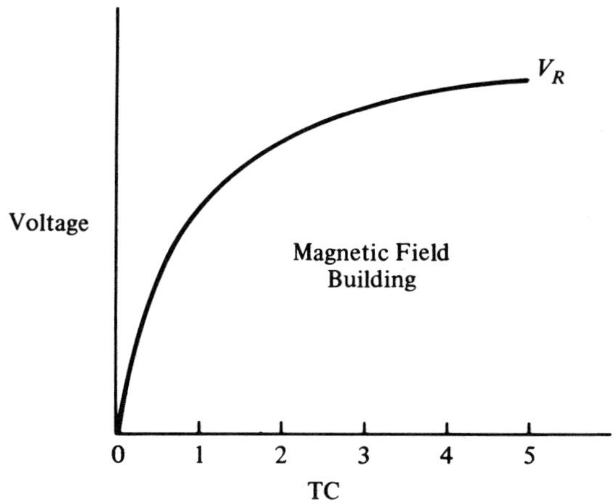

FIGURE 6.23

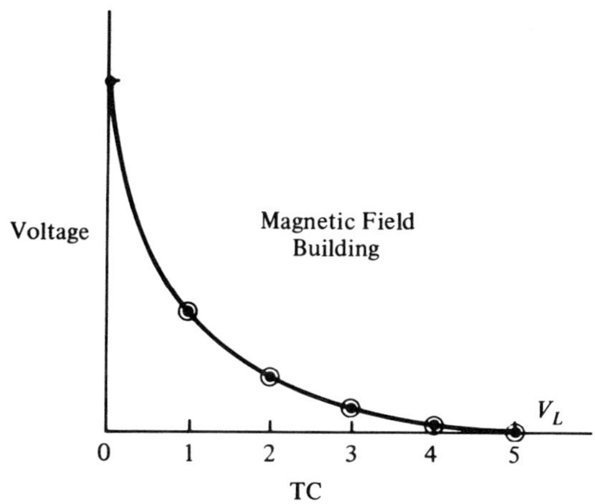

FIGURE 6.24

Continuing our problem, we consider the waveforms for the supply voltage and the voltages across the coil and resistor. Since the period (time for one cycle) of the supply voltage is T = 0.02 second, the time for one-half cycle is $\frac{1}{2}$ T = $\frac{1}{2} \times 0.02$ second = 0.01 second. Recalling that 1 TC = 0.001 second for this circuit and that 5 TC = 5×0.001 second = 0.005 second is needed to build the magnetic field, we see that one-half cycle (0.01 second) provides adequate time for this process. As a matter of fact, one-half cycle is actually twice as much time as is really needed to build the magnetic field (0.01 second = 2×5 TC = 2×0.005 second). Thus, our circuit is a peaking circuit.

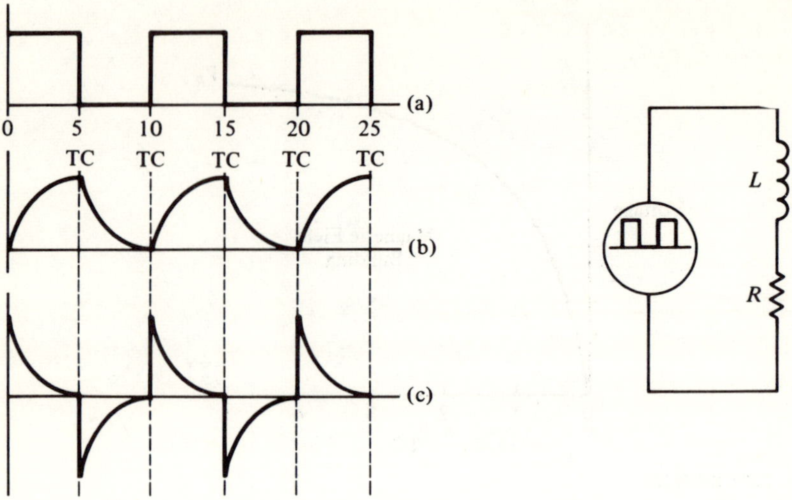

FIGURE 6.25 *(a) Applied voltage; (b) voltage across resistor,* V_R*; (c) voltage across coil,* V_L*.*

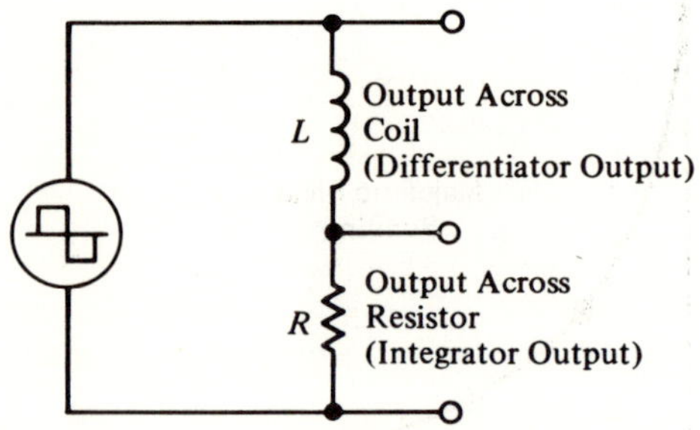

FIGURE 6.26 $E = 5$ *volts;* $R = 10{,}000$ *ohms;* $L = 10$ *henries;* $f = 50$ *Hz;* $T = 1/f = 1/50 = 0.02$ *second;* 1 $TC = L/R = 10$ *henries*/10,000 *ohms* = 0.001 *second.*

These waveforms (Figure 6.27) are derived by reasoning which is so similar to that used for RC circuits that we will not attempt a detailed explanation. You should observe that at any instant, the sum of V_R and V_L equals the supply voltage. Thus, at the beginning of the second half cycle, the supply voltage has dropped to -5 volts, a decrease of 10 volts. The voltage across the resistor (V_R) is 5 volts; thus, the entire voltage drop is seen across the coil and $V_L = -10$ volts. ($E = V_R + V_L$; -5 volts = 5 volts − 10 volts.) Notice also that after the first half cycle, the magnetic field about the coil collapses, then rebuilds with each change in current direction, that is, during each half cycle, just as in the RC cir-

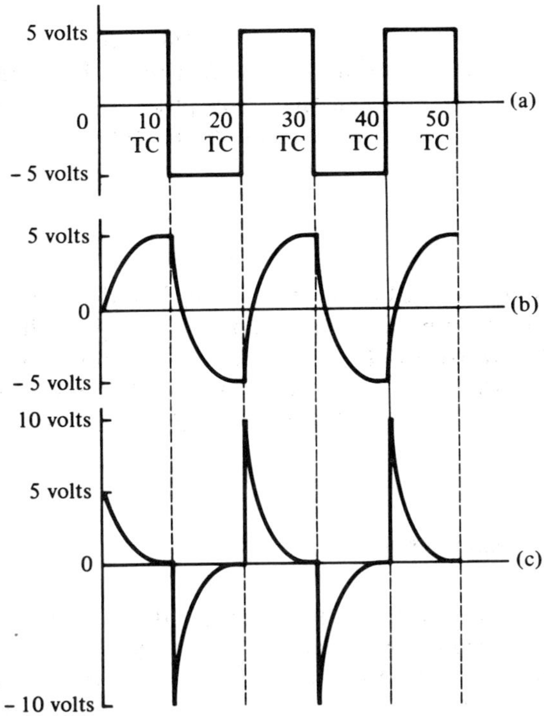

FIGURE 6.27 *(a) Applied voltage; (b) voltage across resistor, V_R; (c) voltage across coil, V_L.*

cuit, the capacitor discharged and charged in the opposite direction during each half cycle.

6.7 SUMMARY

When a signal is passed from one circuit to another, some form of coupling is required. Normally we like to pass the signal with a minimum of distortion. The RC circuit with its output across the resistor can accomplish this over a wide range of frequencies, as we will see when we discuss RC amplifiers. This coupling takes place with minimum distortion when the time constant (1 TC $= R \times C$) is at least ten times as large as the period (T $= 1/f$). In this case, there is little time for the voltage across the resistor to change and thus little distortion. On the other hand, peaking circuits occur when the period is ten times as large as the time constant, allowing ample time for the capacitor to charge and discharge in an RC circuit or for the magnetic field to build and collapse in an RL circuit (1 TC = L/R). For peaking circuits, the output is taken across the resistor in an RC circuit and across the coil in an RL circuit. An integrating circuit, an RC circuit with the output across the capacitor, can be used to sort frequencies for a varying input voltage.

Questions

(1) The time constant for an RC circuit is 1 TC = RC. What is the relation between the time constant and the period of the input, if the output across the resistor does not differ from the input wave?

(2) What is the relation between TC and T in a peaking circuit?

(3) Where is the output taken in an RC peaking circuit? An RL peaking circuit?

Suggested Experiments

(1) The charge and discharge of a capacitor may be observed with the aid of an oscilloscope. When the switch is in position A, the capacitor charges through the resistor R. When switched to B, the capacitor discharges through R. If the oscilloscope just indicates a sudden change when the switch is in position A, the time constant (1 TC = RC) is too short and may be lengthened by increasing either R or C. (See Figure 6.28.)

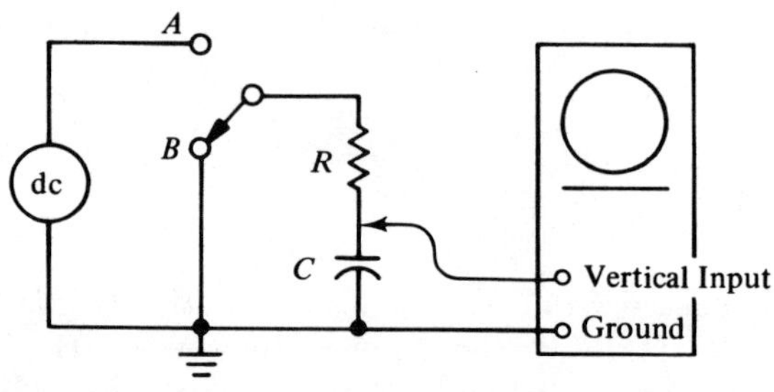

FIGURE 6.28

CHAPTER 7

Transformers

7.1 MUTUAL INDUCTION

We have observed that a coil of wire connected to a generator will induce in itself a current and voltage in opposition to the normal flow of current from the generator. This self-induced current and voltage is the product of the coil's magnetic field, which alternates as the original current and voltage from the generator alternate. If another coil of wire is placed near the first coil, there is the likelihood that the changing magnetic field around the coil attached to the generator will "cut" the coils of the wire in the second coil. Cutting these coils of wire with a changing magnetic field will induce a current and voltage in the second coil. The induction of current and voltage in a nearby coil is referred to as mutual induction.

7.2 TRANSFORMER SCHEMATIC

The device which utilizes this phenomenon, mutual induction, is the transformer. (See Figure 7.1.) It is important to note that there is no electrical connection between the coil attached to the generator (the primary coil) and the coil in which the current is induced (the secondary coil). This electrical isolation of the primary and secondary coils in a transformer is often used to protect the secondary. The principal use of the transformer, as the name implies, is to transform, that is, to increase (step up) or decrease (step down) current and voltage. Later we will show how the transformer is used to match impedances, for example to match the output impedance of an amplifier to the input impedance of a speaker.

The schematic for a transformer is shown in Figure 7.2. The lines between the primary and secondary coils, which appear in all but the first diagram, represent the iron cores on which the two coils are wound. Remember that the iron core tends to concentrate the magnetic lines. The current in the primary is alternating, but at one instant one end of the coil will be positive, the other negative; the same may be said of the

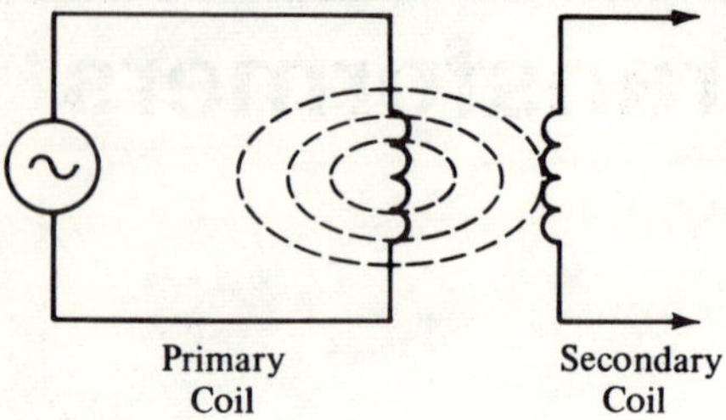

FIGURE 7.1 *Transformer.*

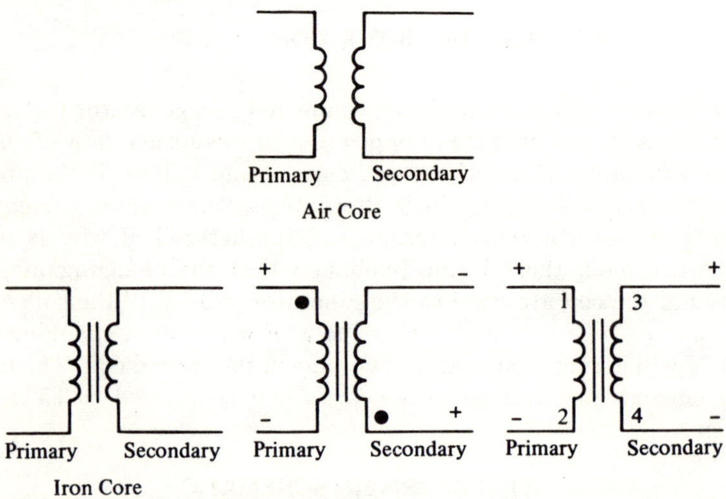

FIGURE 7.2

secondary. The dots in the third schematic are used to indicate the ends of the two coils that are in phase, that is, both positive (or both negative) at the same time; the ends with dots are said to have the same polarity. The numbers seen on the fourth diagram are just another method of indicating phase: odd numbers have the same phase. Notice also that the schematics are read from left to right, that is, the primary is always on the left side regardless of whether it has more or fewer turns than the secondary.

7.3 VOLTAGE IN THE TRANSFORMER

The changing magnetic field induces a voltage in the primary coil (self induction) and in the secondary coil (mutual induction) of the transformer. The magnitude of the induced voltage depends on the voltage supplied by the generator to the primary coil. Since there is no generator connected to the secondary coil, but instead just the magnetic link to the primary, the voltage induced in each turn of the secondary is the same as that induced in each turn of the primary. Thus, without its own voltage generator, the secondary coil has a voltage dependent upon the number of turns of the coil.

Let us follow through with this idea in a problem. An ac generator supplies 200 volts effective to the primary coil of a transformer. The primary has 100 turns and the secondary 60 turns. What voltage is induced in the secondary coil of the transformer? (See Figure 7.3.)

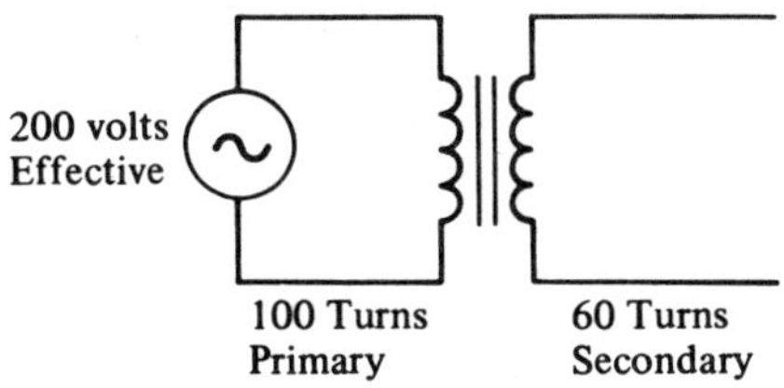

FIGURE 7.3

The voltage induced in just one turn of the primary can be found by dividing the total voltage by the number of turns in the primary coil. Voltage in 1 turn of primary = 200 volts/100 turns = 2 volts in each turn. This is the same voltage induced in each turn of the secondary coil. Since there are 60 turns in the secondary coil and each one has an induced voltage of 2 volts, the total voltage in the secondary would be 120 volts. (2 volts in each turn $\times$ 60 turns = 120 volts.)

Since the induced voltage in one turn of the primary equals the voltage in one turn of the secondary, we can write voltage of the primary (V_P) divided by the number of turns in the primary (N_P) equals voltage of the secondary (V_S) divided by the number of turns in the secondary (N_S). Voltage in one turn of primary $= V_P/N_P =$ voltage in one turn of secondary $= V_S/N_S$, or $V_P/N_P = V_S/N_S$. (In our problem, 200 volts primary/100 turns primary = 120 volts secondary/ 60 turns secondary = 2 volts in each turn.) If we want to rewrite this equation with both voltages on one side and number of turns on the other side, we can do so by first dividing both sides of the equation by V_S, then multiplying both sides by N_P. We will not change the equality as long as we do the same thing to both sides. Dividing by V_S, $V_P/N_PV_S = V_S/N_SV_S$ or $V_P/N_PV_S = 1/N_S$. Multiplying by N_P, $N_PV_P/N_PV_S = N_P(1/N_S)$. Therefore,

$$\frac{V_P}{V_S} = \frac{N_P}{N_S}$$

In the above problem, the voltage in the secondary (V_S = 120 volts) was smaller than the voltage in the primary (V_P = 200 volts). This is a step-down transformer.

Looking at another transformer problem, we will again determine the voltage in the secondary coil, this time by applying the above equation. Since we want to find V_S, we will solve our equation for V_S by multiplying on both sides by N_S. We have $V_S/N_S = V_P/N_P$. Multiplying, $V_SN_S/N_S = V_PN_S/N_P$ or $V_S = V_PN_S/N_P$. Now we are ready to substitute numerical values from the schematic. (See Figure 7.4.) $V_S =$

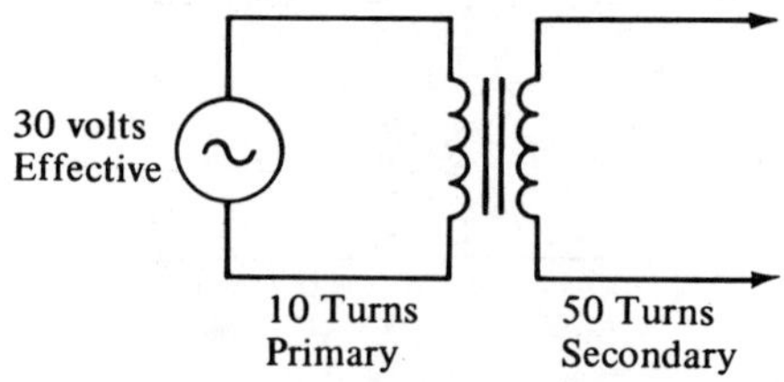

FIGURE 7.4

V_PN_S/N_P = 30 volts × 50 turns/10 turns = 150 volts. Since the voltage in the secondary is larger than the voltage in the primary, this is a step-up transformer. The voltage increased from 30 volts in the primary to 150 volts in the secondary.

Observe that when the number of turns in the secondary is less than the number of turns in the primary, the voltage in the secondary is also less than the voltage in the primary; when there are more turns in the secondary than in the primary, the voltage in the secondary is greater than the voltage in the primary.

7.4 CURRENT IN THE TRANSFORMER

When there is no load connected across the secondary coil, there is no path for current flow through the secondary circuit, and the changing magnetic field about the coils is produced entirely by the extremely small current in the primary coil. Note that we refer to this current as "extremely small." In fact, the inductive reactance of the primary coil is so large that the current flowing in the primary is practically zero; when the secondary circuit is open, there is essentially no current in the primary. If we now connect a load, for simplicity say a resistance, across the secondary coil, there will be a flow of current (I_S) in the secondary circuit. (See Figure 7.5.) With the secondary circuit closed, there is another

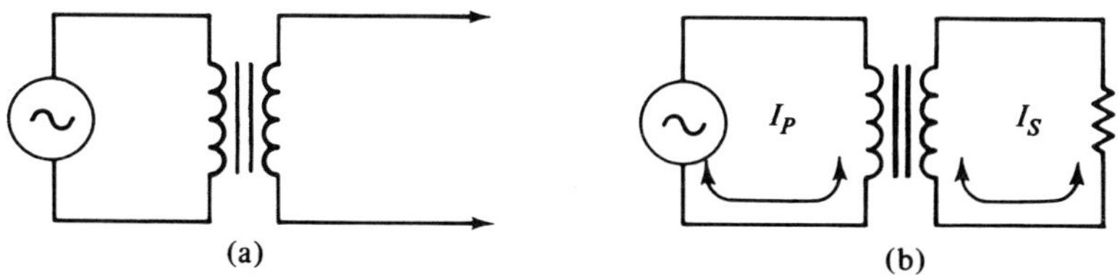

FIGURE 7.5 *(a) Secondary circuit open—no current in secondary, nearly zero current in primary; (b) secondary circuit closed—current in secondary and primary.*

changing magnetic field due to the current in the secondary coil. This field will induce a voltage in the primary coil, which is in the opposite direction to the primary's self-induced voltage and hence lessens the primary's opposition to current flow. Thus, the total effect of closing the secondary circuit is to reduce the inductive reactance of the primary coil and allow a much larger current (I_P) to flow in the primary. I_P is so much larger than the extremely small current mentioned earlier that it is taken to be the total current in the primary.

Now let us calculate these currents (I_P and I_S) in a transformer. In order to find current, we will return to an earlier concept of power. You remember that power was found to be the product of the voltage times the current in the circuit. $P = E \times I$. Power is a measure of how fast energy is used or stored in the circuit. While working with the transformer, we will assume that the power generated in the primary circuit is equal to the power used in the secondary circuit. In other words, we assume that no energy is lost in the transformation. Another way to express this assumption is to say that the transformer is 100 percent efficient. However, we would suspect that not all the magnetic lines are used in inducing the voltage in the secondary coil, and we know that the turns in the coils have resistance, which will dissipate some energy in the form of heat. Thus, the assumption that the transformer is 100 percent efficient is really just an approximation.

In the following examples, we assume that the transformer is 100 percent efficient, that is, we assume that the power of the secondary (P_S) is the same as the power of the primary (P_P). $P_P = P_S$. In the problem shown in Figure 7.6, there is a 25-ohm load resistance in the secondary circuit. Returning to the equation for voltage in the secondary, we find $V_S = V_P N_S / N_P = 200$ volts $\times$ 50 turns/100 turns $= 100$ volts. Applying Ohm's Law, we find current in the secondary $I_S = V_S/R = 100$ volts/25 ohms $= 4$ A. Now knowing both current and voltage in the secondary, we proceed to calculate power $P_S = V_S I_S = 100$ volts $\times$ 4 A $= 400$ watts. Remember that to use this equation for power, we must substitute effective current and voltage rather than peak values.

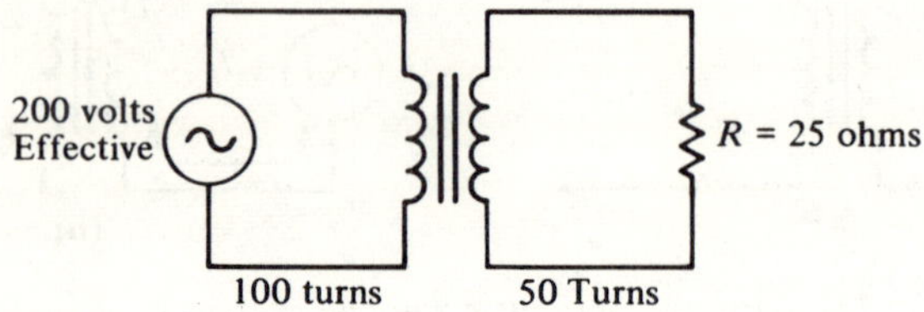

FIGURE 7.6

Using our assumption that the transformer is 100 percent efficient, we say that power in the primary is also 400 watts. $P_P = 400$ watts. Do not lose sight of the original objective to determine current in both the primary and secondary coils. We can now find I_P using the equation $P_P = V_P I_P$. This equation can be solved for I_P by dividing both sides of the equation by V_P to obtain $P_P/V_P = V_P I_P / V_P$ or $I_P = P_P/V_P$. Thus, $I_P = 400$ watts/200 volts $= 2$ A.

If we divide both sides of the equation $V_P I_P = V_S I_S$ by V_P and by I_S, it can also be rewritten as $V_P/V_S = I_S/I_P$. We already have a relation between voltages and number of turns, $V_P/V_S = N_P/N_S$, so we can now combine all three ratios.

$$\frac{N_P}{N_S} = \frac{V_P}{V_S} = \frac{I_S}{I_P}$$

Thus, besides the relation between current and voltage, we have a relation between current and number of turns ($I_S/I_P = N_P/N_S$), which can be written $I_P = N_S I_S / I_P$. Thus, in the same problem we can find I_P by this third method: $I_P = 50$ turns $\times$ 4 A/100 turns $= 2$ A.

Now let us consider a similar problem with a step-up transformer. (See Figure 7.7.) Using the equation relating voltage and number of turns, we find $V_S = V_P N_S / N_P = 10$ volts $\times$ 100 turns/20 turns $=$ 50 volts. We can apply Ohm's Law to determine current in the secondary. $I_S = V_S/R = 50$ volts/20 ohms $= 2.5$ A. Knowing one of the currents in the transformer, we find the other using the relation $I_P = I_S V_S / V_P$, obtained from the equation relating voltages, currents, and number of turns. $I_P = 2.5$ A $\times$ 50 volts/10 volts $= 12.5$ A. Observe that fewer turns in the primary ($N_P = 20$ turns) than in the secondary ($N_S = 100$ turns) implies a larger current in the primary ($I_P = 12.5$ A) than in the secondary ($I_S = 2.5$ A).

When there is a step-down in number of turns from primary to secondary, there is a corresponding step-down in voltage but a step-up in current from primary to secondary. When there is a step-up in number of turns, there is a step-up in voltage but a step-down in current.

This step-down in current in a step-up transformer is one of the principal reasons why the current and voltage supplied throughout the country is ac. The power lost, or dissipated, by heat in a wire due to its

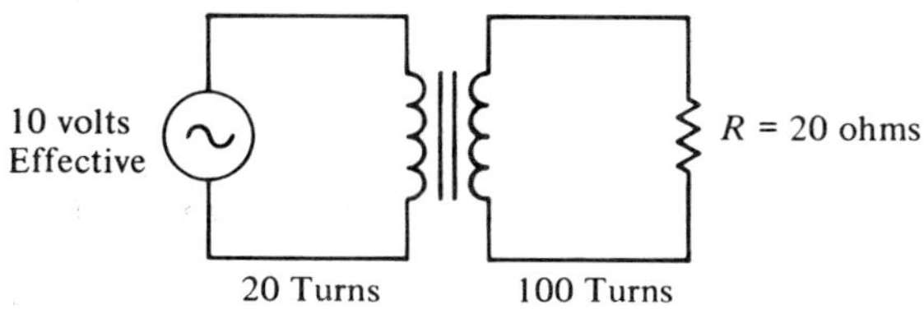

FIGURE 7.7

resistance R is $P = I^2R$. For transfer of electrical current over long distances, the resistance of the wire is necessarily large. Thus, to keep the power loss small and give greater efficiency, the current I in the wire must be small. This is accomplished in the secondary circuit of a step-up transformer. When more current is needed, such as in the home, a step-down transformer is used; this transformer reduces the voltage to 220 volts or 110 volts and, at the same time, increases the current.

7.5 RESISTANCE IN THE TRANSFORMER

Now let us return to the last two problems to illustrate one other idea: the reflection of the resistance of the secondary circuit, sometimes called load resistance or load, from the secondary to the primary circuit.

Return to the first circuit with the values previously calculated. The resistance of the secondary circuit (R_S) is 25 ohms, excluding the inductive reactance of the secondary coil. (See Figure 7.8.) If we apply Ohm's Law to the voltage and current in the primary, we find $R_P = V_P/I_P =$ 200 volts/2 A = 100 ohms. The 25-ohm resistance in the secondary, which increases the flow of current in the primary, also causes a resistance in the primary of 100 ohms; we say it reflects in the primary as 100 ohms. Comparing the ratio of these two resistances to the turns ratio we see they are not equal ($\neq$). 100 turns primary/50 turns secondary $\neq$ 600 ohms primary/125 ohms secondary. The resistance ratio (4 to 1) does not equal the turns ratio (2 to 1); however, if we squared the turns ratio, we

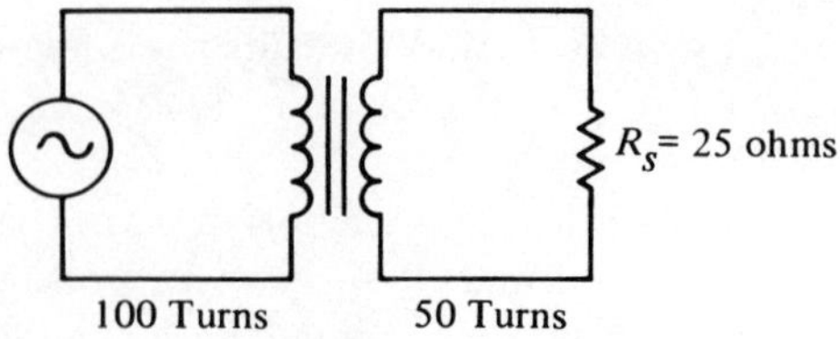

FIGURE 7.8 $I_P = 2$ *amperes;* $I_S = 4$ *amperes;* $V_P = 200$ *volts;* $V_S = 100$ *volts.*

would have $2^2/1^2 = 4/1$, which does equal the resistance ratio. In equation form, the resistance is reflected according to the ratios

$$\frac{N_P{}^2}{N_S{}^2} = \frac{R_P}{R_S}$$

Let us check this relation with the second circuit. (See Figure 7.9.) First, we find R_P using the equation $R_P/R_S = N_P{}^2/N_S{}^2$, which can be

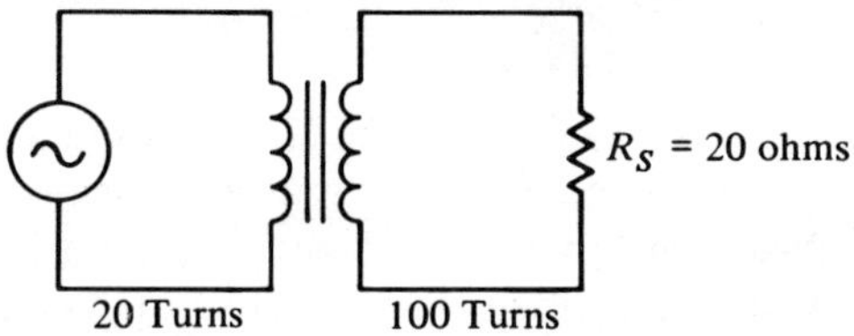

FIGURE 7.9 $I_P = 12.5$ *amperes;* $I_S = 2.5$ *amperes;* $V_P = 10$ *volts;* $V_S = 50$ *volts.*

written $R_P = N_P{}^2R_S/N_S{}^2 = (20 \text{ turns})^2 \times 20 \text{ ohms}/(100 \text{ turns})^2 = 0.8$ ohm. We verify this result for our example using Ohm's Law, $R_P = V_P/I_P = 10 \text{ volts}/12.5 \text{ A} = 0.8$ ohm.

We see that the same load resistance can be made to reflect different resistances to the primary simply by changing the turns ratio. For example, suppose we change the turns ratio in the above circuit from a ratio of 20 turns to 100 turns (1 to 5) to a ratio of 60 turns to 20 turns (3 to 1) and find the reflected resistance. (See Figure 7.10.)

$$R_P = \frac{N_P{}^2R_S}{N_S{}^2} = \frac{(60 \text{ turns})^2 \times 20 \text{ ohms}}{(20 \text{ turns})^2}$$

$$R_P = 180 \text{ ohms}$$

Note that without changing the secondary resistance R_S, the reflected resistance R_P has changed from 0.8 ohm to 180 ohms.

We will use this ability to change R_P by adjusting the turns ratio to match input and output impedances. Our concern will be to match impedances in such a way as to obtain high efficiency, that is, to avoid wasting

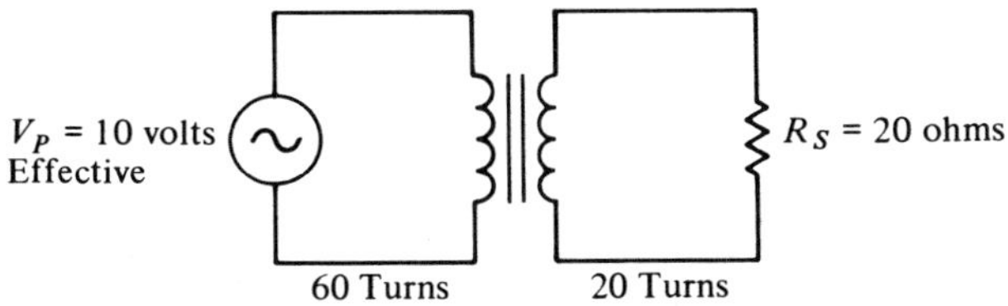

FIGURE 7.10

energy. To these ends we will briefly consider maximum power transfer, then return to study impedance matching in the transformer for maximum power transfer from one circuit to another.

7.6 MAXIMUM POWER TRANSFER

To determine when maximum power transfer occurs, we consider the following experimental circuit, calculations, and graph. The circuit (Figure 7.11) consists of an ac generator, with an effective voltage of

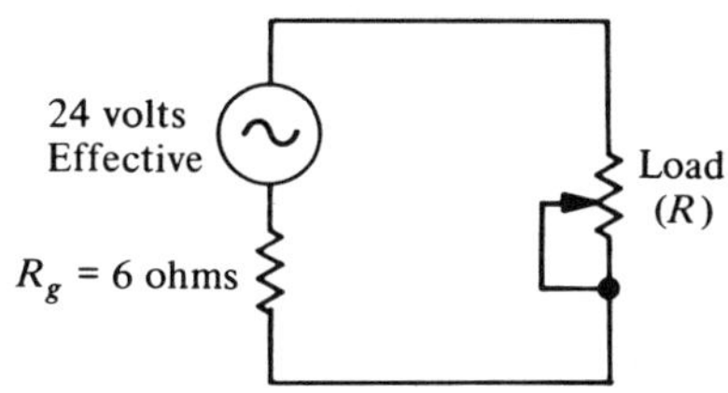

FIGURE 7.11

24 volts but with an internal coil resistance of 6 ohms, which is connected to a variable load resistor.

We now carry out a number of calculations and place them in Table 7.1.

TABLE 7.1

Load R	Total Resistance $R_T = R + R_g$	Current $I = E/R_T$	Current Squared I^2	Power to Load $P = I^2R$
0 ohm	6 ohms	4 A	16 A^2	0
2 ohms	8 ohms	3 A	9 A^2	18 watts
4 ohms	10 ohms	2.4 A	5.76 A^2	23 watts
6 ohms	12 ohms	2 A	4 A^2	24 watts
8 ohms	14 ohms	1.7 A	2.89 A^2	23.1 watts
10 ohms	16 ohms	1.5 A	2.25 A^2	22.5 watts
12 ohms	18 ohms	1.3 A	1.69 A^2	20.3 watts
14 ohms	20 ohms	1.2 A	1.44 A^2	20.2 watts
16 ohms	22 ohms	1.1 A	1.21 A^2	19.4 watts

After the calculations are complete, we plot the values of power (P) for different values of load resistance (R). This is done in Figure 7.12. Notice, looking at the table and at the figure, the largest value of power occurred when the resistance of the load $R = 6$ ohms. The generator supplied maximum power to the load resistance when the load resistance (R) was equal to the resistance of the generator (R_g). $R = R_g = 6$ ohms.

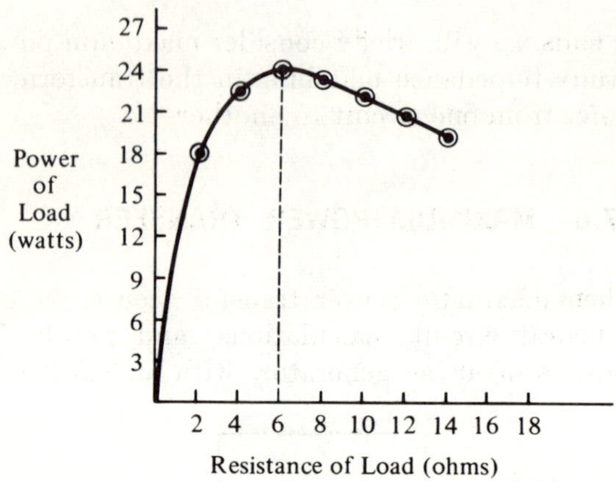

FIGURE 7.12

In this problem, the load resistance was variable. When neither the resistance of the generator nor the load resistance is variable, and yet maximum power transfer is desired, then a transformer is employed to match impedances.

7.7 IMPEDANCE MATCHING

Consider the following situation which requires an impedance matching transformer. (See Figure 7.13.) A speaker with a resistance of 16 ohms

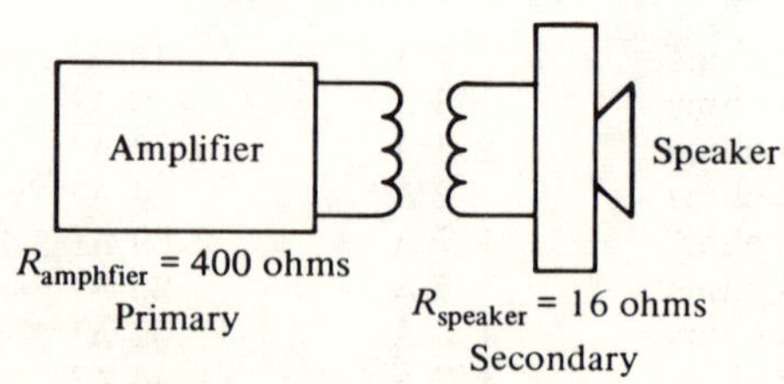

FIGURE 7.13

is to be connected to an amplifier whose output resistance is 400 ohms. The amplifier corresponds to the generator in our last problem, and the resistance of the amplifier corresponds to R_g. The speaker represents the load resistance. To obtain maximum power transfer (resistance of the amplifier, 400 ohms, must equal resistance of the load), we place a transformer between the amplifier and the speaker whose turns ratio is such that the resistance of the speaker (secondary circuit) is reflected to the amplifier (primary circuit) as a resistance of 400 ohms.

We can find the turns ratio of this speaker by using the equation $R_P/R_S = N_P^2/N_S^2$. We know $R_P = 400$ ohms and $R_S = 16$ ohms. From the above equation, $N_P^2/N_S^2 = 400$ ohms$/16$ ohms $= 25/1$. Now we can let the number of turns in the secondary (N_S) be equal to 1 since $N_S^2 = 1^2 = 1$. Thus, $N_P^2/1 = 25/1$. Now we must find N_P. We see that N_P is a number that, when squared, is equal to 25. Since we recall $5 \times 5 = 5^2 = 25$, then $N_P = 5$. ($N_P^2/N_S^2 = 5^2/1^2 = 25/1$.) Thus, the ratio of the number of turns in the primary (amplifier side) to the number of turns in the secondary (speaker side) is N_P to N_S or 5 to 1. For each turn in the secondary, there must be 5 turns in the primary.

If we want to be a little more mathematical in our solution, we can take the square root of both sides of the equation $N_P^2/N_S^2 = R_P/R_S$, which gives $\sqrt{N_P^2}/\sqrt{N_S^2} = \sqrt{R_P}/\sqrt{R_S}$ or $N_P/N_S = \sqrt{R_P}/\sqrt{R_S}$. Substituting resistance in the equation, we have $N_P/N_S = \sqrt{R_P}/\sqrt{R_S} = \sqrt{400}/\sqrt{16}$. To find the square root of 400 ($\sqrt{400}$), we look for a number which when squared (multiplied by itself) gives 400. Since $20^2 = 20 \times 20 = 400$, then $\sqrt{400} = 20$. Similarly, since $4^2 = 16$, then $\sqrt{16} = 4$. Hence $N_P/N_S = 20/4 = 5/1$, as found earlier.

If we wanted to attach a 2500-ohm microphone to the input of our amplifier, which we will take as 90,000 ohms, we would need another impedance matching transformer. (See Figure 7.14.) What would the turns

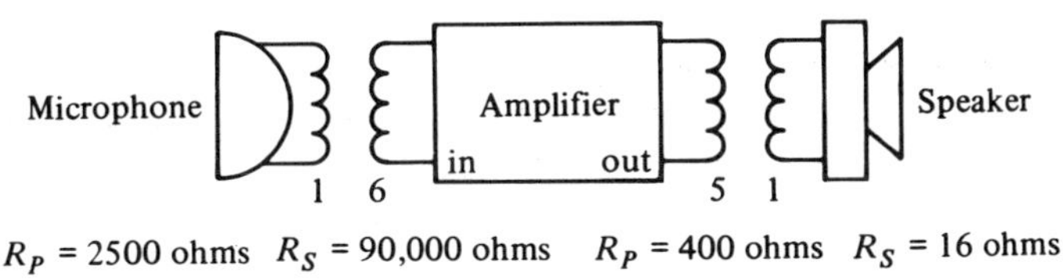

FIGURE 7.14

ratio of this transformer be for maximum power transfer? This time the microphone represents the primary resistance ($R_P = 2500$ ohms) and the amplifier input the secondary resistance ($R_S = 90{,}000$ ohms), in keeping with the convention of reading schematic diagrams from left to right. Applying the equation $N_P/N_S = \sqrt{R_P}/\sqrt{R_S}$, we have $N_P/N_S = \sqrt{2500}/\sqrt{90{,}000} = 50/300 = 1/6$. The ratio for the impedance matching transformer is 1 to 6; for each turn in the primary coil, there are six turns in the secondary coil.

7.8 DC BLOCKING

Note that a transformer is able to pass current from primary coil to secondary only because of the alternating magnetic field caused by an alternating current (ac). Thus, the transformer acts like a capacitor in a circuit in that it blocks the flow of dc. If the ac is at some dc potential in either a capacitor or transformer circuit, the blocking action will eliminate the dc and just pass the ac. (See Figure 7.15.)

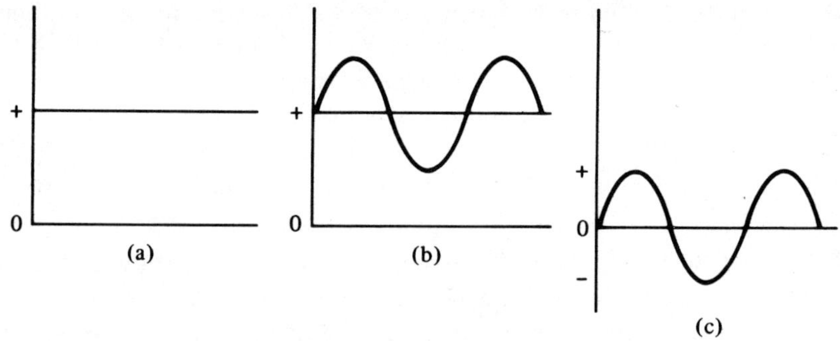

FIGURE 7.15 *(a) dc voltage or current; (b) ac at a dc level; (c) dc level removed by transformer or capacitor.*

Questions

(1) What is the relation between (a) voltage in the primary and secondary (b) current in the primary and secondary and (c) power in the primary and secondary, of a step-up transformer?

CHAPTER 8

Diodes

8.1 THE EDISON EFFECT

Most of you know the story of Thomas A. Edison, the Wizard of Menlo Park, and the invention of the incandescent light bulb. Edison found that when a thin carbon filament was placed in a glass envelope, the bulb evacuated, and a small voltage applied to the filament, the filament glowed and gave off light. An early problem with carbon filaments was that after they were in use for a short while, the inside of the glass envelope became coated with carbon black. Thus, the light bulb gave off less and less light the longer it was in use. One solution to this problem, Edison found, was to place a small metal plate inside the bulb and connect it to the positive terminal of the battery; the carbon black would then coat the metal plate and not the glass. (See Figure 8.1.) Later he discovered

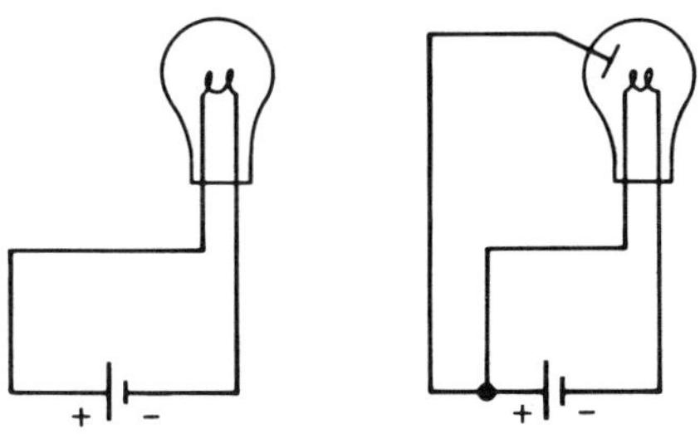

FIGURE 8.1

that the metal tungsten, when drawn into fine wire, made an excellent filament for his lamp and, of course, had the advantage of not giving off carbon black.

While working with his first solution for the carbon black problem, Edison observed by placing an ammeter between the metal plate and the positive terminal of the battery, that a small current flows while the filament glows. On the other hand, if the metal plate is connected through the ammeter to the negative terminal of the battery, no current flows between the metal plate and the negative terminal of the battery.

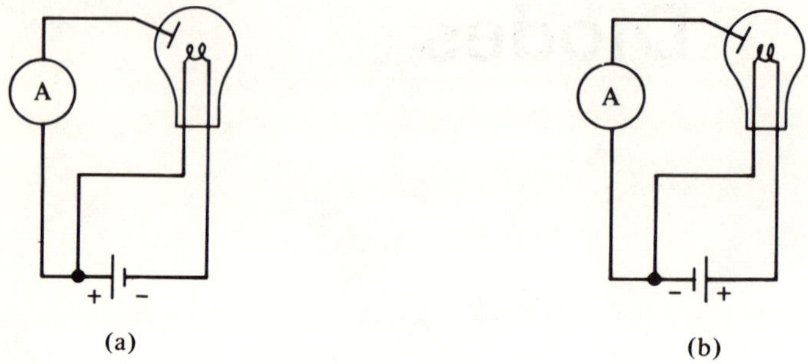

FIGURE 8.2 (*a*) *Current flow;* (*b*) *no current flow.*

(See Figure 8.2.) The observation of current flow with the metal plate connected to the positive terminal is called the Edison effect.

8.2 THERMIONIC EMISSION

The filament inside the evacuated glass bulb, whether it be carbon, tungsten, or some other material, offers enough resistance to the flow of electrons that, when it is attached to a battery, it gets hot enough to glow and hence give off light. It not only gives off light, it also gives off electrons. The filament is heated until it actually "boils off" electrons much like boiling water. This "boiling off" of electrons is referred to, in scientific terms, as thermionic emission and is just one of several ways in which electrons may be obtained. When light strikes certain materials, electrons are given off; this phenomenon is called photoelectric emission. High energy electrons striking a metal surface sometimes transfer enough energy to the metal surface to enable surface electrons to be emitted; this is called secondary emission.

We see then that thermionic emission, the boiling off of electrons from the filament, was taking place in Edison's light bulb. And we should realize that, although carbon black molecules were attracted to the metal plate, the Edison effect was caused by electrons from the filament moving to the positive metal plate and from there to the positive terminal of the battery, for it is the flow of current (the flow of electrons) which the ammeter measures.

8.3 VACUUM TUBE DIODES

The evacuated envelope contains two distinct objects and hence is referred to as a diode or, more precisely, as a vacuum tube diode for there are now many different types of diodes. The source of electrons, which includes the filament, is called the cathode. The metal plate of the vacuum tube diode is properly called the anode but popularly is just referred to as the plate. In the cathode, the filament may be the direct source of electrons, as in the bulb described above, or it may be placed in a metal sleeve or cylinder and be used to heat the sleeve, which then gives off the electrons. For this reason, the filament is commonly called the heater. This metal sleeve is an indirectly heated emitter of electrons, and, because of its larger surface area, it is a better supplier of electrons than a directly heated emitter like the filament. However, it does initially take longer to heat the metal sleeve and obtain electrons from it than to obtain electrons directly from the filament. The schematic of the vacuum tube diode shows the isolation of anode and cathode in the envelope; actually as you see in Figure 8.3, the anode or plate surrounds the cathode. As the cathode

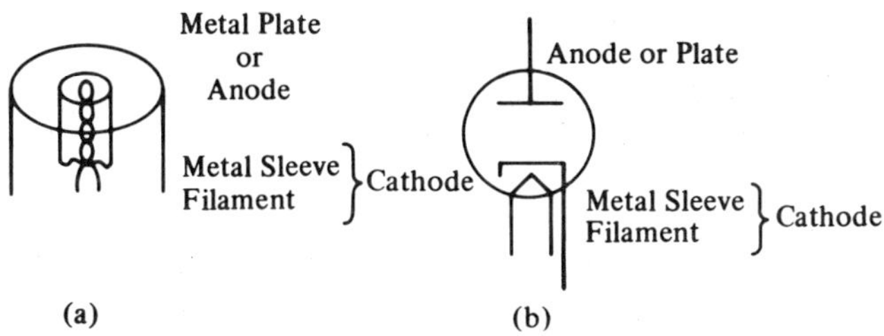

FIGURE 8.3 *(a) Cut-away picture of diode; (b) schematic of diode.*

temperature increases, more electrons are emitted. In order to increase the temperature of the cathode, the filament voltage must be increased.

The flow of thermionically emitted electrons from the indirectly heated metal sleeve depends on the voltage or potential between the anode and cathode. (See Figure 8.4.) If the anode or plate is negative with respect to the cathode, as is the case in Figure 8.5 where the plate is connected to the negative end of a battery, the cathode to the positive end, then the electrons emitted from the cathode will be repelled by the plate (from the law of electrostatics that like charges repel each other). Thus, the thermionic electrons will have nowhere to go; they will build up a negative space charge between the hot cathode and the negative plate.

If we remove the battery between the plate and cathode and replace it with a wire, then we observe a very feeble current flow from cathode to

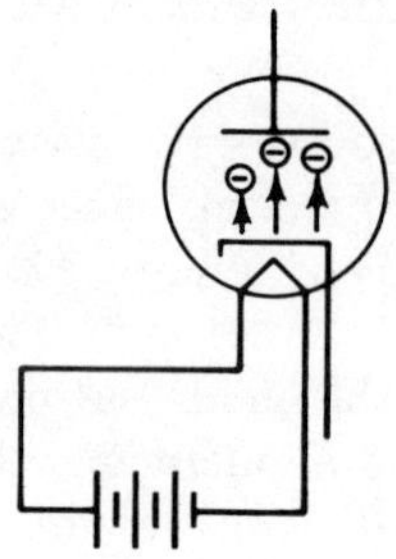

FIGURE 8.4 *Electrons from cathode.*

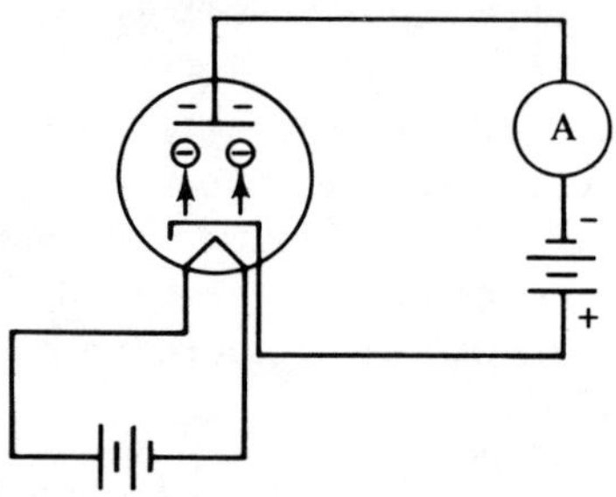

FIGURE 8.5 *Negative potential on plate with no current flow. (Note: Negative plate electrons form negative space charge.)*

plate. The reason the current flow is so weak is that the thermionic emission does not supply the majority of electrons with enough energy to reach the anode; only a few of the "boiled off" electrons have the energy necessary to travel to the anode, thus causing the weak current. (See Figure 8.6.)

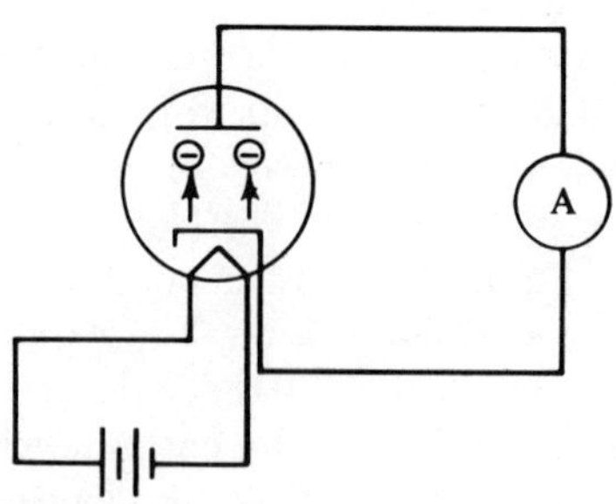

FIGURE 8.6 *Zero potential on plate with little or no current flow.*

If we now place the battery back in the plate circuit but this time in such a way as to make the anode positive with respect to the cathode, we observe a much larger current flow. (See Figure 8.7.) This time, since the plate is positive, the electrons are attracted to it (unlike charges attract).

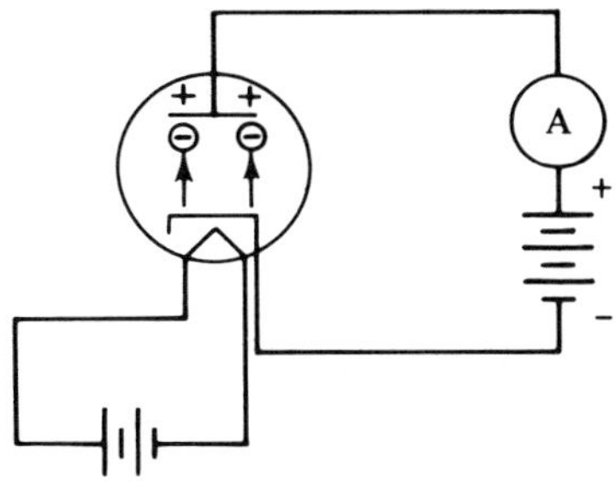

FIGURE 8.7 *Positive potential on plate with current flow.*

If the battery in the plate circuit is increased in size, the current through the ammeter will increase, since the greater plate potential attracts more electrons. There will finally be a point, when the battery in the plate circuit or the plate potential is made large enough, at which all of the electrons emitted at the cathode are collected at the plate. This point is called saturation. Any increase in plate voltage beyond saturation will not increase plate current. The relation between plate voltage and plate current for different filament voltages is shown in Figure 8.8.

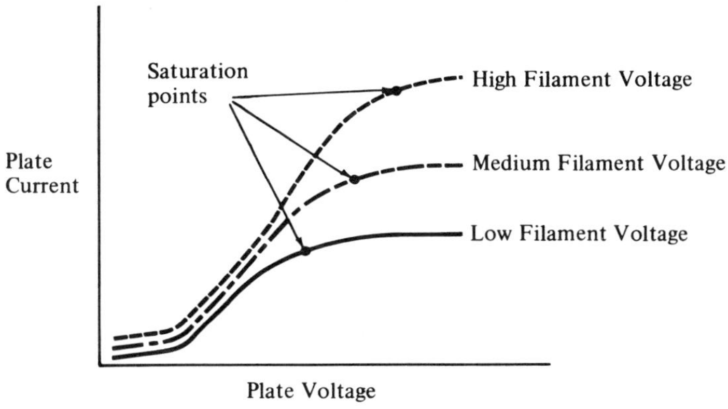

FIGURE 8.8 *(Note: Fewer electrons emitted from cathode at low temperature.)*

The saturation point can be found on the graph by looking for the point at which the curve ceases to rise (the current stops increasing). The diode normally operates well below saturation. We can understand why filament voltage affects the graph of plate voltage and plate current, if we recall that the higher the filament voltage, the more electrons are "boiled off."

With more electrons available to the plate at higher filament voltages, more current can flow before all the electrons are being used (that is, before saturation).

In the vacuum tube diode, the flow of electrons is from cathode to plate, never from plate to cathode. Thus, the diode allows the flow of electrons, or current, in only one direction. The flow of electrons occurs when the plate is positive with respect to the cathode, never when the plate is negative. The current or flow of electrons then can be turned on or off by changing the polarity of the plate voltage. For this reason, the two-element diode is sometimes referred to as a valve. Because of this switching on and off, the diode, or valve, also acts as a variable resistor. When there is current flow, the resistance of the diode is small; when the plate becomes negative and there is no current flow, the resistance of the diode can be thought of as being very, very large. In addition, there is an extremely large resistance to the flow of current from plate to cathode. Before we look at some applications of diodes, such as rectifiers and detectors, we will consider a few other types of diodes besides the vacuum tube diode just mentioned.

8.4 METALLIC DIODES

All materials may be roughly divided into three categories with respect to their conductivity of electrons: There are good conductors, semiconductors, and poor conductors or insulators. Metallic diodes consist of two different materials placed side by side. One of the materials is a better conductor of electrons than the other. In the metallic diode, the flow of electrons, or current, is from the better conductor to the poorer conductor; current will not flow in the other direction, so the metallic diode has many of the properties of the vacuum tube diode. The schematic of the metallic diode is deceiving in that the arrow (-▷) representing the anode points in the opposite direction to the flow of electrons. (See Figure 8.9.) To briefly

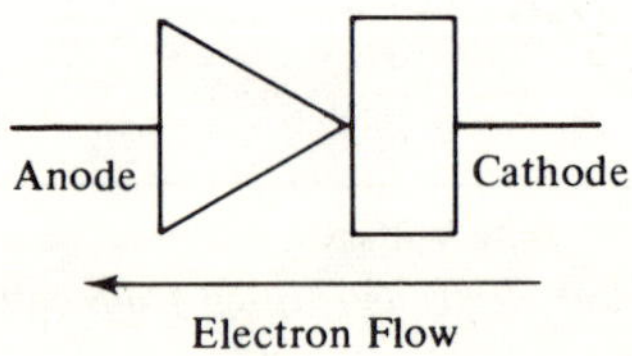

FIGURE 8.9

explain this, in the past current was thought to flow from positive to negative. Thus, the anode arrow pointed in the direction of conventional current flow rather than electron flow. In our discussion, we always consider current to be in the same direction as electron flow, from negative to positive.

Disc type diodes, often referred to as rectifiers, consist of alternating discs of good and poor conductors. The discs often have a hole drilled through their centers so they may be held together by an insulated bolt. Two common types of disc diodes are: (1) the copperoxide rectifier consisting of copper and copperoxide discs, and (2) the selenium diode made up of selenium and iron discs. (See Figure 8.10.) Metallic diodes, such as

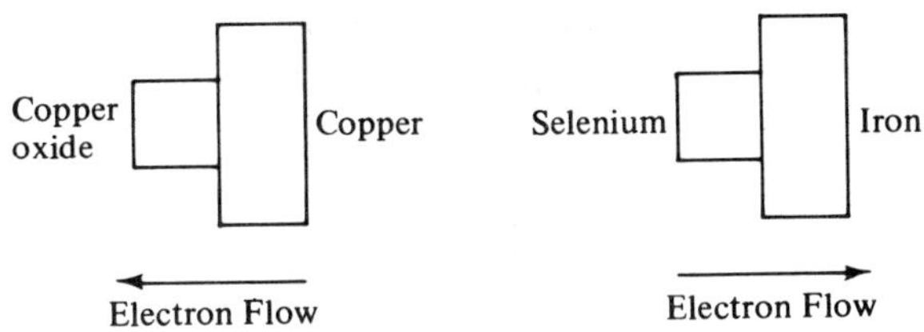

FIGURE 8.10

these, are employed in half-wave, full-wave, and bridge rectifiers, in battery chargers, electroplating, and arc welding, not to mention their use as rectifiers in electronic instruments from radio to radar. The basic disc type diode, shown above as consisting of two unlike discs, is called a cell and may be combined with others to increase the voltage rating of the rectifier. Some of the discs are larger; these act as cooling fins in order to allow larger currents to flow.

Diffused junction diodes are another form of metallic diode. The germanium–indium junction diode consists of pellets of germanium and indium melted together. The germanium, being a better conductor than the indium, again allows electrons to flow in only one direction. The indium may be replaced by a pointed wire touching the surface of the germanium. This is called a point contact diode. (See Figure 8.11.)

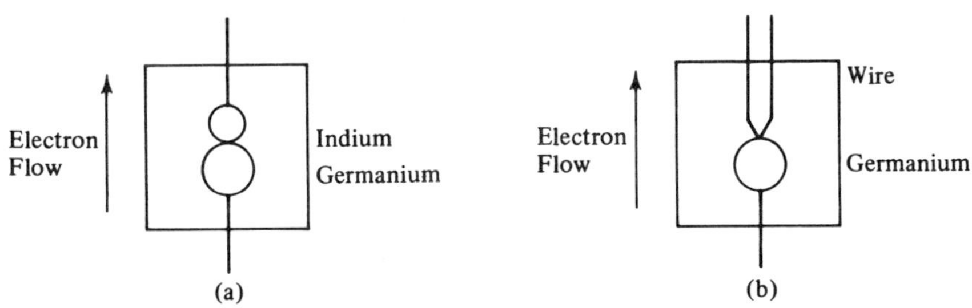

FIGURE 8.11 *(a) Diffused junction diode; (b) point contact diode.*

Each of the metallic diodes has a preferred operating range. The point contact diode is limited to currents as small as 60 mA (0.06 A). The diffused junction diodes, such as the germanium–indium diode, can handle currents up to 1 A; and silicon–iron junction diodes up to 250 A, when properly cooled. In the design of an electronic circuit to do a specific job, the current and voltage requirements of the components are calculated, then the proper component is chosen. Books containing listings of vacuum tubes and transistors presently available, along with descriptions of their characteristics, can be obtained from the manufacturer. Lists of transformers, coils, capacitors, resistors, and so on, may also be obtained from the manufacturer or from a retail catalog.

8.5 COVALENT BONDING

Earlier when we briefly discussed atomic theory, we indicated that electrons travel about the nucleus of an atom in rather well-defined orbits, much as planets travel about the sun. The number of electrons preferred in each orbit is well established, and an atom which does not have the desired number may share electrons with other atoms. This sharing does not affect the charge of the atom, but it does, in a sense, give the atom a complete outer orbit and thus more stability.

Germanium (Ge) has four electrons in its outer orbit, as does selenium, and needs four more electrons for a complete outer orbit containing eight electrons. In order to obtain this stable state, a germanium atom will share one of the outer electrons of each of four surrounding germanium atoms in a germanium crystal. This sharing of outer, or valence, electrons is called covalent bonding and can be shown in Figure 8.12. Notice

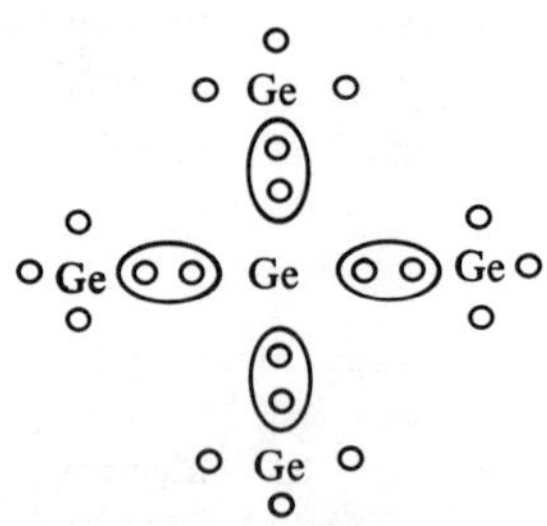

FIGURE 8.12 *o—outer electrons; covalent bonds are circled.*

that the center germanium atom shares a pair of electrons with each of the four surrounding atoms; thus the central atom essentially has eight electrons in its outer orbit.

You may find this covalent bonding somewhat disturbing in that we have studied and applied the law of electrostatics, which tells us that like charges repel each other. How then can two negatively charged electrons form a stable bond? The electrons orbiting the nucleus also rotate or spin on their own axis, just as the earth spins on an axis through its poles. The spinning electron with its negative charge produces a magnetic field. In the covalent bond, the magnetic attraction between the two electrons in the pair is greater than their electrostatic repulsion.

8.6 *N*-TYPE SEMICONDUCTOR

If the central germanium atom is replaced by phosphorus (P), arsenic (As), or antimony (Sb), each of which has five valence electrons, the covalent bond can be pictured as in Figure 8.13. Notice that four of the

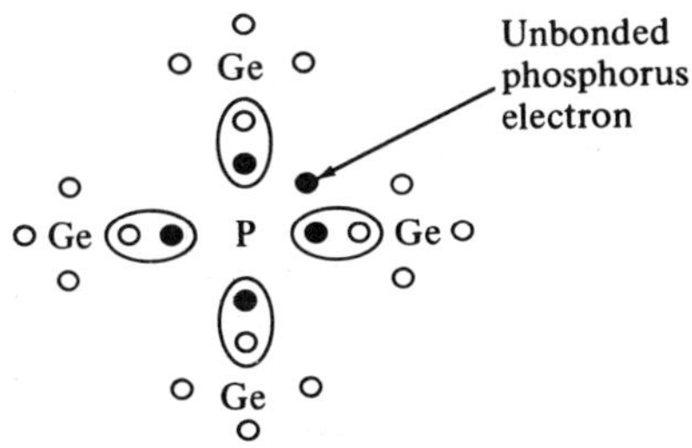

FIGURE 8.13 *N-type semiconductor.*

phosphorus valence electrons form covalent bonds with the four surrounding germanium atoms, but the fifth phosphorus electron does not have a valence germanium electron with which to bond. Because of this unbonded electron, the phosphorus atom is called a donor atom. At room temperatures, the unbonded electron possesses enough energy to break away from the donor atom. Thus, the donor atom contributes, or donates, a free electron to the germanium crystal. This leaves the donor atom positively ionized since it has one more proton than electrons. In comparison to the large number of germanium atoms, the donor atoms are few in number. In fact, they are thought of as an impurity in the germanium crystal. The entire germanium crystal, containing its donor impurities, has a net charge of zero; however, since the unbonded electron is relatively free to leave its parent atom, the material is called *N*-type semiconductor, the letter *N* representing the negatively charged electron supplied by the donor atom.

8.7 *P*-TYPE SEMICONDUCTOR

Now let us replace the donor atom, which had five valence electrons, with an atom such as boron (B), gallium (Ga), or indium (In), each of which has three valence electrons. (See Figure 8.14.) Notice that each of

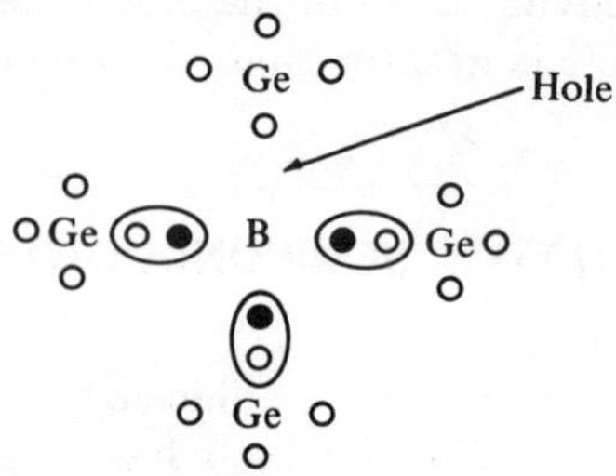

FIGURE 8.14 *P-type semiconductor.*

the three valence electrons forms an electron pair bond with three of the four surrounding germanium atoms; there is no electron available for the fourth germanium atom. Remember that four covalent bonds are needed to make the outer orbit complete with eight electrons. The "missing" electron creates what we call a hole or an electron–hole pair. Since this hole is essentially a space where another electron could go or be accepted, these atoms with three valence electrons are called acceptor atoms. The number of acceptor atoms is small and they, like the donor atoms, are said to be an impurity. A germanium crystal with acceptor atom impurities has a net charge of zero. If one of the electrons around an acceptor obtains enough energy to break away, leaving another hole, it positively ionizes its parent acceptor; this free electron may fill the hole in another acceptor atom, negatively ionizing it. Note that although the electron actually moves from the first acceptor atom to the second, the hole also appears to move from the second atom to the first. (See Figure 8.15.)

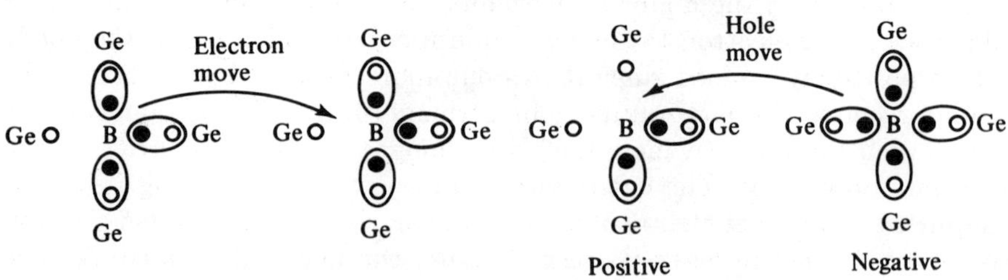

FIGURE 8.15

The positive charge of the ionized parent acceptor and the negative charge of the nearby acceptor still leave a net charge of zero. The positively ionized acceptor is the source of the name *P*-type for the germanium acceptor semiconductor.

8.8 *PN* JUNCTION DIODES

Fusing the two types of semiconductors just discussed, *N*-type and *P*-type, forms a *PN* junction diode. This diode has the same schematic as the metallic diode. (See Figure 8.16.)

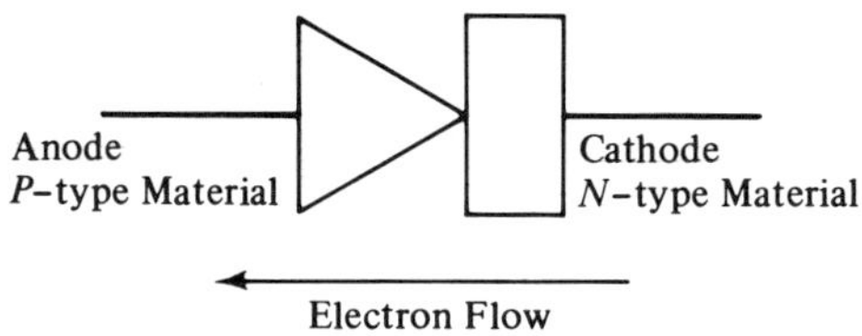

FIGURE 8.16

When the *PN* junction is fused, some of the higher energy free electrons from the donor atoms in the *N*-type semiconductor will flow toward the holes in the acceptor atom of the *P*-type semiconductor. This leaves part of the *N*-type positive (positive charge barrier) and makes part of the *P*-type negative (negative charge barrier). The two barriers form the so-called depletion zone. (See Figure 8.17.)

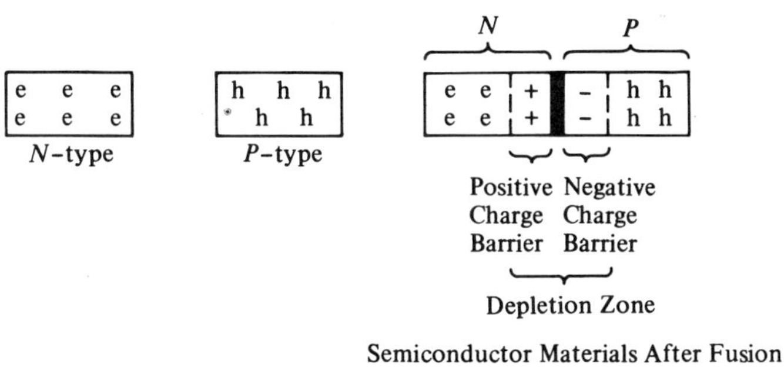

FIGURE 8.17 (*Note: e's represent free electrons, not a net negative charge; h's represent holes.*)

If we connect leads from a battery to the ends of the *PN* junction diode in such a way that the negative terminal of the battery is connected to the *N*-type semiconductor and the positive terminal to the *P*-type semiconductor, we have what is called forward bias. (See Figure 8.18.)

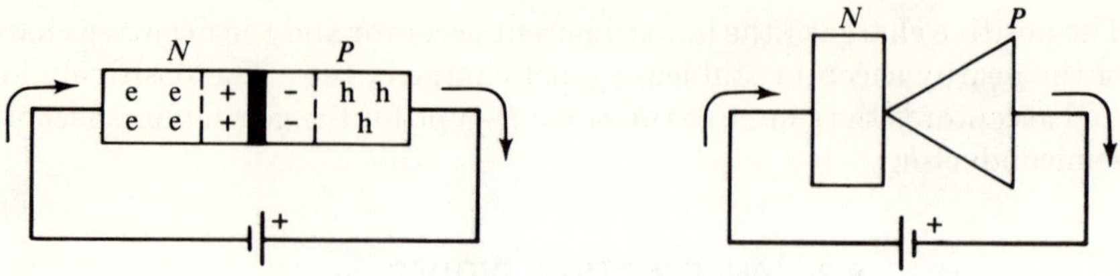

FIGURE 8.18 *Forward biased PN junction diode.*

The battery will inject an electron into the *N*-type semiconductor. This extra electron repels *N*-type free electrons, already present in the diode, toward the positive charge barrier of the depletion zone. Meanwhile the battery is also removing an electron from the *P*-type semiconductor. This leaves a positively ionized acceptor atom (with two holes) in the *P*-type semiconductor. The extra hole "moves" (by the mechanism discussed earlier) toward the negative charge barrier of the depletion zone. Thus, the extra electron in the *N*-type semiconductor and the extra hole in the *P*-type semiconductor both are forced toward the junction, where they combine. One electron has entered the *N*-type, one electron has left the *P*-type; thus the net charge of the diode is still zero, and current has flowed from *N* to *P*.

As the forward bias voltage from the battery increases, the flow of current increases as shown in Figure 8.19. Before any voltage is applied,

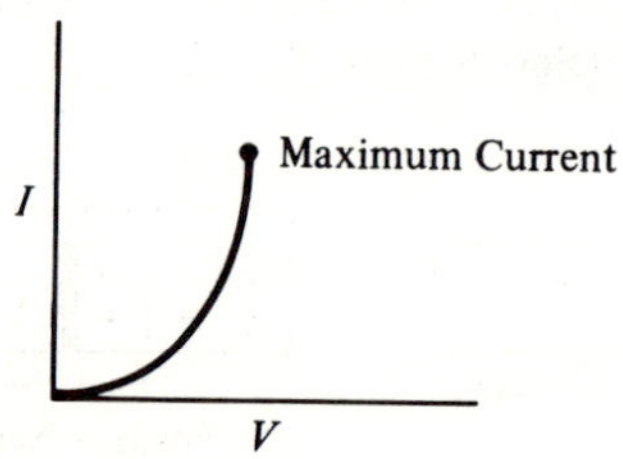

FIGURE 8.19 *Voltage-current characteristic of PN diode.*

there is no flow of current in the diode. As the voltage increases, more electrons receive enough energy to cross the junction and combine with a hole. There is a maximum current flow indicated on the graph. The amount of impurity in the germanium limits the flow of current in the *PN* diode, that is, the total number of acceptors or donors present in the crystal limits the rate of combining at the junction and hence limits the current flow.

If we reverse the leads from the battery, the bias is reversed. (See Figure 8.20.) The battery now attracts electrons in the *N*-type semicon-

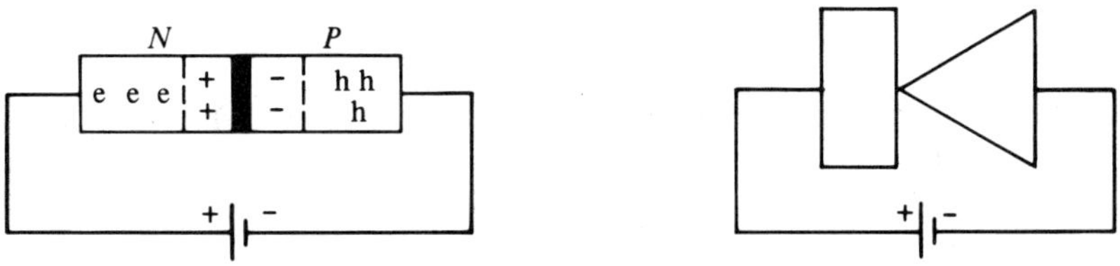

FIGURE 8.20 *Reverse biased PN junction diode.*

ductor away from the junction, thus making the positive charge barrier larger. At the same time, the battery injects electrons into the *P*-type semiconductor; these electrons fill holes in the acceptor atoms, thus making the negative charge barrier larger. Thus, with reverse bias the depletion zone increases in size. (See Figure 8.21.)

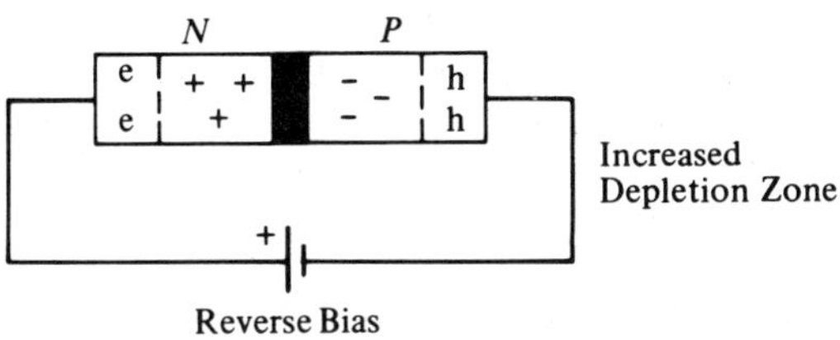

FIGURE 8.21

Therefore, when the bias is reversed in the *PN* junction diode, the net effect is an increase in the size of the depletion zone until all the holes in the *P*-type semiconductor are filled with free electrons taken from the *N*-type semiconductor. This takes place in a relatively short period of time and is analogous to charging a capacitor. It would appear that, once this process was complete, there would be no current flow. However, there will continue to be a small current flow. At room temperature (70°F), many loosely bound electrons, and a few more tightly bound electrons, are freed by heat energy. Heat energy will produce a few holes in the *N*-type material, and it will liberate a few electrons in the *P*-type material. These *N*-type holes and *P*-type electrons are called minority carriers. (See Figure 8.22.) With the injection of an electron in the *P*-type, the heat-liberated minority electron will flow toward the junction. The removal of an electron from the *N*-type will cause the heat-generated minority hole to move toward the junction, by the mechanism already discussed, where it combines with the minority electron. Thus, the minority *P*-type electrons and *N*-type holes allow a small current to flow under reverse bias by the same process as the *N*-type electrons and *P*-type holes allowed the larger current to flow under forward bias. The very small reverse bias current in the *PN* junction diode is limited by the small number of heat liberated minority carriers (electrons and holes).

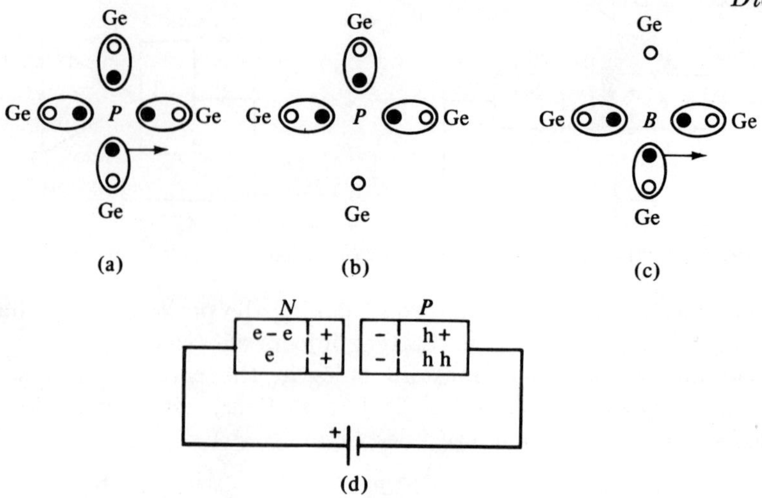

FIGURE 8.22 *(a) Electron liberated by heat energy produces hole in N-type; (b) hole in N-type; (c) electron liberated by heat energy produces free electron in P-type; (d) reversed bias showing minority carriers.*

The current–voltage graph of the *PN* diode, showing both reverse and forward bias, is given in Figure 8.23. Notice that both current and

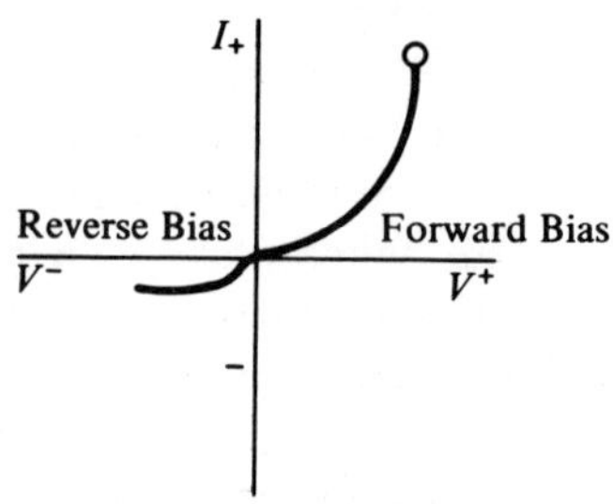

FIGURE 8.23

voltage are negative for reverse bias, since they are both in the opposite direction to current and voltage for forward bias. Observe that making the reverse voltage larger (in the negative direction) does not increase the reverse current. This is because the small number of minority carriers limits the reverse current.

8.9 ZENER DIODE

If we continue to increase the reverse voltage, or potential, across the *PN* junction diode, this increased reverse bias will finally break down more covalent bonds, thus liberating more minority carriers (electrons and holes) and allowing a large reverse current to flow. The reverse voltage at which this occurs is called the Zener potential. Besides breaking down the covalent bond, this large reverse voltage also gives the free electrons enough energy to liberate additional electrons, with which they collide, thus further increasing the current. This reverse biased *PN* junction diode is called a Zener diode; its schematic and its current–voltage graph are shown in Figure 8.24.

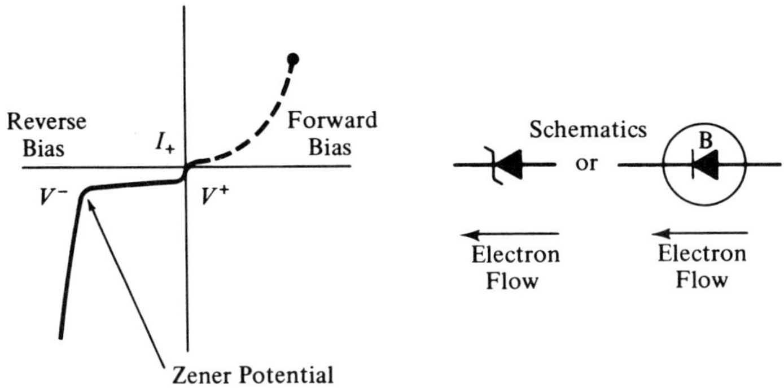

FIUGRE 8.24 (*Note: The B indicates the breakdown of covalent bonds in the Zener diode.*)

The Zener diode, although mentioned at this point, is not used as a rectifier or a detector, as are the other diodes. Its principal use, we will see later, is as a voltage regulator.

8.10 SUMMARY

To compare briefly the types of diodes discussed, we recall that the vacuum tube diode requires a filament voltage to liberate electrons from the cathode and is therefore inherently hotter than the metallic diodes. In addition, the current in the vacuum tube is limited by the number of electrons emitted by the cathode; in the metallic diode, the junction temperature is the principal factor limiting current flow (too high a temperature affects the materials in such a way as to increase their resistance to current flow). By cooling the junction, the metallic diodes will have higher current ratings than the vacuum tube diode. In addition, the forward opposition to current flow is less in the metallic diodes than in the vacuum tube. We will continue to refer to the vacuum tube diode in circuits throughout this book, but we will also try to show an equivalent circuit using one of the metallic diodes, such as copperoxide, selenium, point contact, or *PN* junction.

Questions

(1) What is the normal direction of electron flow in the diode?

(2) After an electron enters *N*-type material in a *PN* junction diode, describe what happens to return the diode to normal (net charge of zero).

Suggested Experiments

(1) Diode Tube Characteristic. The curve of plate current versus the plate voltage at a specific filament voltage may be drawn by first establishing a specific filament voltage, then taking readings of plate current for each new setting of the plate voltage. (See Figure 8.25.)

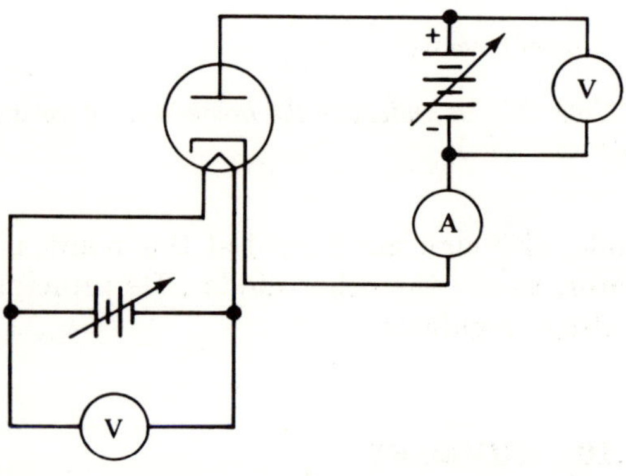

FIGURE 8.25

CHAPTER 9

Voltage Rectifiers and Multipliers

9.1 HALF-WAVE RECTIFIERS

Each type of diode has its limitations, but all have the basic ability to change ac to dc. Whether employed as a rectifier or as a detector, the diode allows current to flow principally in just one direction. If we connect a diode to an ac supply, the circuit is a half-wave rectifier. (See Figure 9.1.) The input ac voltage is stepped up by the transformer, and this stepped-up voltage is applied across the diode. A separate secondary winding supplies the filament voltage. Looking at the secondary coil of the transformer, we recall that, as the current alternates, *A* is positive when *B* is negative and vice versa. Remember that current flows in the tube when the anode, or plate, is positive with respect to the cathode. Consider first the half cycle for which *A* is positive. When *A* is positive, the anode is positive, and current flows from cathode to plate, through the secondary winding, through the load resistance, and back to the cathode. The load resistance provides the circuit with a sufficient voltage drop so that the dc output may be observed across it. When *A* is negative on the second half cycle, the plate is negative, and no current flows. (See Figure 9.2.)

If the voltage across the diode on the negative half cycle (*A* negative, *B* positive) were quite large, there is the possibility that a few electrons on the plate, or anode, might arc across to the cathode. The voltage, at which this arcing could occur, is called the peak inverse voltage. For our half-wave rectifier, the voltage from the transformer must be such that the peak inverse voltage is not reached, so the diode never conducts on the negative half cycle.

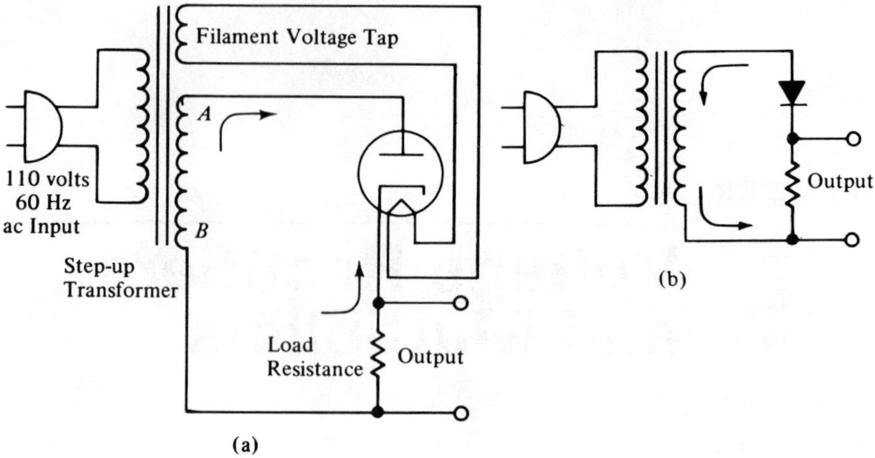

FIGURE 9.1 *(a) Vacuum tube half-wave rectifier; (b) half-wave rectifier with metallic diode.*

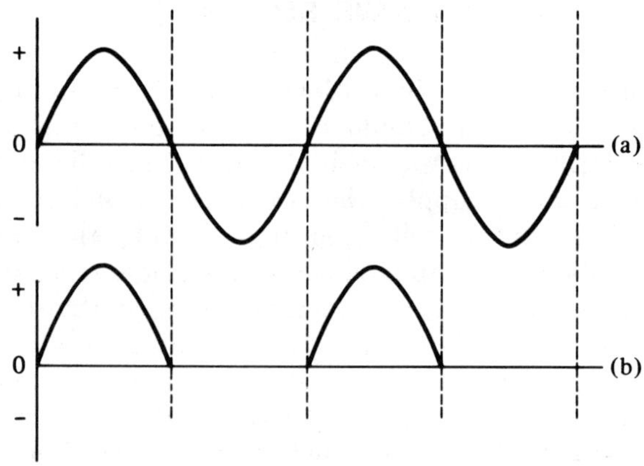

FIGURE 9.2 *Voltage waveforms for a half-wave rectifier. (a) Voltage across secondary coil of transformer (input voltage); (b) voltage across load resistor (output voltage).*

The above circuit is referred to as a half-wave rectifier, because it rectifies every other half cycle. Its dc output voltage is unlike the dc voltage from a battery. The output observed across the load resistance in the half-wave rectifier is called dc ripple; it is dc, and not ac, in that it does not go negative (the current does not change direction). (See Figure 9.3.)

Remember that effective voltage for ac was found by the equation $V_{\text{Effective ac}} = 0.707 V_{\text{Peak ac}}$. For the dc ripple voltage, since it occurs

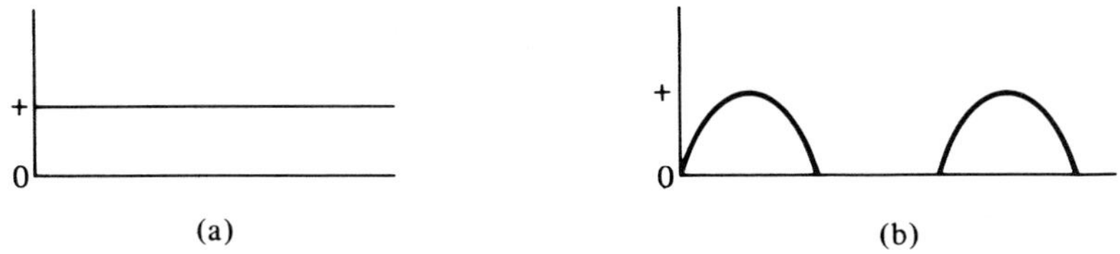

FIGURE 9.3 *(a) Steady dc from a battery; (b) dc ripple from a half-wave rectifier.*

every other half cycle, the effective value can be found by the equation $V_{\text{Effective dc ripple}} = \frac{1}{2} \times 0.707 V_{\text{Peak dc ripple}}$.

If we think of the dc ripple wave as sand and smooth it out by knocking the tops of the ripples into the valleys, the level of the smooth sand would represent the average value of the dc ripple and would be found to be $V_{\text{Average dc ripple}} = 0.318 V_{\text{Peak dc ripple}}$. (See Figure 9.4.) Notice that

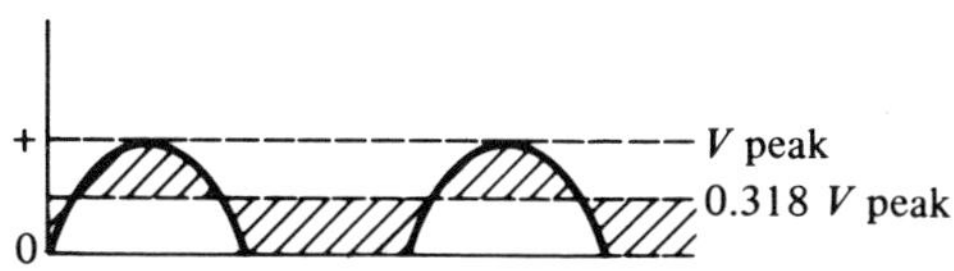

FIGURE 9.4

this average value is a little smaller than the effective value found above. Later, with the application of a filter network consisting of a capacitor and a coil or resistor, we will be able to smooth out the dc ripple voltage, just as we smoothed out the sand, to a steady dc voltage (equal to $V_{\text{Average dc ripple}}$) like that produced by a battery.

9.2 FULL-WAVE RECTIFIERS

The half-wave rectifier is limited to applications which do not require much current. The full-wave rectifier, although it does employ two diodes, utilizes both half cycles of the input from the transformer. The half-wave rectifier only conducts every other half cycle, whereas, with the full-wave rectifier, current will flow across the load for the entire period. Thus, the output of the full-wave rectifier, in terms of effective and average values, will be twice that of the half-wave rectifier.

The full-wave rectifier utilizes a step-up transformer, as does the half-wave rectifier. (See Figure 9.5.) The secondary of the transformer has a center tap at point B. If we want the peak output of the full-wave rectifier to be the same as that of the half-wave rectifier, then the turns

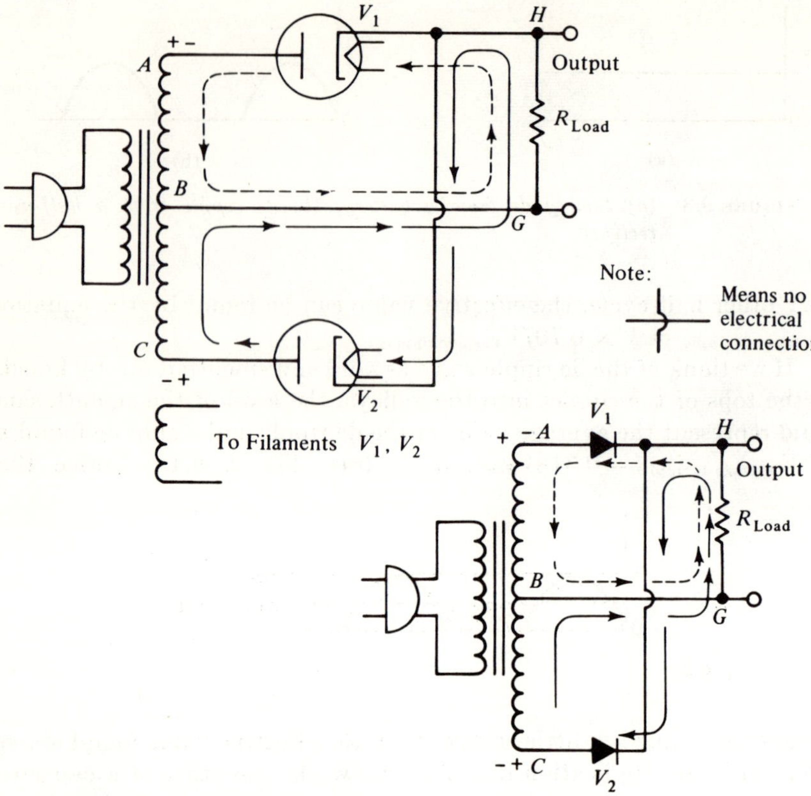

FIGURE 9.5 *Full-wave rectifier.*

ratio must be twice as great because of this center tap. Doubling the turns ratio gives the same voltage across AB (or BC) as across the entire secondary coil in the half-wave rectifier; thus we obtain the same voltage across each diode in the full-wave as across the single diode in the half-wave rectifier.

When A is positive, B is negative and C even more negative. With C negative, the anode of diode V_2 is negative with respect to the cathode, and V_2 does not conduct. With A positive, the anode of diode V_1 is positive with respect to the cathode, and V_1 conducts. Thus, on the first half cycle, the path of the electrons is from the cathode of V_1 to the plate, through half of the secondary from A to B, through the load resistance, and back to the cathode. Notice that current through the load flows from G to H.

On the second half cycle point C is positive, point B less positive, point A is negative. With A negative, diode V_1 will not conduct. Since

C is positive with respect to B, the anode of V_2 is positive with respect to the cathode, and V_2 will conduct. The flow of electrons is from cathode to anode of V_2, through CB of the secondary, from G to H in the load resistance, then back to the cathode. Notice that the current flow through the load was in the same direction (from G to H) on both half cycles. The waveforms in Figure 9.6 show the input and output and also show

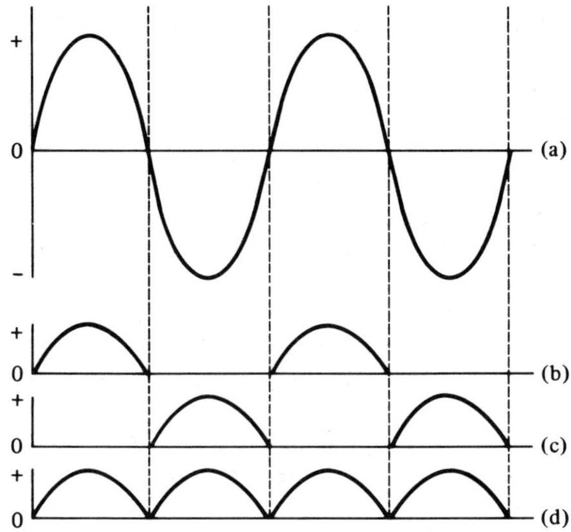

FIGURE 9.6 *(a) Voltage across entire secondary of transformer (input); (b) voltage across load when V_1 conducts; (c) voltage across load when V_2 conducts; (d) entire voltage output across load resistance.*

individual outputs when each diode conducts. The peak values of the output voltages are only half the peak value of the voltage across the secondary due to the center tap. The voltage across the load reflects the fact that just half of the secondary coil is used on each half cycle.

As we have mentioned, the effective and average voltages for the full-wave rectifier are twice those of the half-wave rectifier when the peak values are the same. $V_{\text{Effective full-wave}} = 2V_{\text{Effective half-wave}} = 2 \times \frac{1}{2} \times 0.707V_{\text{Peak dc ripple}} = 0.707V_{\text{Peak dc ripple}}$. $V_{\text{Average full-wave}} = 2V_{\text{Average half-wave}} = 2 \times 0.318 \times V_{\text{Peak dc ripple}} \doteq 0.636V_{\text{Peak dc ripple}}$. Summarizing,

$$V_{\text{Effective full-wave}} = 0.707V_{\text{Peak dc ripple}}$$

which incidentally is the same value as that for $V_{\text{Effective ac}}$, and

$$V_{\text{Average full-wave}} = 0.636V_{\text{Peak dc ripple}}$$

With capacitor and coil or resistor, we will be able to filter this dc ripple output of a full-wave rectifier to a steady dc output.

9.3 BRIDGE RECTIFIERS

The full-wave rectifier doubled the output of the half-wave rectifier. We will now look at the bridge rectifier whose output is nearly twice that of the full-wave rectifier when the same transformer is used for both. This is because the bridge rectifier does not have the center tap on the secondary coil that was present in the full-wave rectifier. Another advantage of the bridge rectifier, in comparison to the full-wave rectifier, is that, because it has two tubes in series each half cycle, its inverse voltage is one half that of the full-wave rectifier. Remember that by inverse voltage, we mean the voltage across the diode when it is not conducting; it is important that the inverse voltage be small enough that the diode never arcs, or conducts, from anode to cathode. The vacuum tube bridge rectifier requires filament voltages for V_1, V_2, V_3, and V_4; other than that the two schematics for the bridge rectifiers are the same.

On the first half cycle, A is positive and B is negative. (See Figure 9.7.) Thus, the cathode of V_4 is positive, so it does not conduct; the plate of V_2 is negative, and it does not conduct. The cathode of V_1 is negative and the plate of V_3 is positive, so V_1 and V_3 both conduct. If we follow the electrons leaving the cathode of V_3, they travel to the anode of V_3, then around to A of the secondary coil (remember they cannot go through V_4 because it does not conduct this half cycle), through the coil to B, from there to the cathode of V_1. They pass through V_1, across the load resistance from G to H, and back to the cathode of V_3. Notice that the cathode voltages of vacuum tubes V_1 and V_3 are different and thus require separate filament supplies. Normally, all the filaments are tied together at one transformer winding; however, since the cathodes of V_1 and V_3 are at different potentials and since arcing from filament to cathode is common, their filaments are not connected in order to prevent discharging of one cathode by the other.

On the second half cycle when B is positive and A is negative, the plate of V_2 is positive and the cathode of V_4 is negative, so V_2 and V_4 both conduct. V_1 will not conduct (cathode is positive), nor will V_3 (anode is negative). Electrons leaving the cathode of V_2 flow to the plate of V_2, through the secondary of the transformer from B to A, to the cathode of V_4. They then move to the plate of V_4, pass through the load moving in the same direction as on the previous half cycle from G to H, then back to the cathode of V_2. Again, we see that V_2 and V_4 have

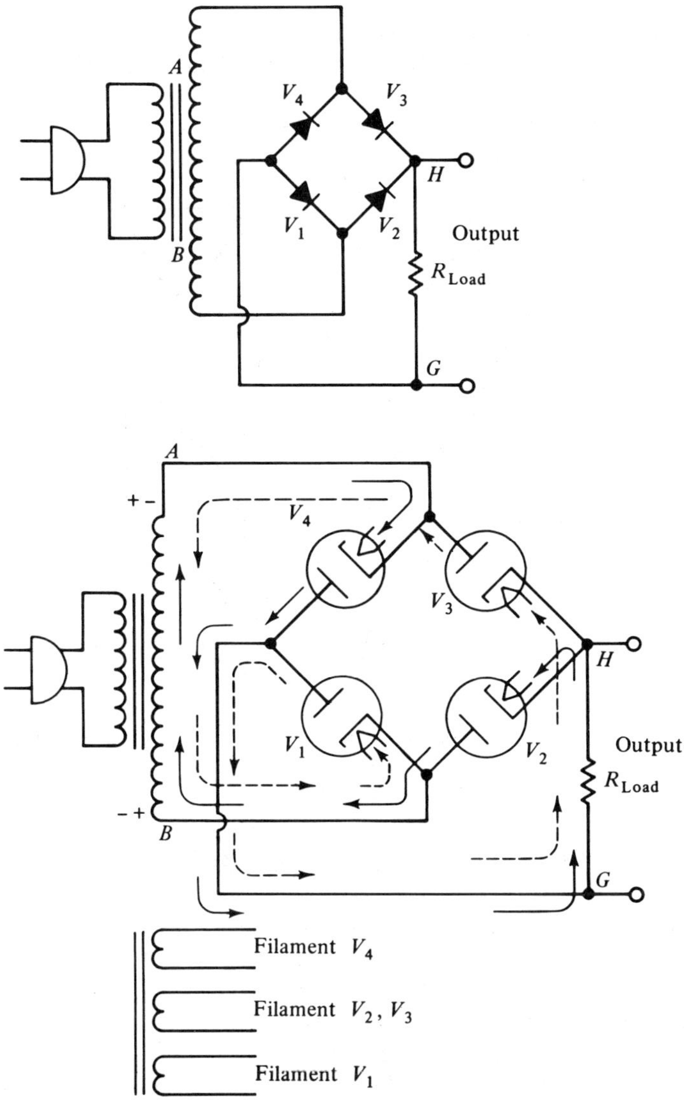

FIGURE 9.7 *Bridge rectifiers.*

separate filament voltage supplies since their cathodes are at different potentials. Notice that since the cathodes of V_2 and V_3 are connected, they are at the same potential and can use the same filament voltage supply. This requirement for separate filament voltages, in fact the necessity for any filament voltage supply at all, is, of course, eliminated when metallic diodes are used.

The input and output voltages for the bridge rectifier are shown in Figure 9.8, along with the outputs across the load when V_1, V_3 conduct and when V_2, V_4 conduct.

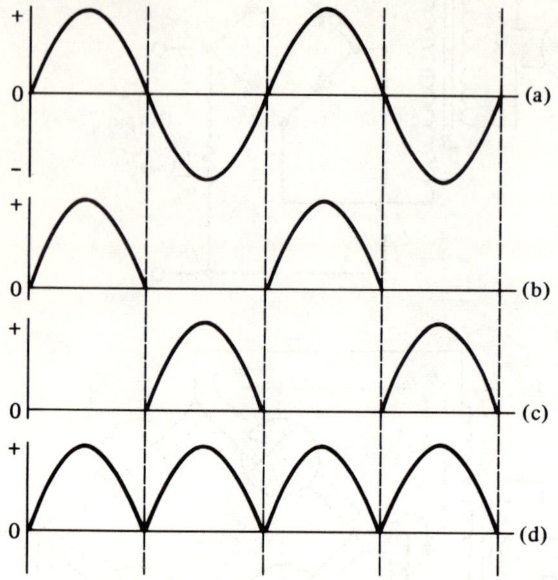

FIGURE 9.8 *(a) Voltage across secondary (input); (b) voltage when V_1 and V_3 conduct; (c) voltage when V_2 and V_4 conduct; (d) voltage across load resistance (output).*

9.4 CASCADE VOLTAGE DOUBLERS

If capacitors are charged every other half cycle and their voltages added, the output will be nearly twice the input. The voltage doubler is a circuit which utilizes this procedure of alternately charging capacitors. A voltage doubler circuit has two forms: the half-wave, or cascade, voltage doubler and the full-wave voltage doubler. Later, we will see that the output of the full-wave voltage doubler is easier to filter, or smooth out, than that of the cascade type. In doubling the voltage with one of these circuits, the current output will be rather low, which limits its applications. The peak voltage output from a voltage doubler is twice that from a half-wave or bridge rectifier using the same transformer. We will consider the circuit for a cascade voltage doubler first.

In Figure 9.9, when A is negative and B is positive, the plate of V_1 will be positive, so V_1 will conduct. V_2 will not conduct, since its plate will be negative. Electrons flow from the cathode of V_1 to the plate, through the secondary coil from B to A, charging the left plate of C_1

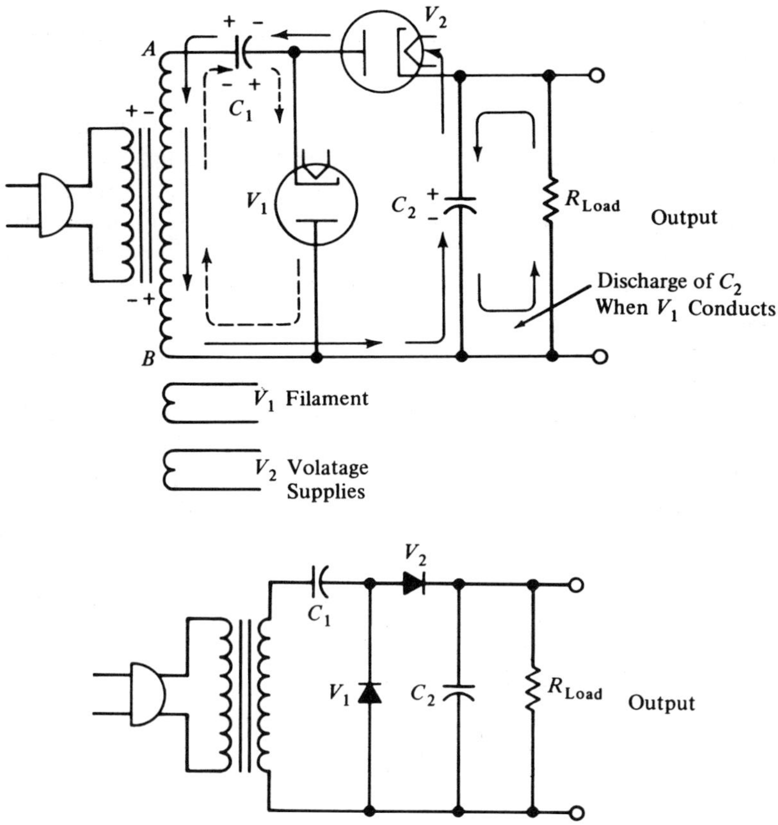

FIGURE 9.9 *Cascade voltage doublers.*

negatively. Electrons are repelled from the right plate of C_1, returning to the cathode of V_1 and leaving the right plate positive.

On the next half cycle, when A is positive and B is negative, the plate of V_2 will be positive and V_2 will conduct. Electrons flow from the cathode to the plate of V_2; they build up on the right plate of C_1, discharging it, then charging it in the opposite direction, the plate on the right becoming negative and the plate on the left positive. As C_1 discharges, the electrons already on the left plate of C_1 from the previous half cycle are repelled, then as C_1 charges in the opposite direction, additional electrons are repelled to make the left plate positive. Thus, the number of electrons leaving C_1 is nearly twice the number originally deposited by V_1. These electrons then flow through the secondary from A to B and charge C_2, its bottom plate negative, top plate positive. Electrons repelled from the top plate of C_2 return to the cathode of V_2.

Essentially, all of the electrons go to C_2 rather than passing through the load resistance, because R_{Load} is very large in comparison to the capacitive reactance of C_2.

Notice that C_2 is charged only when V_2 conducts. On the next half cycle, when V_1 conducts, C_2 tends to discharge through the load resistance. The time constant ($1\text{ TC} = R \times C$) for the RC circuit involved in the discharge (C_2 and R_{Load}) will be large, thus C_2 will not have time to completely discharge during the half cycle when V_1 conducts. As we have seen, when V_2 conducts, C_1 is first discharged through C_2, then charged in the opposite direction through C_2; thus, C_2 charges to nearly twice the normal charge produced by the input voltage. Recalling the relation for capacitance $C_2 = Q/V$ and solving for voltage $V = Q/C_2$, we see that doubling the charge doubles the voltage. (Voltage $= 2Q/C_2 = 2 \times Q/C_2 = 2V =$ twice the original voltage.) The capacitance C_2 is constant in that it depends only on the physical size of the plates, the distance between the plates, and the dielectric used. Thus, the output voltage across C_2 when V_2 conducts is nearly twice the input voltage.

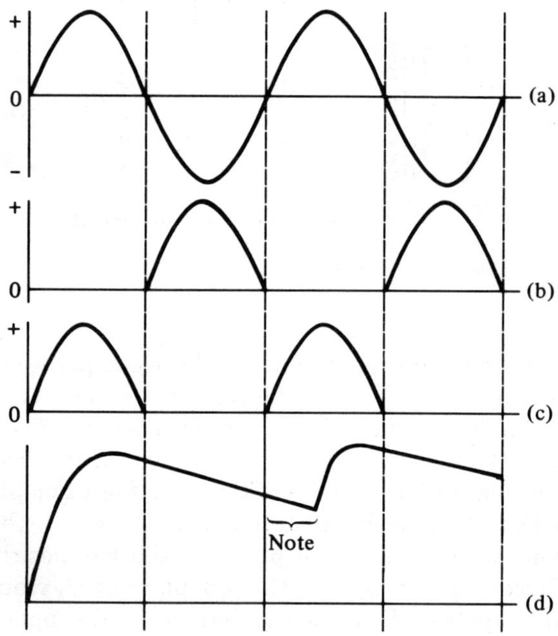

FIGURE 9.10 *(a) Voltage across secondary (input); (b) voltage across V_1 when V_1 conducts (capacitor discharging); (c) voltage across V_2 when V_2 conducts (capacitor charging to twice input voltage); (d) voltage output across load resistor (output). (Note: C_2 continues to discharge into the third half cycle until voltage across V_2 increases to a point that it can begin to build the charge on C_2 again.)*

When V_1 conducts, C_2 discharges slowly and hence the output voltage drops slightly. (See Figure 9.10.)

As we observe in Figure 9.10, the output voltage across the load is partly filtered, or smoothed, by the slow discharge of C_2. The term cascade is used for this voltage doubler, because the output of V_1 flows into the output of V_2. When the circuit is not operating (there is no input), C_2 will have time to completely discharge through the load resistance; thus, the load resistor also acts as a bleeder resistor in that it bleeds off any latent charge on C_2.

9.5 FULL-WAVE VOLTAGE DOUBLERS

The full-wave voltage doubler adds to the output voltage every half cycle, rather than allowing the output to decrease (while the capacitor discharges) through every other half cycle, as was the case with the half-wave voltage doubler. The circuit for the full-wave voltage doubler uses the same components as the half-wave voltage doubler; again, if vacuum tube diodes are used, separate filament voltage supplies are required.

In Figure 9.11, when point A is positive and B is negative, the anode of V_1 will be positive, and V_1 will conduct. Electrons flow from cathode to plate of V_1, through the secondary from A to B and charge C_1. The lower plate of C_1 is charged negatively; the electrons repelled from the upper plate of C_1, leaving it positively charged, return to the cathode of V_2. Again, practically all the electrons build up on C_1, rather than passing through R_1, because the resistance of R_1 is much larger than the capacitive reactance of C_1.

On the next half cycle, A is negative and B is positive. The cathode of V_2 will then be negative, and V_2 will conduct. Electrons flow from cathode to anode of V_2 and charge the lower plate of C_2 negatively. Electrons repelled from the upper plate of C_2 cannot move up to C_1, since its lower plate is negatively charged from the last half cycle, so they travel to the secondary, through it from B to A, and back to the cathode of V_2.

Notice that while C_2 is charging quickly, C_1 is discharging slowly through R_1, and vice versa; on the next half cycle, C_2 discharges slowly while C_1 charges. In order to make the capacitors discharge very slowly, the time constants for the two RC circuits, through which the capacitors discharge (C_1 through R_1 and C_2 through R_2), must be quite large. Note that the top plate of C_1 is positive and the bottom plate of C_2 is negative at all times. This entire voltage drop over C_1 and C_2 is felt across the load resistance ($R_1 + R_2$) and represents the output voltage. Since C_1 charges to the input voltage on one-half cycle and C_2 charges to the input voltage on the next, the entire voltage drop, or output, will be nearly twice the input voltage.

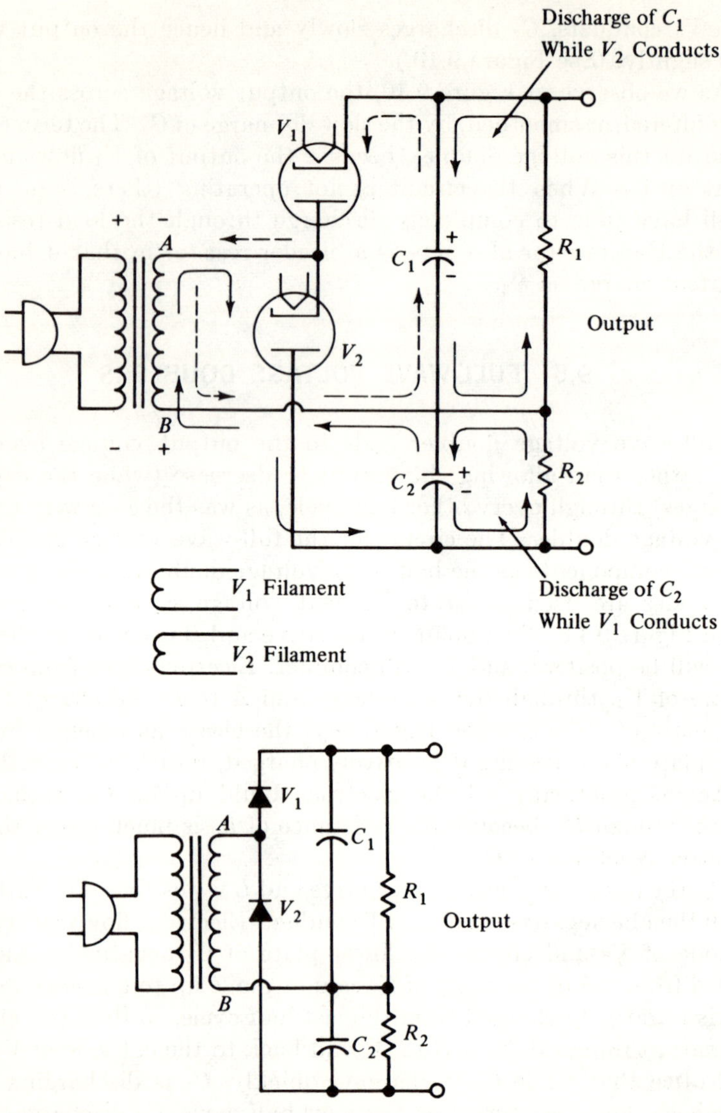

FIGURE 9.11 Full-wave voltage doublers.

In comparing the outputs of the full-wave and half-wave voltage doublers, we observe that the full-wave has twice as many ripples as the half-wave voltage doubler and, in addition, the full-wave output is smoother (valleys are not as deep). (See Figure 9.12.) The resistors R_1 and R_2 serve as the load resistance when the circuit is in operation and as bleeder resistors, discharging the capacitors, when the circuit is turned off.

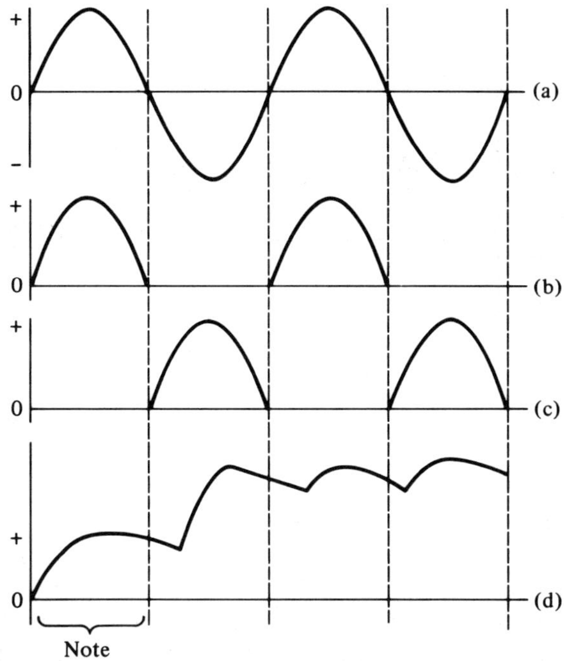

FIGURE 9.12 *(a) Voltage across secondary (input); (b) voltage across V_1; (c) voltage across V_2; (d) voltage across load resistance (output). (Note: On the first half cycle when input first connected, voltage will just be that across C_1 since C_2 has not had a chance to charge.)*

9.6 VOLTAGE TRIPLERS AND QUADRUPLERS

It is possible to triple or quadruple voltage as well as double it. Practical considerations limit the multiplication factor to four or five. Voltage multipliers, because of their many stages, normally use metallic diodes rather than vacuum tube diodes in order to eliminate the requirement for multiple filament sources. Voltage multipliers frequently consist of cascade voltage doublers with their outputs summed. Let us consider voltage tripler and quadrupler circuits.

The voltage tripler is made up of the cascade voltage doubler, whose output has already been discussed, and a half-wave rectifier with a capacitor. (See Figure 9.13.) When A of the secondary is negative, V_3 will conduct, charging C_3 to the voltage across the secondary. During the next half cycle, when A is positive, V_3 will not conduct and C_3 will discharge slowly through the resistance R_3. Recall that C_2 will charge to double the voltage across the secondary when A is positive. Thus, C_2 and C_3 charge on alternate half cycles. However, the time constants for their discharge are so long that practically their entire voltage is felt across

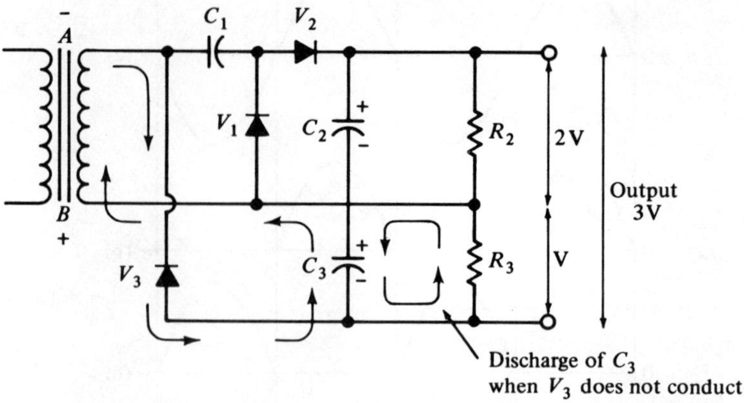

FIGURE 9.13 *Voltage tripler.*

R_2 and R_3 all the time. The output voltage is taken across the load resistance ($R_2 + R_3$), so it is the sum of the voltage drops across R_2 and R_3. Since the voltage drop due to the charge on C_2 is twice the voltage across the secondary, and the voltage drop due to the charge on C_3 equals the voltage across the secondary, the output voltage will be nearly three times the voltage across the secondary, that is, the voltage is tripled.

The voltage quadrupler consists of two cascade voltage doublers with their output voltages in series. (See Figure 9.14.) On the half cycle

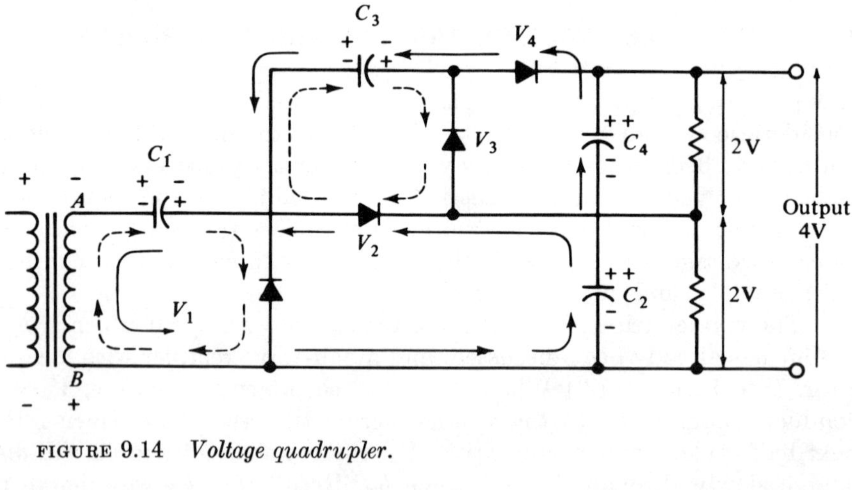

FIGURE 9.14 *Voltage quadrupler.*

when A is negative, C_1 and C_3 both charge to the voltage across the secondary. On the next half cycle, when A is positive, both C_1 and C_3 will be discharged and charged in the opposite direction; this means that

C_2 and C_4 will each have twice the charge that was on C_1 and C_3, respectively, and thus will each be charged to twice the voltage across the secondary. When the total voltage drop is taken, we thus have an output of four times the voltage across the secondary.

Questions

(1) Rectification is the process of changing ac to dc. How can you justify this statement when you compare the output of a half-wave or full-wave rectifier to the output of a battery?

(2) Why would a voltage doubler not be used to supply dc for the purpose of charging an automobile battery?

Suggested Experiments

(1) Rectifier Waveforms. The input and output waveforms of half-wave, full-wave, and bridge rectifiers and voltage doublers can be observed with an oscilloscope.

CHAPTER 10

Filters

10.1 CAPACITOR FILTER

The dc output of the voltage doublers and multipliers just discussed was partially filtered, or smoothed, due to the long time constants and consequent slow discharge of the capacitors through the load resistance. The dc output from half-wave, full-wave, and bridge rectifiers was unfiltered and, you recall, was referred to as dc ripple voltage. We will be concerned in this section with methods of smoothing, or filtering, the dc voltages from these sources. As we noted in the doubler circuits, a capacitor and resistor, connected in parallel and placed across the output, tend to slow the decrease of the output voltage; this slower decrease in the output voltage is due to the long time constant of the RC circuit. The waveforms illustrate the capacitor filter action in the half-wave and full-wave rectifier circuits. (See Figure 10.1.) The capacitor in each rectifier charges to the peak value of the voltage quickly because of the small resistance of the charging path. When the voltage from the diode drops below that across the capacitor, the capacitor begins to discharge slowly through the load resistor. Because of the long time constant, the capacitor does not lose much charge before the voltage from the diode has increased again to a point where it begins to recharge the capacitor.

Since the filtered voltage output never reaches zero, the average dc output for a filtered voltage is greater than for the same unfiltered output. As you can see from the graphs, there will still be a ripple to the dc even after capacitor filtering, so additional filtering action would be in order.

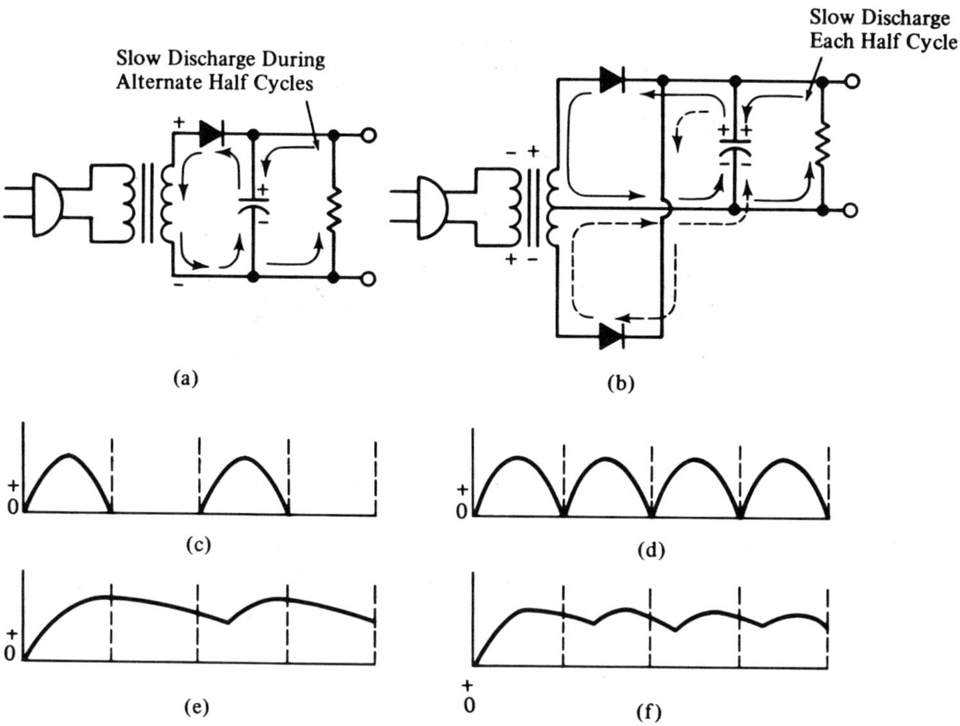

FIGURE 10.1 *(a) Half-wave rectifier with capacitor filter; (b) full-wave rectifier with capacitor filter; (c) output: half-wave rectifier; (d) output: full-wave rectifier; (e) output: half-wave rectifier with capacitor filter; (f) output: full-wave rectifier with capacitor filter.*

10.2 INDUCTOR FILTER

An inductor, or coil, tends to oppose any change in current flow with a self-induced current. Remember the changing current in the coil has a changing magnetic field associated with it; it is this changing magnetic field which produces a current in the coil in the opposite direction to the original current. A dc ripple can be reduced by placing an inductor in series with the output. The inductor is called a choke coil, or choke, because it "chokes" the ripple. The choke opposes current (and voltage) change and thus keeps the output from dropping to zero as quickly as it would without this filter. The peak value for the filtered voltage is less than that for nonfiltered due to the resistance of the coil. (See Figure 10.2.)

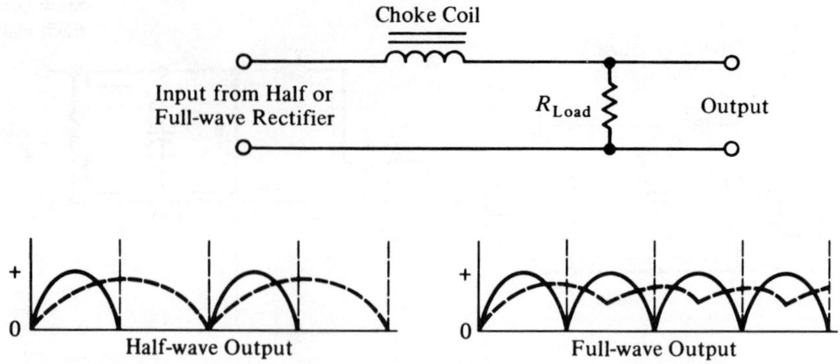

FIGURE 10.2 *(Note: Solid wave is unfiltered output; dotted wave is output with inductor filter.)*

The advantage of the choke filter is that, since it is a coil, it acts as a regulator in that it opposes any sudden change in current with a self-induced current in the opposite direction. Therefore, it would oppose any sudden large current drain by the load and, by thus maintaining a relatively steady current, would prevent the output voltage from changing drastically.

10.3 π FILTER

The most common type of filter combines two capacitors and a coil in a configuration that resembles the Greek letter π (pi). The π filter is shown in Figure 10.3. Note that L_2 and C_2 are in series, and this series combina-

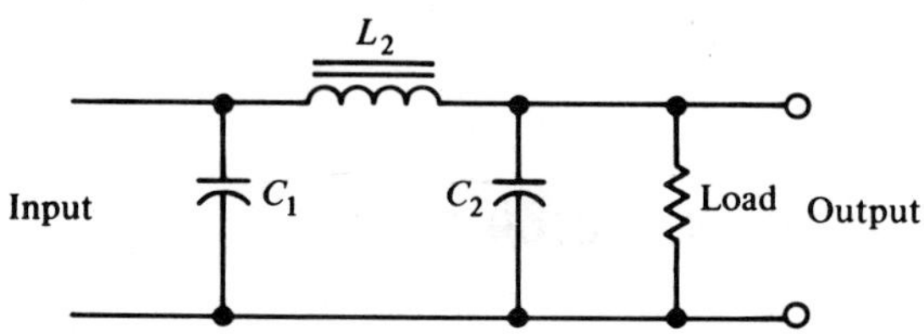

FIGURE 10.3 *Capacitor input π (pi) filter.*

tion is in parallel with C_1. Practically all the electrons from the input will go to charge C_1 because of the much higher impedance of the other path through L_2 and C_2. The capacitor C_1 thus acts as a reservoir storing dc ripple voltage and discharging through L_2 and C_2. The voltage across C_1 for half-wave and full-wave inputs is shown in Figure 10.4.

The same voltage will be felt across the parallel branch (L_2 and C_2) as across C_1. This is still a dc ripple voltage, as shown above. We can think of this voltage as a ripple superimposed on a pure (or constant) dc voltage.

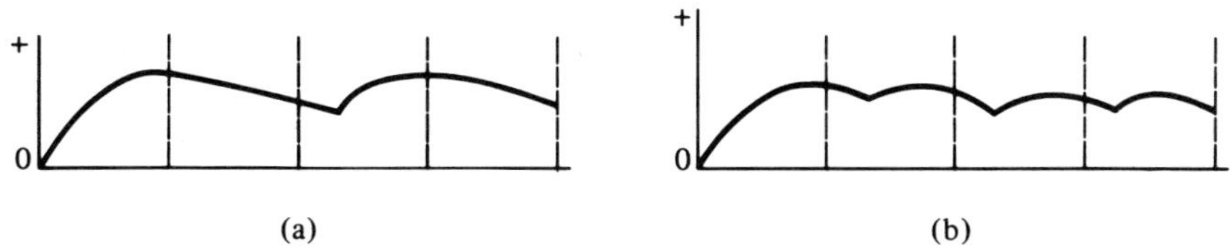

FIGURE 10.4 *Voltage across C_1. (a) Half-wave input; (b) full-wave input.*

This voltage will be divided between the coil L_2 and the capacitor C_2. Since the inductive reactance of the coil ($X_L = 2\pi fL$) is directly proportional to the frequency, the coil will offer a high opposition to the ripple, that is, to changes in the dc voltage. The capacitive reactance of C_2 ($X_C = 1/2\pi fC$) is inversely proportional to frequency and thus offers little opposition to the ripple. Therefore, most of the ripple will be dropped, or smoothed out, by the choke coil L_1; L_1 is also called a smoothing coil when preceded by a capacitor like this. Most of the ripple is removed by L_1 and very little by C_2; this is fortunate because any ripple across C_2 is felt across the output. The pure dc voltage is also divided between coil and capacitor. Since this part of the voltage is constant, it does not have a changing magnetic field associated with it, and hence the coil offers very little impedance to the pure dc voltage. We already know that a capacitor blocks the flow of dc, so practically all of the pure dc voltage is felt across the output, and the filter has changed ripple dc voltage to nearly constant dc voltage. (See Figure 10.5.) Because the capacitor input

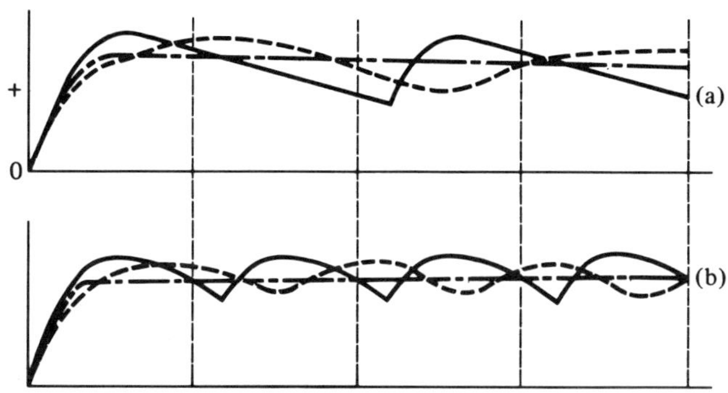

FIGURE 10.5 *(a) π filter with half-wave input; (b) π filter with full-wave input. (Note: Voltage across C_1 is shown in solid wave; voltage across coil is in dotted wave; voltage across load (output) is in dashed-and-dotted wave.)*

(C_1) charges quite rapidly, the π filter is limited to small currents. If a large pulse of current from C_1 were felt across the inductor, it would damage the coil.

Filtering action by coil ($X_L = 2\pi fL$) and capacitor ($X_C = 1/2\pi fC$) to remove ripple voltage is better at higher frequencies. At high frequencies, the coil offers large inductive reactance to the ripple, thus removing most of it, while the capacitor offers little reactance to the ripple. The frequency of the half-wave rectifier output is the same as the input voltage (they have the same number of positive half cycles in a given length of time). But the frequency of the full-wave rectifier output is twice that of the input (the full-wave has twice as many positive half cycles as the input in a given length of time). We can then conclude that the voltage from a full-wave rectifier, since it has a higher frequency, is easier to filter than that from a half-wave rectifier. (See Figure 10.6.)

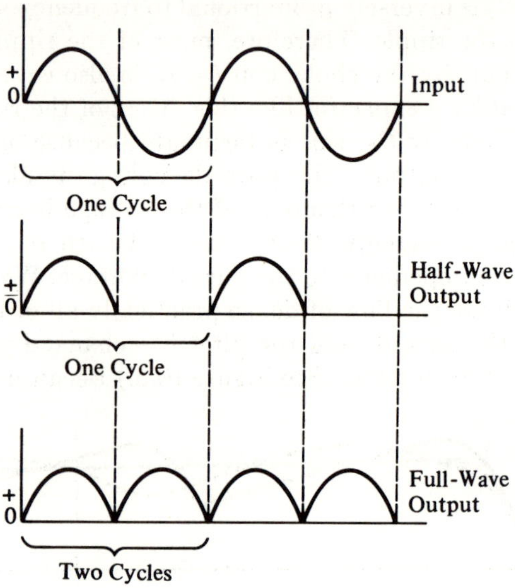

FIGURE 10.6

In all of these filters, the load resistor also serves as a bleeder resistor in that it removes the charge from the capacitor when the circuit is not in operation. This protects anyone handling the circuit, when it is not in operation, from being shocked.

A coil preceding a π filter, although it does reduce the voltage output of the filter, also maintains a more uniform current output. Since the advantage of this choke input is its ability to allow a continuous flow of current from a rectifier, it is normally not used with a half-wave rectifier where current only flows every other half cycle anyway. The continuous flow of current from the choke input (L_1) protects the other coil (L_2) from the pulsating current of the rectifier. The coil L_1, called a swinging choke, by smoothing out the current input, allows C_1 to charge at a steadier rate, which, of course, reduces the ripple across C_1; as we have seen, a coil also reduces the input voltage somewhat. L_2 and C_2 act as a voltage divider across C_1 and, as we have just discussed with the capacitor input π filter, further reduce the ripple. (See Figures 10.7 and 10.8.)

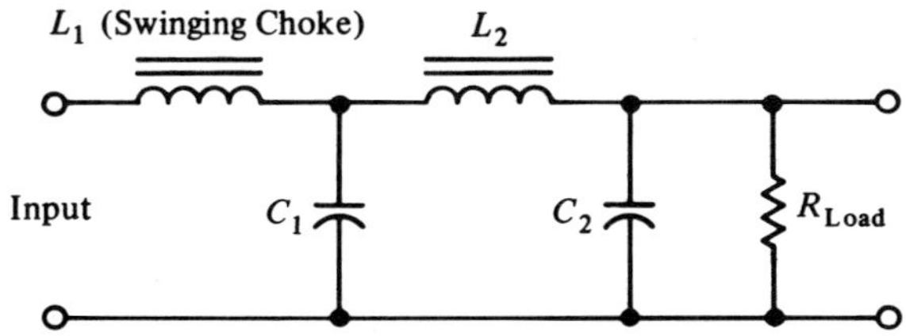

FIGURE 10.7 *π filter with swinging choke.*

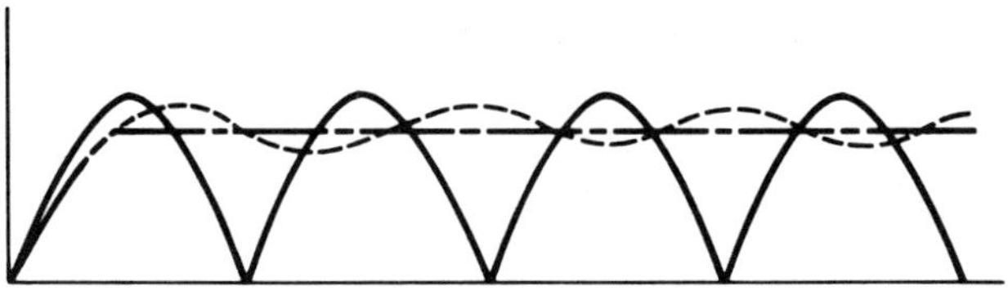

FIGURE 10.8 *Choke input π filter. (Note: Full-wave input is in solid wave; voltage across C_1 in dotted wave; output voltage in dashed-and-dotted wave.)*

This swinging choke (L_1) is designed in such a way that when the output requires more current, its inductive reactance decreases, allowing C_1 to charge to a higher value and thus deliver more current without affecting the output voltage. Correspondingly, the reactance of L_1 will increase when the output requires less current. In this way, the swinging choke tends to regulate the voltage output even when the current requirements are variable. We will see how this is done when we take a more careful look at voltage regulators in Chapter 11.

A graphic comparison of the capacitor input π filter and the choke input π filter can be made if we consider the current output of each one at different voltages. The voltage output of the capacitor input π filter is larger than that of the choke input π filter. This is understandable when we recall that the choke coil, because of its resistance, does not allow the capacitor to charge to the peak value of the input voltage. The choke input filter does a better job in regulating the voltage output than the capacitor input filter does. As we see in Figure 10.9, the voltage from the choke input filter remains nearly constant over a wide range of currents.

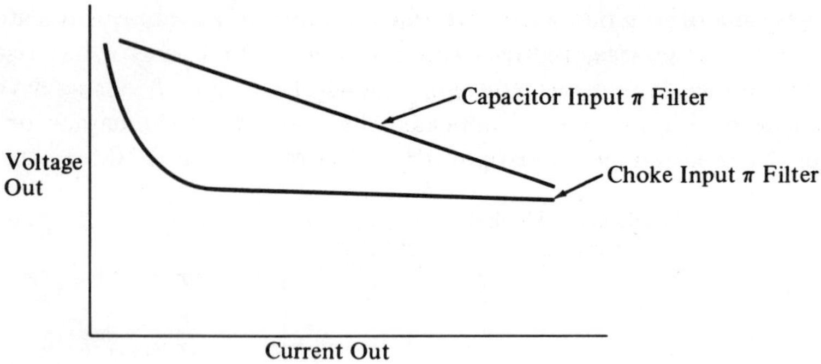

FIGURE 10.9 *(Note: Capacitor input filter delivers larger voltage at smaller currents; voltage is nearly constant for choke input filter.)*

Thus, the choke input filter is preferable, when the current required by the output is variable. If there is little current drain, the capacitor input filter delivers more voltage and is better suited for this output requirement.

10.4 RC FILTER

Because coils cost more than resistors, a resistor often replaces the coil in the capacitor input π filter. The resulting resistance–capacitance filter offers poorer regulation than the capacitor input π filter using a coil; in addition, some of the steady dc voltage will be dropped over the resistor which, of course, reduces the dc voltage output. (See Figure 10.10.)

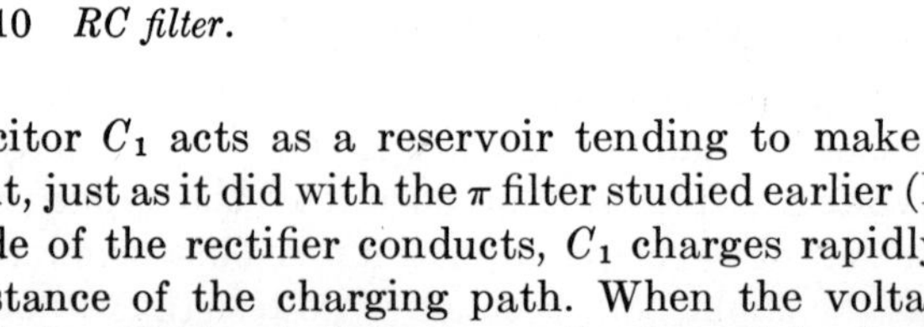

FIGURE 10.10 *RC filter.*

The capacitor C_1 acts as a reservoir tending to make the voltage output constant, just as it did with the π filter studied earlier (Figure 10.3). When the diode of the rectifier conducts, C_1 charges rapidly because of the small resistance of the charging path. When the voltage from the rectifier drops below the level of charge on C_1, then C_1 begins to discharge slowly through R and C_2. R and C_2 act as a voltage divider for the dc ripple from C_1, R offering some resistance to steady dc but C_2 offering a much higher reactance to steady dc. With respect to the ripple in the voltage, the capacitor is such that it offers a lower capacitive reactance (X_C) to the ripple than the resistor offers resistance (R). Hence the capacitor C_2 passes the ripple and stores most of the steady dc voltage for discharge through the load; the resistor R drops the ripple voltage and some of the steady dc voltage. When the output does not require large currents and there is little or no variation in the current drain, the resistance–capacitance filter can be used as an inexpensive substitute for the capacitor input π filter with choke coil.

Questions

(1) dc filters ordinarily consist of capacitors in parallel and coils in series with the output of a rectifier. What difference does it make if the filter has a capacitor input or a choke input?

Suggested Experiments

(1) The performance of various filters when applied to half-wave and full-wave rectifiers may be observed using an oscilloscope.

CHAPTER 11

Voltage Regulators

11.1 MANUALLY OPERATED SERIES VOLTAGE REGULATOR

The removal of dc ripple voltage in rectifier or voltage multiplier circuits is accomplished by a filter. Filtering action thus provides a steady dc output not unlike that from a battery. However, even this output may vary and, in addition, the load resistance to which it is connected may change, causing a corresponding change in the output voltage across the load. Voltage regulators are needed to provide a dependably steady voltage output.

A simple, manually operated voltage regulator is shown in Figure 11.1. The series voltage regulator consists of a variable resistor connected

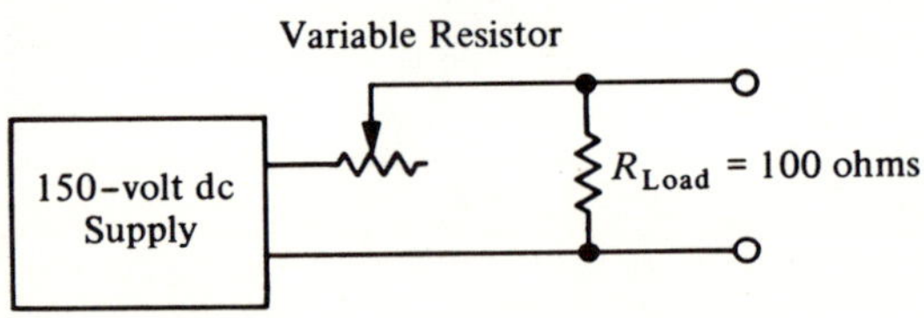

FIGURE 11.1

to the load. The voltage from the supply (rectifier–filter) is divided between the load and the variable resistor. $V_{Supply} = V_{Load} + V_{Variable}$. If the output across the load must be maintained at 100 volts, we have 150 volts $=$ 100 volts $+ V_{Variable}$ and hence $V_{Variable} = 50$ volts.

Suppose the output of the supply increased to 160 volts. In order to maintain the 100-volt output, the variable resistor would have to have a voltage drop of 60 volts. 160 volts = 100 volts + V_{Variable} or V_{Variable} = 60 volts. To increase the voltage drop across the variable resistor (from 50 volts to 60 volts), we had to increase its resistance. Thus, an increase in supply voltage requires an increase in variable resistance in order to maintain a steady output voltage.

If the rectifier–filter supplied something less than 150 volts, for example, the supply voltage dropped to 140 volts, the voltage drop across the variable resistor would have to decrease to maintain the same output. 140 volts = 100 volts + V_{Variable}; V_{Variable} = 40 volts. This decrease in V_{Variable} is brought about by a decrease in the resistance of the variable resistor. A decrease in supply voltage requires a decrease in variable resistance.

Now let us suppose the supply voltage remains constant, but the load resistance changes. In Figure 11.2, the load resistance increases from 100 ohms to 250 ohms. At present, we take the variable resistance to be 50 ohms.

$$\begin{aligned} R_T &= R_{\text{Load}} + R_{\text{Variable}} \\ &= 100 \text{ ohms} + 50 \text{ ohms} \\ &= 150 \text{ ohms} \\ I &= \frac{E}{R_T} \\ &= \frac{150 \text{ volts}}{150 \text{ ohms}} \\ &= 1 \text{ A} \\ I_T &= I_{\text{Load}} = I_{\text{Variable}} = 1 \text{ A} \\ V_{\text{Load}} &= I_{\text{Load}} \times R_{\text{Load}} \\ &= 1 \text{ A} \times 100 \text{ ohms} \\ &= 100 \text{ volts} \end{aligned}$$

$$\begin{aligned} R_T &= R_{\text{Load}} + R_{\text{Variable}} \\ &= 250 \text{ ohms} + 50 \text{ ohms} \\ &= 300 \text{ ohms} \\ I_T &= \frac{E}{R_T} \\ &= \frac{150 \text{ volts}}{300 \text{ ohms}} \\ &= 0.5 \text{ A} \\ I_T &= I_{\text{Load}} = I_{\text{Variable}} = 0.5 \text{ A} \\ V_{\text{Load}} &= I_{\text{Load}} \times R_{\text{Load}} \\ &= 0.5 \text{ A} \times 250 \text{ ohms} \\ &= 125 \text{ volts} \end{aligned}$$

We first find the total resistance of the series circuit before and after the increase in load resistance. Using Ohm's Law ($I = E/R$), we can calculate the total current in the circuit. We observe that, with an increase in load resistance, the current decreases (an inverse proportion). In a series circuit, the current in each component is the same as the total current, so we can use Ohm's Law ($E = IR$) to find the voltage across the load. We see that the load voltage increased with an increase in load resistance.

Originally we were interested in adjusting the variable resistance in such a way as to keep the load voltage constant even when the load resistance varied. How do we adjust the variable resistance in the unbalanced state (Figure 11.2) in order to make $V_{Load} = 100$ volts? We

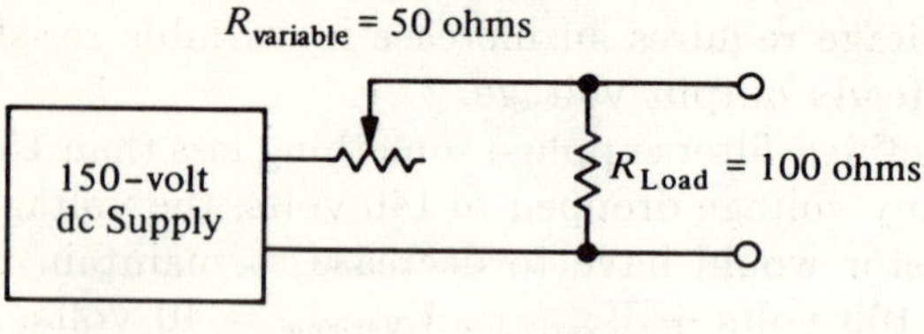

(a)

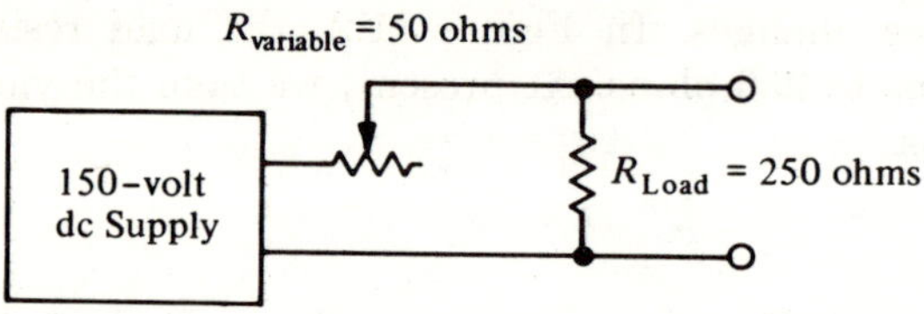

(b)

FIGURE 11.2 *Series voltage regulator (increase in load resistance). (a) Balanced state; (b) unbalanced state: change $R_{Variable}$ to return output to 100 volts.*

have to find a way of decreasing V_{Load} from 125 volts to 100 volts; this can be done by decreasing the current through R_{Load}. The decreased current can be found as follows: $I_{Load} = V_{Load}/R_{Load} = 100$ volts/250 ohms $= 0.4$ A. If the load current is 0.4 A, then the total current in the series circuit is also 0.4 A. $I_T = I_{Load} = 0.4$ A. Knowing the total current desired and total voltage, we can find the necessary total resistance for the circuit: $R_T = E/I_T = 150$ volts/0.4 A $= 375$ ohms. Since $R_T = R_{Load} + R_{Variable}$, 375 ohms $= 250$ ohms $+ R_{Variable}$, and $R_{Variable} =$ 125 ohms. When the load resistance increased, the variable resistance also had to increase (from 50 ohms to 125 ohms) in order to maintain a constant output voltage of 100 volts.

To summarize: When the load resistance increases, there will be a corresponding decrease in load current and an increase in load voltage. In order to maintain load voltage at a constant level, an increase in the variable resistance is necessary. In the reverse situation, when the load resistance decreases, there will be a current increase and a load voltage decrease. To offset this decrease in load voltage, the variable resistance must be decreased. This can be written in a shorter form using ↑ to represent an increase and ↓ to represent a decrease.

When $R_{\text{Load}}\uparrow$, then $\begin{cases} I_T\downarrow \\ \text{and} \\ V_{\text{Load}}\uparrow \end{cases}$ (To decrease V_{Load} to its original level: If $R_{\text{Variable}}\uparrow$, then $V_{\text{Variable}}\uparrow$ and $V_{\text{Load}}\downarrow$ to desired level.)

When $R_{\text{Load}}\downarrow$, then $\begin{cases} I_T\uparrow \\ \text{and} \\ V_{\text{Load}}\downarrow \end{cases}$ (To increase V_{Load} to its original level: If $R_{\text{Variable}}\downarrow$, then $V_{\text{Variable}}\downarrow$ and $V_{\text{Load}}\uparrow$ to desired level.)

Combining these observations with our earlier comments about a variable supply voltage, we can say that, for a series voltage regulator, (1) when the voltage across the load increases (whether this increase is due to an increase in the supply voltage or an increase in the load resistance), the variable resistance must be increased to return the load voltage to the desired level; (2) a decrease in load voltage (due to a decrease in supply voltage or load resistance) requires a decrease in the variable resistance for steady output.

11.2 BALLAST TUBE REGULATOR

A partial solution to the problem of manually adjusting the variable resistor is provided by the ballast tube, a tube containing a fine iron wire in a bath of hydrogen gas. Although the ballast tube does automate the adjustment of the variable resistor in a series voltage regulator, its application is limited to counteracting variations in supply voltage only. (See Figure 11.3.)

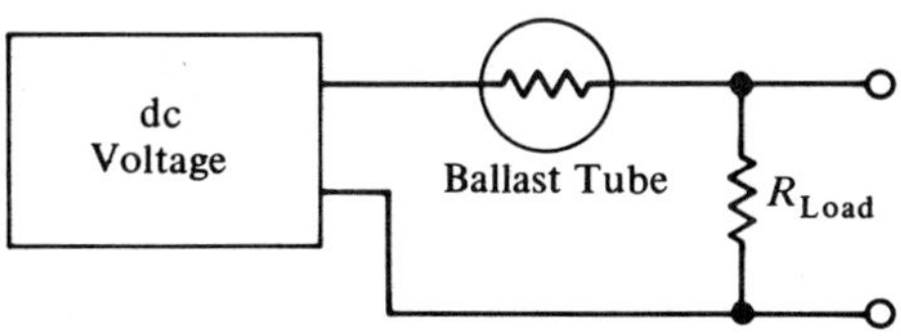

FIGURE 11.3

When the voltage from the rectifier–filter increases, the current in the series circuit increases. This increase in current causes the iron wire in the ballast tube to get hot. But the iron wire is such that its resistance increases when it gets hot. This increase in resistance means that a larger voltage drop occurs across the ballast tube, and the voltage across the load is maintained at a constant level. When the supply voltage decreases, there is less current, and the iron wire cools, thus decreasing its resistance and the voltage drop across it. With a smaller part of the total voltage appearing across the ballast tube, more appears across the load, and the load voltage remains steady even with the decrease in supply voltage.

The reason the ballast tube cannot be used as a voltage regulator when the load resistance changes becomes apparent when we recall from our example that an increase in load resistance caused a decrease in total current and an increase in load voltage which had to be offset by an increase in variable resistance. But the only way to increase the resistance of the ballast tube is by increasing the current. Since the current actually decreases, the ballast tube operates in the opposite manner to that in which it should when the load resistance changes.

11.3 MANUALLY OPERATED PARALLEL VOLTAGE REGULATOR

Another form of voltage regulator is the parallel voltage regulator. We will first consider a manually operated parallel voltage regulator before installing an automatically adjusted variable resistor. The circuit is a series–parallel resistive network. We will attempt to maintain 100 volts across the load by adjusting the variable resistor. We first take R_{Variable} to be 200 ohms. (See Figure 11.4.)

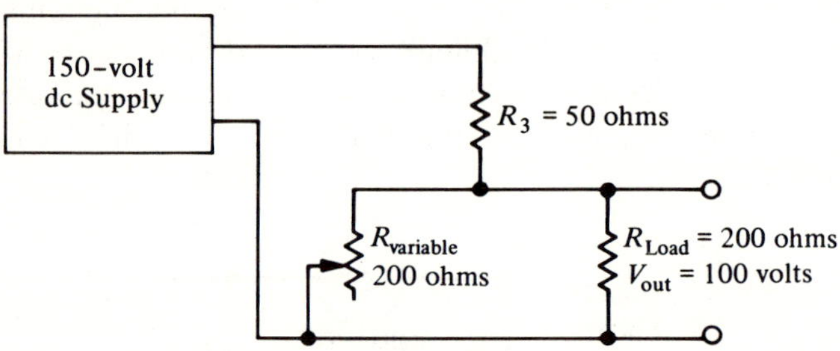

FIGURE 11.4 *Parallel voltage regulator.*

Notice that the load and variable resistance are in parallel. We can find the resistance of this parallel combination (R_{Parallel}) using the "product divided by the sum" technique. $R_{\text{Parallel}} = (R_{\text{Variable}} \times R_{\text{Load}})/(R_{\text{Variable}} + R_{\text{Load}}) = (200 \times 200)/(200 + 200) = 100$ ohms. The equivalent circuit then is shown in Figure 11.5. To find the total resistance of the circuit, since R_3 and R_{Parallel} are in series, we use $R_T = R_3 + R_{\text{Parallel}} =$

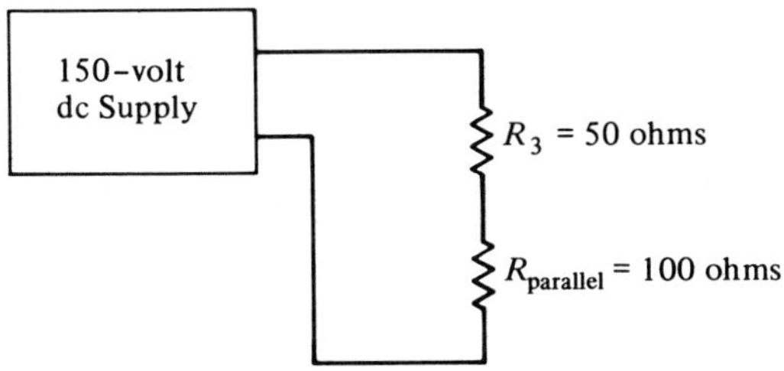

FIGURE 11.5

50 ohms + 100 ohms = 150 ohms. Since the applied voltage from the rectifier filter is 150 volts, we use Ohm's Law to find total current: $I_T = E/R_T$ = 150 volts/150 ohms = 1 A. This total current divides at the parallel branch, part of it going through the load and part through the variable resistance. Since we know the load resistance and load voltage, we can find current through the load. $I_{Load} = V_{Load}/R_{Load}$ = 100 volts/200 ohms = 0.5 A.

Now let us consider what action must be taken to maintain 100 volts across the load when the rectifier–filter begins to generate a larger voltage, for example, E = 200 volts. Before adjusting the variable resistor, we determine how the increase in supply voltage affects the voltage across the load. Since none of the resistances has been changed, the total opposition to current flow remains the same, R_T = 150 ohms. The total current can be calculated $I_T = E/R_T$ = 200 volts/150 ohms = 1.33 A. This total current passes through R_3, then divides between $R_{Variable}$ and R_{Load}. The voltage drop across R_3 is $V_3 = I_3 \times R_3$ = 1.33 A × 50 ohms = 66.5 volts. The remaining voltage from the supply is dropped across the parallel branch. $E = V_3 + V_{Parallel}$; 200 volts = 66.5 volts + $V_{Parallel}$; $V_{Parallel}$ = 133.5 volts. In a parallel circuit, the voltage drop across each branch is the same, so $V_{Parallel} = V_{Variable} = V_{Load}$ = 133.5 volts. The current through the load is $I_{Load} = V_{Load}/R_{Load}$ = 133.5 volts/200 ohms = 0.667 A.

Thus, we have observed that an increase in supply voltage (from 150 volts to 200 volts) causes an increase in voltage across the load (from 100 volts to 133.5 volts) and an increase in load current (from 0.5 A to 0.667 A). To return the load voltage to the steady 100 volts desired, we must adjust the variable resistor. To decrease the voltage drop across the parallel branch to 100 volts, more of the supply voltage must be dropped across R_3. In fact, since $V_{Supply} = V_3 + V_{Parallel}$ and 200 volts = V_3 + 100 volts, we see that the voltage drop across R_3 must be increased to 100 volts. Knowing V_3 and R_3, we find I_3 (which is equal to I_T); $I_T = I_3 = V_3/R_3$ = 100 volts/50 ohms = 2 A. This total current will divide at the parallel combination, part of the current going through the load and part through the variable resistance. We use Ohm's Law to find the current through the load for the desired load voltage of 100 volts. $I_{\text{Load}} = V_{\text{Load}}/R_{\text{Load}} = 100$ volts/200 ohms $= 0.5$ A. The rest of the current must pass through the variable resistor. $I_T = I_{\text{Variable}} + I_{\text{Load}}$; 2 A $= I_{\text{Variable}} + 0.5$ A; $I_{\text{Variable}} = 1.5$ A. What value of the variable resistance will allow a current of 1.5 A to flow? We know the voltage across the variable resistance will be 100 volts, the same as the voltage across the load, because the two resistances are in parallel. Thus, $R_{\text{Variable}} = V_{\text{Variable}}/I_{\text{Variable}} = 100$ volts/1.5 A $= 66.6$ ohms. To review, when the voltage from the supply increases (from 150 volts to 200 volts), the variable resistance in a parallel voltage regulator must decrease (from 200 ohms to 66.6 ohms) in order to maintain a steady load voltage (100 volts).

Without repeating the mathematics, we will state the consequences of a decrease in supply voltage. If the voltage from the rectifier–filter decreased, a decrease in voltage across the load would be observed along with a decrease in load current. In order to offset this decrease and maintain a constant output voltage across the load, the variable resistance must increase.

Variations in load resistance will also require adjustments of the variable resistor in the parallel voltage regulator to maintain a steady output voltage. Suppose the load resistance increases from 200 ohms to 400 ohms. We have already found currents and voltage drops for the circuit in its original balanced state. We will now compute output and current in the load for the unbalanced circuit and then determine what adjustments must be made in the variable resistance to return the circuit to its balanced state, that is, to return the output to 100 volts. (See Figure 11.6.) First, the resistance of the parallel combination is found [$R_{\text{Parallel}} = (R_{\text{Variable}} \times R_{\text{Load}})/(R_{\text{Variable}} + R_{\text{Load}}) = (200 \times 400)/(200 + 400) = 133.3$ ohms] and the equivalent circuit drawn. (See Figure 11.7.) From the equivalent circuit we can find the total resistance: $R_T = R_3 + R_{\text{Parallel}} = 50$ ohms $+ 133.3$ ohms $= 183.3$ ohms. Knowing the total voltage supplied and the total resistance of the circuit, we can find total current. $I_T = E/R_T = 150$ volts/183.3 ohms $= 0.818$ A. Since the equivalent circuit is a series circuit, $I_T = I_3 = I_{\text{Parallel}} = 0.818$ A. From Ohm's Law, the voltage drop across R_3 is $V_3 = I_3 \times R_3 = 0.818$ A $\times$ 50 ohms $= 41$ volts. In the series circuit, $V_T = V_3 + V_{\text{Parallel}}$; 150 volts $= 41$ volts $+ V_{\text{Parallel}}$; $V_{\text{Parallel}} = 109$ volts. Thus, increasing the load resistance has increased the voltage across the load from 100 volts to 109 volts. The load current is now $I_{\text{Load}} = V_{\text{Load}}/R_{\text{Load}} = 109$ volts/400 ohms $= 0.273$ A, a decrease in load current from 0.5 A.

Now we will return V_{Load} to 100 volts by adjusting R_{Variable}. The current in the load, when it produces the desired voltage drop, must

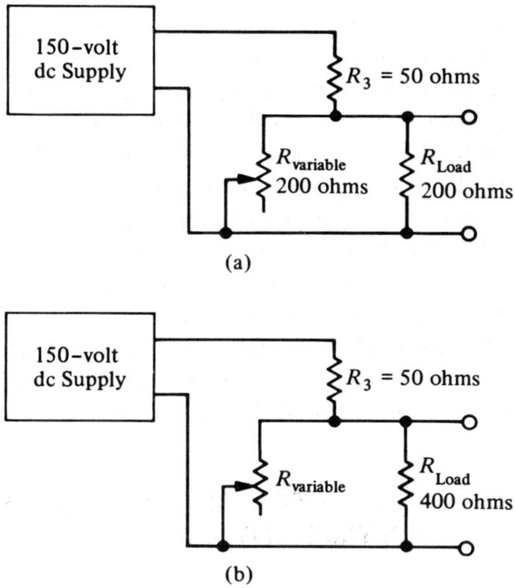

FIGURE 11.6 *(a) Original balanced state; (b) unbalanced state: R_{Load} increased (change $R_{Variable}$ to return output to* 100 *volts).*

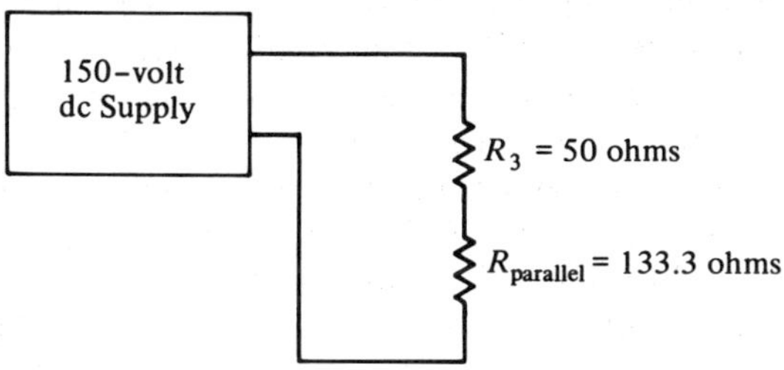

FIGURE 11.7

be $I_{Load} = V_{Load}/R_{Load} = 100$ volts/400 ohms $= 0.25$ A. Since the current in the parallel combination ($I_{Parallel} = 0.818$ A) is divided between the two branches, we have $I_{Parallel} = I_{Variable} + I_{Load}$; 0.818 A $= I_{Variable} + 0.25$ A; $I_{Variable} = 0.568$ A. The voltage across the two branches must be the same, so $V_{Variable} = V_{Load} = 100$ volts. We can now use Ohm's Law to find the variable resistance necessary to allow a current of 0.568 A with a voltage drop of 100 volts. $R_{Variable} = V_{Variable}/I_{Variable} = 100$ volts/ 0.568 A $= 176$ ohms. This represents a decrease in variable resistance.

To review, when the load resistance was increased (from 200 ohms to 400 ohms), the variable resistance had to be decreased (from 200 ohms to 176 ohms) to maintain a steady 100 volts across the load. Without repeating the mathematical procedures, we will state the conclusion which would be reached by calculating the effect of a decrease in load resistance. We invite you to practice the series and parallel relations and check these conclusions. If the load resistance is decreased, the load voltage will also decrease. To return load voltage to the desired constant level, the variable resistance of a parallel voltage regulator must be increased.

Summarizing all our observations concerning parallel voltage regulators, we see: (1) when the voltage across the load increases (due to either an increase in supply voltage or an increase in load resistance), the variable resistance must be decreased to return the load voltage to the desired level and (2) when the voltage across the load tends to decrease (due to a decrease in either supply voltage or load resistance), the variable resistance must be increased to maintain a steady load voltage.

11.4 COMPARISON OF SERIES AND PARALLEL VOLTAGE REGULATORS

At the risk of being repetitious, let us compare the series and parallel voltage regulators. In general, the action of the variable resistance in one is just the opposite of its action in the other. When the supply voltage increases, thus increasing the voltage across the load, the variable resistance in the series circuit is increased, whereas the variable resistance in the parallel circuit is decreased to return the voltage to the desired level. Correspondingly, when the voltage across the load decreases, the series variable resistance must be decreased, but the parallel variable resistance must be increased. If the load resistance increases, then the variable resistance in the series circuit must be increased, or the variable resistance in the parallel circuit decreased, to keep the load voltage at a constant level. On the other hand, if the load resistance decreases, then the series variable resistance must decrease, while the parallel variable resistance must increase. Therefore, (1) an increase in supply voltage or load resistance requires an increase in variable resistance in a series circuit voltage regulator but a decrease in variable resistance in a parallel circuit voltage regulator, and (2) a decrease in supply voltage or load resistance requires a decrease in series variable resistance but an increase in parallel variable resistance.

11.5 VR TUBE

We will automate the parallel voltage regulator by replacing the variable resistor with a VR (voltage regulator) tube. The VR tube is a two-element tube, a diode. However, in contrast to other diodes we have studied, the cathode in the VR tube is not heated, and the tube is not evacuated. The gas in this cold cathode tube is usually neon or argon. These gases can be ionized (stripped of one or more outer electrons) at specific voltages. When the VR tube reaches its ionizing, or firing, voltage, the ionized gas offers a low resistance path for the flow of current through the tube. Below the ionization voltage, the VR tube offers a large resistance to current flow. The schematic of the VR tube shows the cathode as a circle; there is no filament, since the cathode is not heated and the dot inside the envelope indicates that the tube is gas filled. (See Figure 11.8.)

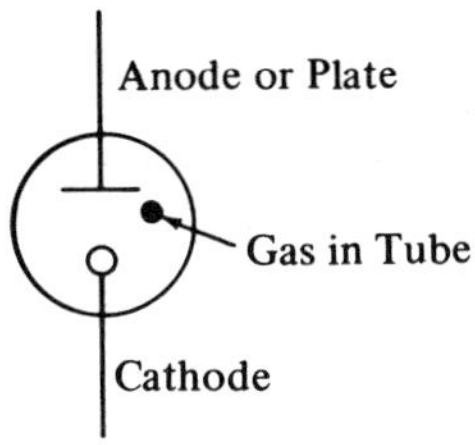

FIGURE 11.8 *VR tube.*

Let us consider a parallel voltage regulator circuit. A VR tube, which ionizes just above the desired load voltage, is selected. (See Figure 11.9.)

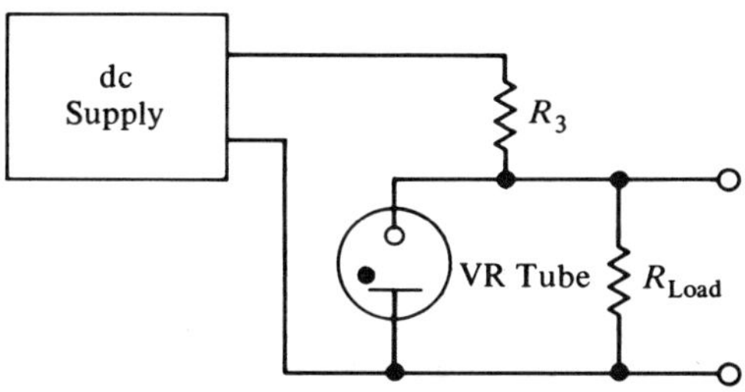

FIGURE 11.9

When the supply voltage increases, the voltage across the load and across the VR tube will increase. If the voltage goes above the desired level, the VR tube will ionize and thus provide a lower resistance path for current than the load provides. With more current traveling through the VR tube, less current travels through the load and, according to Ohm's Law ($V_{Load} = I_{Load}R_{Load}$), this will cause V_{Load} to drop back to the desired level. If the supply voltage decreases, the voltage across the load will decrease. V_{Load} can be brought back up to the desired level by increasing current in the load. This is done automatically by the VR tube as follows: When the voltage drops below the desired level, the ionization level for this VR tube, fewer gas molecules are ionized and the tube offers a high resistance to current; thus, less current goes through the VR tube branch of the parallel circuit and more through the load. This increases V_{Load} to the desired level.

You should recall that when the load resistance increases, the voltage across the parallel combination increases. To counteract this increase in V_{Load}, the excess voltage will ionize more gas molecules in the VR tube allowing more current to flow. With more current through the VR tube and less through the load, the voltage drop across the load returns to normal. Decreasing the load resistance causes the voltage across the parallel combination to decrease. This decrease in VR tube voltage will reduce the number of ionized gas molecules in the tube and increase its resistance. An increased tube resistance means less current in the tube, more current in the load (the other branch of the parallel circuit), and an increased load voltage, bringing it back up to the desired level.

We said that the ionizing voltage of the VR tube should be the same as the desired load voltage. If the desired output voltage is greater than the rating of the VR tube, several tubes may be used in series in such a way that the sum of the ionizing voltages equals the load voltage as shown in Figure 11.10.

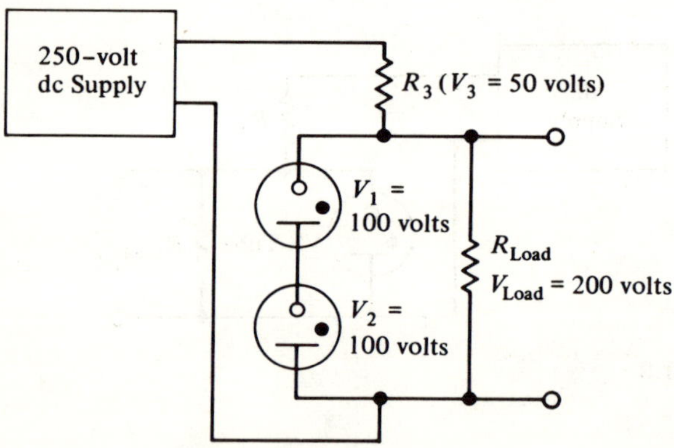

FIGURE 11.10

The VR tube operates over a rather limited current range, approximately from 0 to 40 mA. Thus, any extreme change in supply voltage or load resistance, which would require large VR tube current changes to counteract it, is beyond the capacity of the cold cathode VR tube.

11.6 ZENER DIODE VOLTAGE REGULATOR

Another diode which is gaining wide acceptance as a voltage regulator is the Zener diode. The Zener diode, you remember, is a reverse biased *PN* junction diode, whose covalent bonds have been broken by a large reverse voltage (called the Zener potential), thus allowing a current flow in the reverse direction. (See Figure 11.11.)

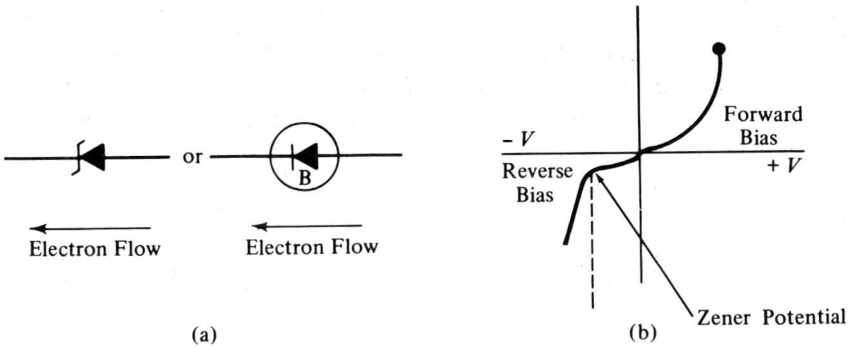

FIGURE 11.11 *(a) Zener-diode schematic; (b) Zener potential.*

In order to use the Zener diode as a voltage regulator, the heat generated by the large reverse current must be dissipated. If the junction is not kept cool, the diode will be destroyed. Thus the Zener diode, like all large current carrying semiconductor devices, requires a heat sink to remove heat from the junction. Usually the heat is simply dissipated through the metal chassis.

Placing the Zener diode in the parallel voltage regulator circuit, we have the schematic in Figure 11.12. Choosing a Zener diode whose Zener potential is the same as the desired load voltage, the action of the diode will be the same as that of a VR tube. We will repeat the four adjustments that might be required using our shorthand notation.

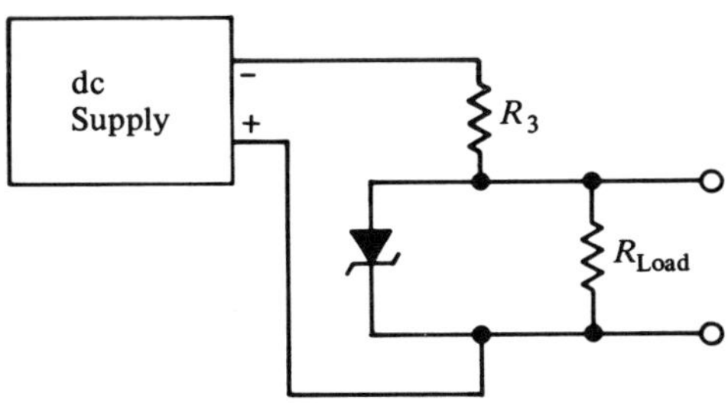

FIGURE 11.12

$$\text{When } E\uparrow \text{ then } \left\{\begin{array}{l} V_{\text{Load}}\uparrow \\ \quad \| \\ V_{\text{Zener}}\uparrow, \text{ then } I_{\text{Zener}}\uparrow, \text{ then } \left\{\begin{array}{l} I_{\text{Load}}\downarrow \\ \text{and} \\ V_{\text{Load}}\downarrow \text{ to desired level}\end{array}\right.\end{array}\right.$$

Since these are rather involved statements, we will read through this first one: "When the supply voltage increases ($E\uparrow$), then the voltage across the load and the equal voltage across the Zener diode both increase $\begin{pmatrix} V_{\text{Load}}\uparrow \\ V_{\text{Zener}}\uparrow \end{pmatrix}$. But an increase in V_{Zener} beyond the Zener potential causes an increase in current through the diode ($I_{\text{Zener}}\uparrow$) and a corresponding decrease in current through the load ($I_{\text{Load}}\downarrow$). This decrease in load current causes the load voltage to decrease ($V_{\text{Load}}\downarrow$), returning it to the desired constant level."

$$\text{When } E\downarrow, \text{ then } \left\{\begin{array}{l} V_{\text{Load}}\downarrow \\ \quad \| \\ V_{\text{Zener}}\downarrow, \text{ then } I_{\text{Zener}}\downarrow, \text{ then } \left\{\begin{array}{l} I_{\text{Load}}\uparrow \\ V_{\text{Load}}\uparrow \text{ to desired level}\end{array}\right.\end{array}\right.$$

The last two statements involve changes in R_{Load}.

$$\text{When } R_{\text{Load}}\uparrow, \text{ then } \left\{\begin{array}{l} V_{\text{Load}}\uparrow = V_{\text{Zener}}\uparrow, \text{ then } I_{\text{Zener}}\uparrow \\ \qquad\qquad \text{and} \qquad\qquad\qquad\qquad \text{and} \\ R_{\text{Parallel}}\uparrow, \text{ then } R_{\text{Total}}\uparrow, \text{ then } I_T\downarrow = I_{\text{Parallel}}\downarrow \end{array}\right\}$$

$$\text{then } \left\{\begin{array}{l} I_{\text{Load}}\downarrow \\ \text{and} \\ V_{\text{Load}}\downarrow \\ \text{to desired} \\ \text{level}\end{array}\right.$$

$$\text{When } R_{\text{Load}}\downarrow, \text{ then } \left\{\begin{array}{l} V_{\text{Load}}\downarrow = V_{\text{Zener}}\downarrow, \text{ then } I_{\text{Zener}}\downarrow \\ \text{and} \qquad\qquad\qquad\qquad\qquad\qquad \text{and} \\ R_{\text{Parallel}}\downarrow, \text{ then } R_{\text{Total}}\downarrow, \text{ then } I_T\uparrow = I_{\text{Parallel}}\uparrow \end{array}\right\}$$

$$\text{then } \left\{\begin{array}{l} I_{\text{Load}}\uparrow \\ \text{and} \\ V_{\text{Load}}\uparrow \\ \text{to desired} \\ \text{level}\end{array}\right.$$

Questions

(1) What are the two types of variations which voltage regulators try to correct?

(2) Explain the action of a VR tube, a Zener diode, and a ballast tube.

CHAPTER 12

Voltage Dividers

12.1 A SIMPLE VOLTAGE DIVIDER

A single electronic circuit often requires several different dc voltages. It would be expensive and bulky to have a separate rectifier–filter–voltage regulator for each needed voltage. However, we can obtain any dc voltage from 0 to the maximum voltage supplied by placing taps across the load resistor of just one power supply. In Figure 12.1, we show the divi-

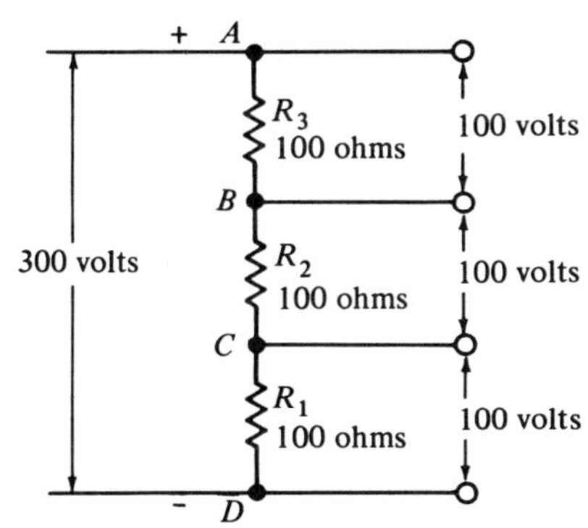

FIGURE 12.1 *Voltage divider.*

sion of a 300-volt supply into three 100-volt sections. The total resistance of the series load is $R_T = R_1 + R_2 + R_3 = 100$ ohms $+ 100$ ohms $+ 100$ ohms $= 300$ ohms. Finding total current, $I_T = E/R_T = 300$ volts/300 ohms $= 1$ A. We can now calculate the voltage drop across each resistor. $V_1 = I_1R_1 = 1 \text{ A} \times 100 \text{ ohms} = 100$ volts; $V_2 = I_2R_2 = 100$ volts; $V_3 = I_3R_3 = 100$ volts. The three voltage drops are equal as expected, since the series resistances are also equal.

12.2 VOLTAGE DIVIDER WITH EXTERNAL CIRCUITS

This circuit is a simple voltage divider. If the resistance of an external circuit, attached across part of the voltage divider, varies, this changes the resistance between the terminals, and we would expect the voltage across the terminals to vary. The voltage regulator will keep the total voltage constant in spite of changes in total load resistance, so the sum of the voltage drops must equal this steady supply voltage.

We will now take this voltage divider and place several external resistance circuits across its terminals.

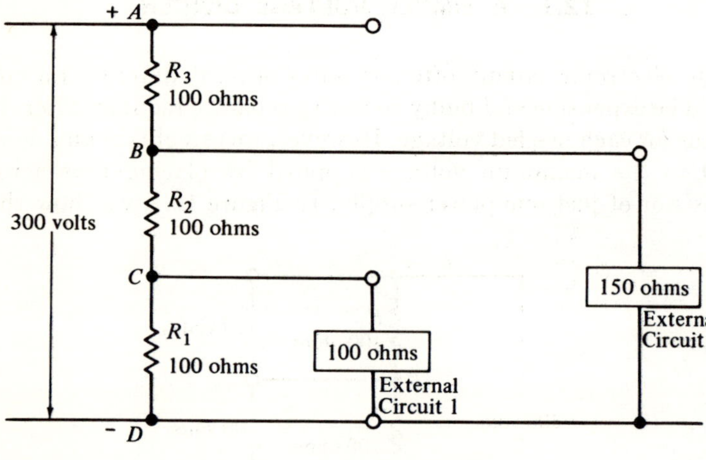

FIGURE 12.2 *Voltage divider with external circuits.*

In Figure 12.2, the resistance between C and D is reduced by the external circuit resistance of 100 ohms (external circuit 1). Since R_1 and the external resistance are in parallel, the resistance between C and D is

$$R_{CD} = \frac{R_1 \times \text{external resistance 1}}{R_1 + \text{external resistance 1}} = \frac{100 \times 100}{100 + 100} = 50 \text{ ohms}$$

R_{CD} is in series with R_2, so their combined resistance is $R = R_{CD} + R_2 = 50 \text{ ohms} + 100 \text{ ohms} = 150$ ohms. This resistance R (due to R_1, R_2, and external circuit 1) is in parallel with the resistance of external circuit 2. Thus, the total resistance between B and D is

$$R_{BD} = \frac{R \times \text{external resistance 2}}{R + \text{external resistance 2}} = \frac{150 \times 150}{150 + 150} = 75 \text{ ohms}$$

Since R_{BD} and R_3 are in series, the total resistance between A and D is $R_{AD} = R_{BD} + R_2 = 75$ ohms $+ 100$ ohms $= 175$ ohms. If the resistance from B to D is 75 ohms and from C to D 50 ohms, then the resistance from B to C would be 25 ohms ($R_{BD} = R_{BC} + R_{CD}$; 75 ohms $=$ $R_{BC} + 50$ ohms; $R_{BC} = 25$ ohms). An equivalent circuit could then be drawn. (See Figure 12.3.)

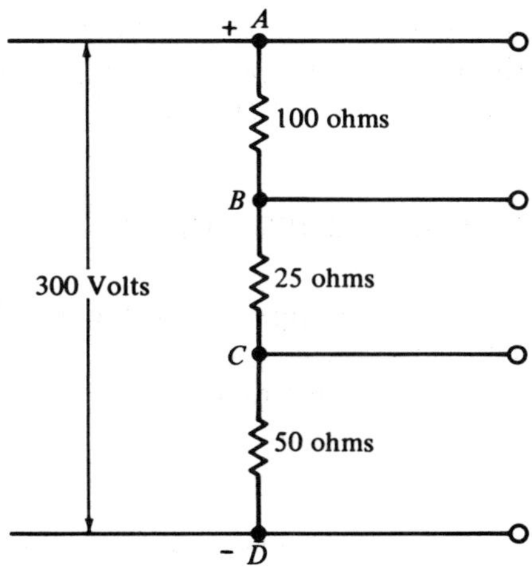

FIGURE 12.3

Knowing the total resistance ($R_{AD} = 175$ ohms) and total voltage, we can find the current in this equivalent circuit. $I_T = E/R_T = 300$ volts/175 ohms $= 1.714$ A. In the series circuit, the current is the same in each resistor; we use this current to find the voltage drop across each resistor. $V_{AB} = I_T R_{AB} = 1.714$ A $\times$ 100 ohms $= 171.4$ volts; $V_{BC} = I_T R_{BC} =$ 1.714 A $\times$ 25 ohms $= 42.9$ volts; $V_{CD} = I_T R_{CD} = 1.714$ A $\times$ 50 ohms $=$ 85.7 volts. Note that the total voltage is still 300 volts: $E = V_{AB} + V_{BC} + V_{CD} = 171.4$ volts $+ 42.9$ volts $+ 85.7$ volts $= 300$ volts. But the voltages across each set of terminals are no longer equal as they were for the same voltage divider before external circuits were attached (Figure 12.1).

The voltage divider does have its limitations. As we have seen in this last example, when the resistances of the external circuits are the same size as, or smaller than, the load resistance divisions to which they are attached, a portion of the current flows through each branch of the parallel circuit (load division and external circuit), and the voltage across each load division is changed significantly. However, if the resistance of the external circuit is extremely large in comparison to the resistance between the terminals, most of the current will continue to flow through the load resistance rather than through the external circuit, and the voltage remains almost unchanged. In other words, if the external circuits require large currents, the voltage divider does not maintain the same voltage drops that were observed across its divisions when no external circuit was attached. On the other hand, if the external circuits do not draw much current (that is, they have a high resistance), the voltage drops remain virtually unchanged.

We might mention one other limitation: Since the voltage divider is a resistor, some of the dc from the power supply will be dissipated as heat before it can be used by the external circuit.

12.3 A DC POWER SUPPLY

An entire dc power supply might appear as in Figure 12.4. This dc power supply consists of a full-wave rectifier, a capacitor input π filter, a parallel voltage regulator utilizing a Zener diode, and a voltage divider with three voltage taps for external circuits.

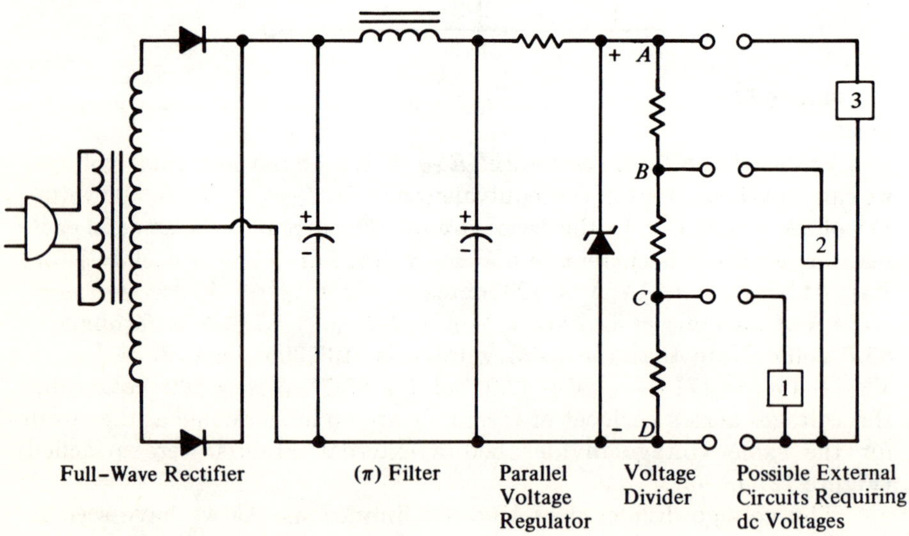

FIGURE 12.4 *dc power supply.*

12.4 GROUND

In the above dc power supply, the negative terminal (D) is common to each of the external circuits. This common point, called ground, does not affect the independent operation of the external circuits. Ground is given the schematic representation ⏚ . Using the ground schematic, our voltage divider and possible external circuits are drawn as shown in Figure 12.5. The term ground does not necessarily imply that this common

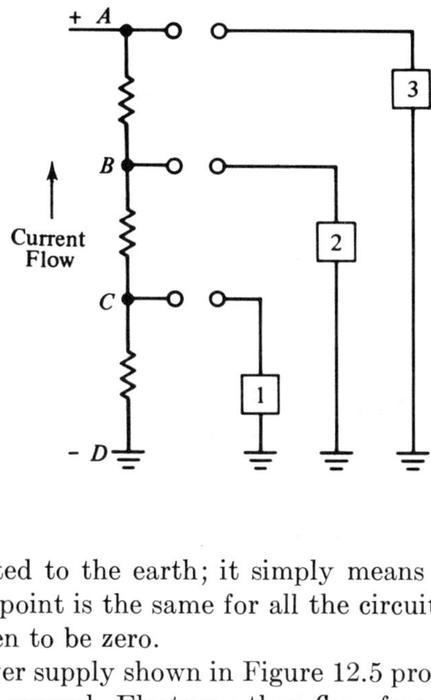

FIGURE 12.5

point is connected to the earth; it simply means that the potential, or voltage, at this point is the same for all the circuits. Usually the ground potential is taken to be zero.

The dc power supply shown in Figure 12.5 provides positive voltages with respect to ground. Electrons thus flow from ground (direction of current flow is indicated on the diagram). We can obtain negative voltages with respect to ground by relocating the common ground as shown in Figure 12.6. (Note that the center tap from the rectifier has been changed from Figure 12.4 in order to reverse the polarity.) Since point A is now negative, and the ground, or common point, is positive, the direction of electron flow is toward ground. When current flows to ground, the voltages are said to be negative.

If we move the common point (ground) to some terminal other than D, we can obtain both positive and negative voltages as shown in Figure 12.7. In this circuit, ground is at point C. Since C is positive with respect to D, current flows through external circuit 1 from D to ground (C). Thus, the voltage across circuit 1 is negative. Point B is positive

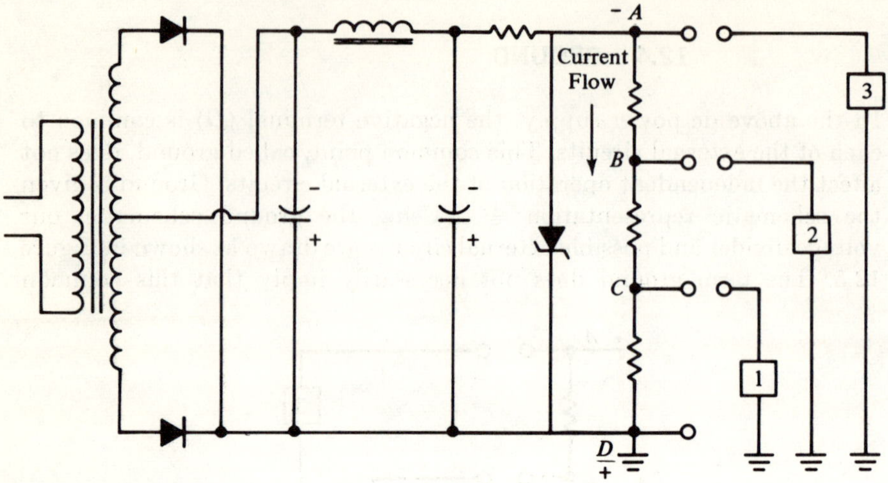

FIGURE 12.6

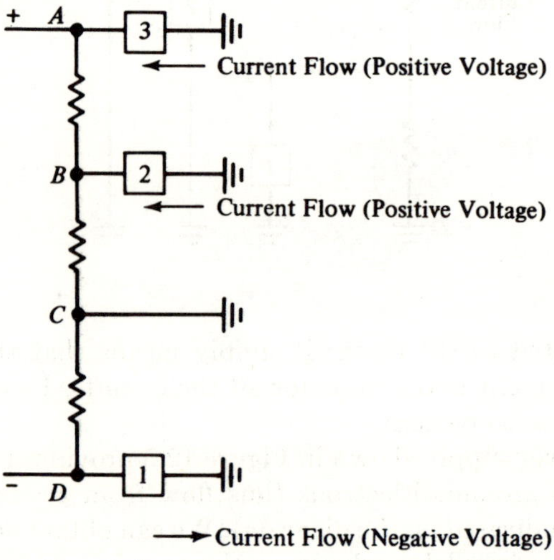

FIGURE 12.7

with respect to *C*. Therefore, current flows from ground (*C*) through external circuit 2 to *B*, and this circuit has a positive voltage. Point *A* is also positive with respect to ground, so external circuit 3 has a positive voltage drop.

12.5 VARIABLE VOLTAGE DIVIDER

A variable voltage divider may be employed in place of the fixed load resistances we have been discussing. The variable voltage divider may be

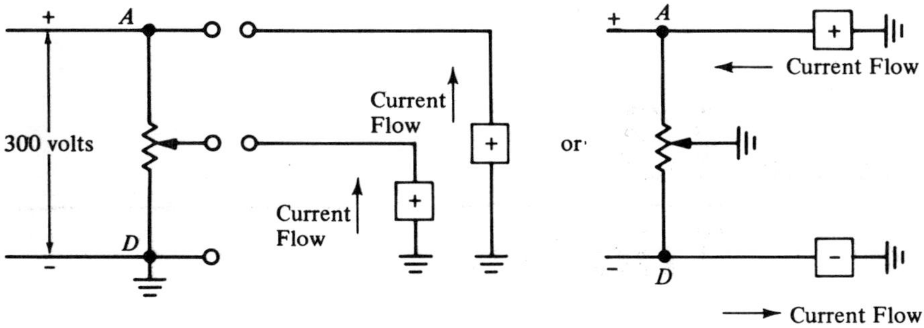

FIGURE 12.8

called a variable resistor, a potentiometer, or a rheostat. (See Figure 12.8.)

Questions

(1) What function does a voltage divider perform in addition to providing a variety of voltages that are less than the terminal voltage of the power supply?

CHAPTER 13

Triodes

13.1 TRIODE SCHEMATIC

The audion, or triode, patented by Lee DeForest in 1907, like the diode, consists of a directly heated cathode and an anode in an evacuated envelope. But placed between the anode and cathode is a coil of wire called the grid. A drawing and schematic of the modern triode are shown in Figure 13.1.

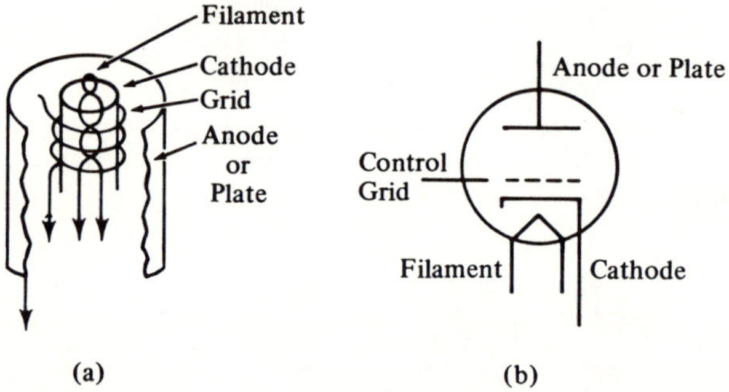

FIGURE 13.1 *(a) Indirectly heated triode; (b) triode schematic.*

The triode, which is a thermionic device (electrons are boiled off the heated cathode), utilizes the law of electrostatics to control the flow of electrons to the plate. If the grid is made negative, it will repel the negative electrons coming from the cathode and thus make it more difficult for them to reach the positive plate. As the grid is made more negative with respect to the cathode, fewer electrons reach the plate.

Making the plate more positive will overcome this somewhat. But the distance between plate and grid is greater than between grid and cathode, and hence the plate does not have as much effect on the electrons as the grid.

13.2 TRIODE CIRCUIT

In the triode circuit, the plate is made positive and the grid negative with respect to the cathode as shown in Figure 13.2. Current flow in

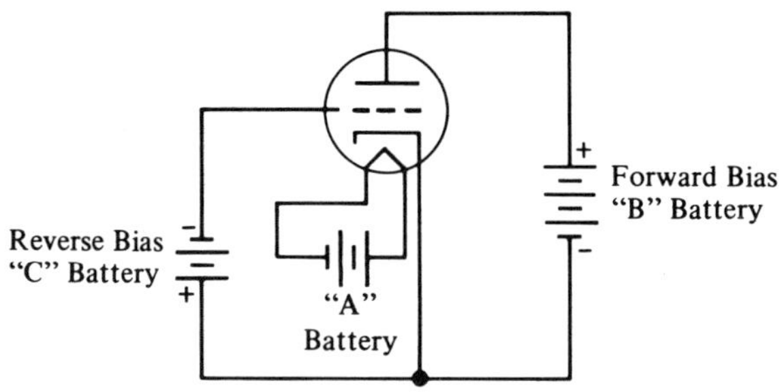

FIGURE 13.2 *Triode circuit.*

the plate circuit is indicated. Note that electrons from the "B" battery do not move to the positive terminal of the "C" battery and through it to the grid, because there is no closed circuit (the electrons cannot flow from grid to plate). As the grid is made more and more negative, the flow of electrons from cathode to plate decreases and finally stops altogether even though the plate remains positive. Thus, the grid voltage at a particular value cuts off the flow of electrons. This cutoff voltage, or bias at cutoff, is, of course, negative with respect to the cathode. If the grid is made positive, it no longer has any control over the flow of electrons. Therefore the less negative (or more positive) the grid becomes, the more electrons reach the plate and the greater the plate current. On the other hand, if the grid voltage is held at some constant value above cutoff, the plate current can also be increased by increasing the plate voltage, thus causing it to attract more electrons.

13.3 TRIODE AMPLIFIER

It would require quite a large change in plate voltage to produce the same change in plate current as that caused by a relatively small change in grid voltage. Since small changes in grid voltage produce quite large changes in plate current, the change in grid voltage, or the signal, is said to be amplified in the plate circuit. A triode amplifier is shown in Figure 13.3.

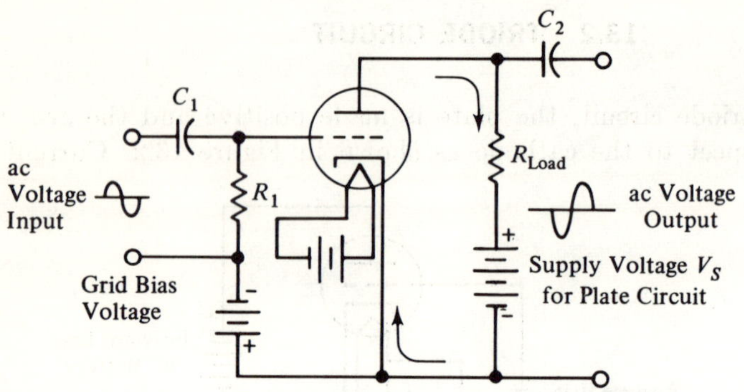

FIGURE 13.3 *Triode amplifier.*

The ac signal is coupled to the amplifier with an RC circuit. The capacitor C_1 removes any previous dc from the signal; the bias voltage establishes a desired dc level for the incoming signal. This small changing signal, when it appears on the grid of the triode, produces large changes in plate current. Since the same current flows in all components of the series plate circuit, these large changes in plate current are also seen as changes in current in the load resistance and cause large changes in load voltage ($V_L = I_P R_L$). The supply voltage is constant and is equal to the sum of the voltage drops about the plate circuit. Current flows through the triode, which has internal resistance and thus produces a voltage drop (V_P) between cathode and plate, and through the load resistor, which produces voltage drop (V_L). Therefore, in equation form, $E = V_P + V_L$. Since E is constant, the voltage drop across the resistor (V_L) must increase as the voltage drop across the tube (V_P) decreases. In order for the voltage across the triode to change, the triode must act as a variable resistor. As the grid goes positive, or less negative, on the positive half cycle of the input signal, there is less opposition to current flow in the triode. The resistance of the triode decreases, so the voltage across the triode decreases. Therefore, the output voltage, taken across the triode, goes negative when the input signal goes positive; there is a polarity inversion in the triode amplifier. Note that coupling capacitor C_2 removes dc from the output voltage. The voltage input on the grid (voltage drop from grid to cathode), the voltage across the triode, and the output voltage across capacitor C_2 are shown in Figure 13.4.

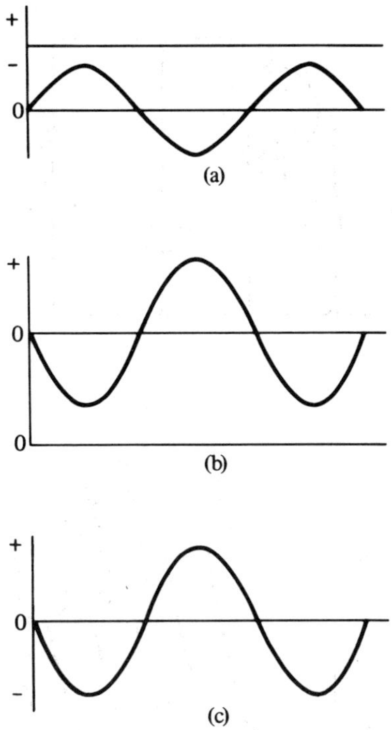

FIGURE 13.4 *(a) Voltage input on grid V_g; (b) voltage across tube V_p; (c) output voltage.*

13.4 TRIODE CHARACTERISTIC CURVES

A circuit with variable voltage supplies to grid and plate can be used to experimentally determine characteristics for an individual triode. (See Figure 13.5.). Notice that voltmeters are placed across portions of the circuit to measure the voltage drop from grid to cathode (V_g) and from plate to cathode across the triode (V_P). There is also an ammeter to measure current in the plate circuit (I_P).

If the grid voltage is held at a constant value and the plate voltage gradually increased, we will observe an increasing plate current. Plotting the value of plate current at each value of plate voltage, we obtain a curve which shows the changes in I_P that occur with corresponding changes in V_P, when V_g is held constant. If a different constant value of V_g were chosen, we could obtain a second curve for I_P and V_P. The graph showing curves obtained at various constant values of V_g is called the plate characteristic curve. (See Figure 13.6.)

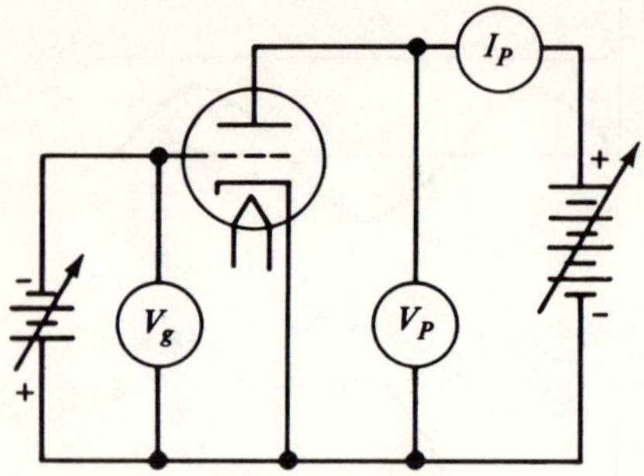

FIGURE 13.5

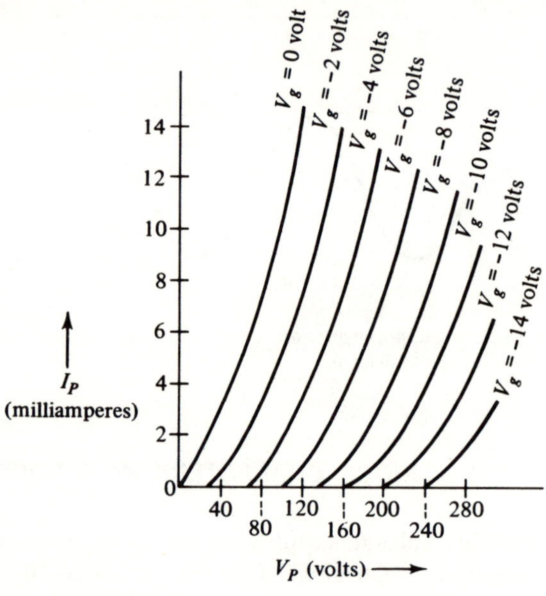

FIGURE 13.6 *Typical plate characteristic graph.*

A second set of curves can be obtained by keeping the plate voltage constant and determining how changes in grid voltage affect plate current. This graph is called the transfer characteristic or mutual characteristic. (See Figure 13.7.)

Finally, if we keep the plate current constant and vary the grid voltage, which varies the resistance of the triode, the plate voltage will vary. This gives us a third set of curves called the constant current characteristic. (See Figure 13.8.)

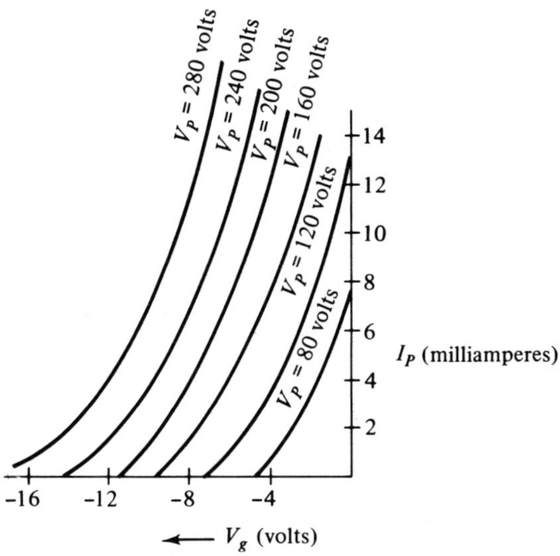

FIGURE 13.7 *Typical transfer characteristic graph.*

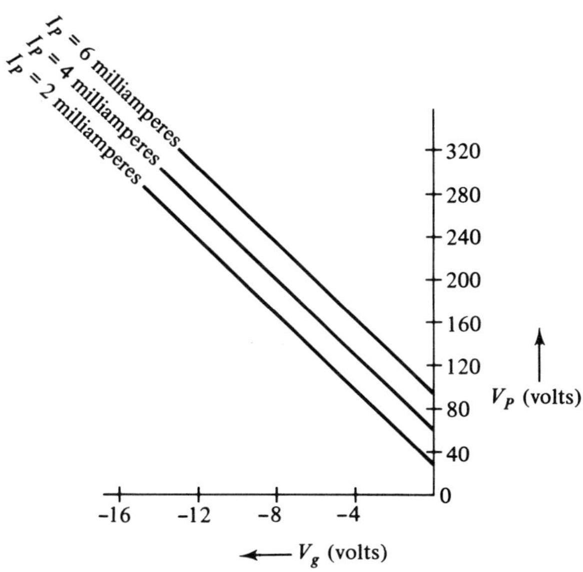

FIGURE 13.8 *Typical constant current characteristic graph.*

13.5 TRIODE CHARACTERISTICS

We can obtain three constants which are characteristic of the individual triode. The first of these constants gives the voltage amplification of the triode. It is called the amplification factor and given the symbol μ (pronounced mu). The amplification factor is simply the ratio of the change in plate voltage to the corresponding change in grid voltage, when the current is held constant. A change in any quantity can be represented by preceding the symbol for the quantity by the Greek letter delta (Δ). Thus, ΔV_P (read delta V_P) means a change in plate voltage, ΔV_g a change in grid voltage. Using this notation, we write

$$\mu = \frac{\Delta V_P}{\Delta V_g} \qquad (I_P \text{ constant})$$

We can find the amplification factor using any one of the three characteristic graphs. Since the plate characteristic is the most common of the three, we will use it. First, choose a constant plate current, for example, $I_P = 6$ mA. Next, drawing a line at the 6-mA level, we see that this line crosses all the curves on our graph. (See Figure 13.9.) If we

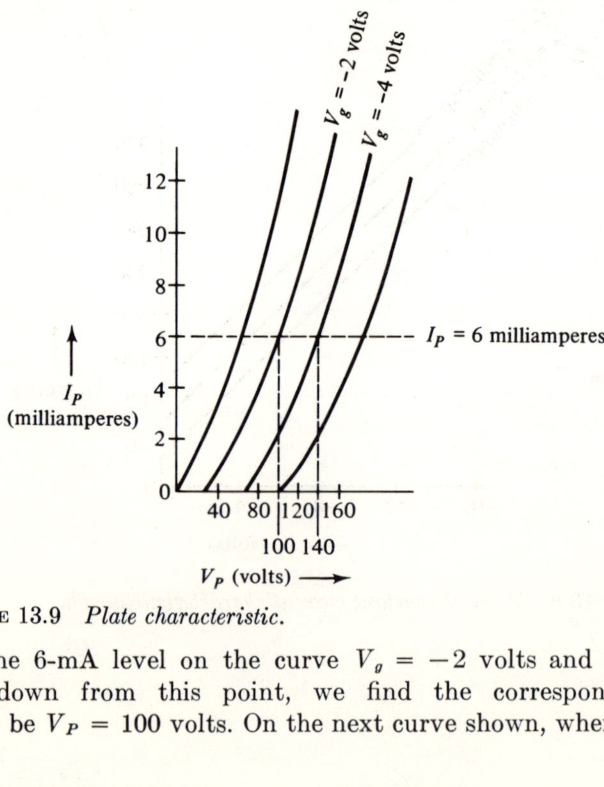

FIGURE 13.9 *Plate characteristic.*

look at the 6-mA level on the curve $V_g = -2$ volts and drop a line straight down from this point, we find the corresponding plate voltage to be $V_P = 100$ volts. On the next curve shown, when $V_g = -4$

volts and I_P is still a constant 6 mA, the corresponding value of V_P is 140 volts. For a constant plate current of I_P = 6 mA, V_P = 100 volts when $V_g = -2$ volts, and V_P = 140 volts when $V_g = -4$ volts. The change in plate voltage, ΔV_P, is from 100 volts to 140 volts, a change of 40 volts; the corresponding change in grid voltage is from -2 volts to -4 volts, a change of 2 volts. Substituting these values in the above equation for the amplification factor, we find $\mu = \Delta V_P/\Delta V_g$ = 40 volts/2 volts = 20. This amplification factor of 20 means that ideally the input voltage will be increased, or amplified, 20 times by the triode.

Earlier, we mentioned that in the plate circuit, the supply voltage was equal to the sum of the voltage drops about the series circuit and that the triode offered resistance to the flow of current, thus causing a voltage drop. The resistance of the triode is referred to as plate resistance (r_P). Referring to this opposition to current flow as resistance may initially seem incorrect, since we recall that the cathode and plate are separated by a dielectric (air or vacuum) and we might then expect the triode to act like a capacitor and have a large reactance. However, when in operation, the triode passes electrons from cathode to plate, it does not charge like a capacitor, so the capacitor analogy breaks down.

Ohm's Law tells us that resistance is voltage divided by current. Therefore, the plate resistance (r_P) can be found by dividing the change in plate voltage (ΔV_P) by the change in plate current (ΔI_P).

$$r_P = \frac{\Delta V_P}{\Delta I_P} \qquad (V_g \text{ held constant})$$

Since we want to find plate resistance, we naturally keep the grid voltage constant and observe changes in plate voltage and plate current. Suppose we choose a constant grid voltage of -4 volts ($V_g = -4$ volts). In the plate characteristic graph, $V_g = -4$ volts along one of the curves. (See Figure 13.10.) Now we look at two points on this curve: one at the 6-mA level (I_P = 6 mA) and the other at the 8-mA level (I_P = 8 mA). Dropping lines straight down from these points, we find that V_P = 140 volts when I_P = 6 mA, and V_P = 160 volts when I_P = 8 mA. The change in plate voltage is from 140 volts to 160 volts (ΔV_P = 20 volts); the change in plate current is from 6 mA to 8 mA (ΔI_P = 2 mA). Using our equation, the plate resistance of the triode is $r_P = \Delta V_P/\Delta I_P$ = 20 volts/2 mA = 20 volts/0.002 A = 10,000 ohms. Thus, when the grid voltage is $V_g = -4$ volts, the resistance of the triode is 10,000 ohms or 10 kiloohms. (The kilo stands for one thousand, 1000, just as milli stands for one thousandth, 0.001.)

Transconductance, in industry usually called mutual conductance, is the "goodness" factor of the triode; it is a quantitative description of

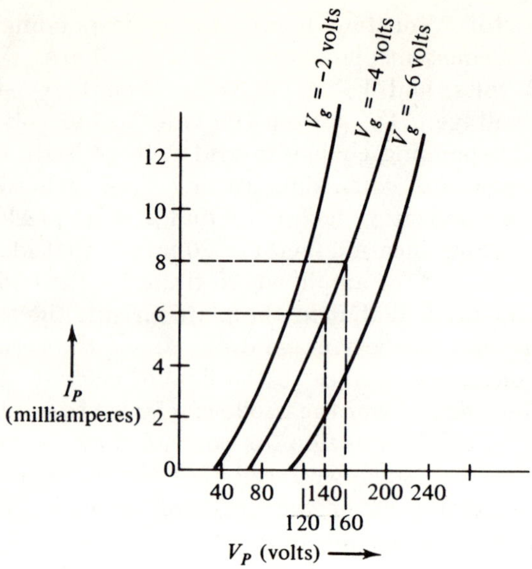

FIGURE 13.10

how well the triode conducts. Transconductance (g_m) is, by definition, the change in plate current (ΔI_P) divided by the change in grid voltage (ΔV_g), when the plate voltage is kept constant.

$$g_m = \frac{\Delta I_P}{\Delta V_g} \qquad (V_P \text{ constant})$$

Notice that the definition is current divided by voltage, not voltage divided by current as in Ohm's Law. Since transconductance (current/voltage) is just the reverse (reciprocal) of V/I, which gives resistance in ohms, the unit of transconductance is called mho (ohm spelled backwards). This is consistent, since transconductance is a measure of how well the tube conducts, not of how "well" it resists.

To find g_m we first choose a constant value for V_P, for example, $V_P = 160$ volts, and draw a vertical line on the plate characteristic graph in Figure 13.11 at 160 volts. Notice that V_P has the same value (160 volts) all along this line. This vertical line intersects curves with values of the grid voltage (V_g) equal to −4 volts and −6 volts. Looking horizontally across from these points of intersection, we can read the corresponding values of plate current. The level of plate current (I_P) is 4 mA when $V_g = -6$ volts and $I_P = 8$ mA when $V_g = -4$ volts. Thus, I_P changes from 4 mA to 8 mA, a change of 4 mA; correspondingly, V_g changes from −6 volts to −4 volts, a change of 2 volts. Substituting

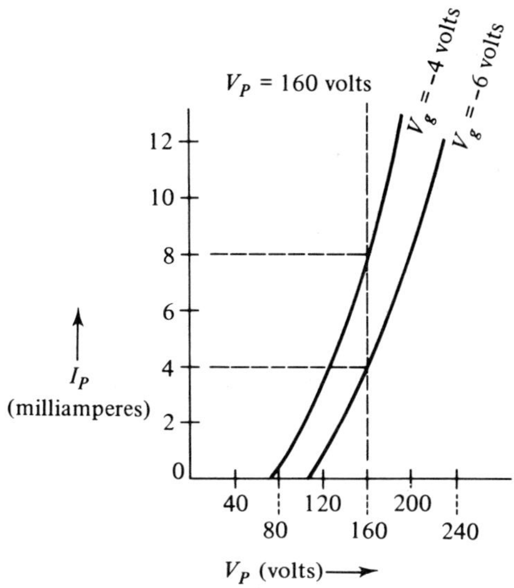

FIGURE 13.11

in the equation for transconductance, $g_m = \Delta I_P/\Delta V_g = 4$ mA/2 volts = 0.004 A/2 volts = 0.002 mho.

The three constants (amplification factor $\mu = \Delta V_P/\Delta V_g$, plate resistance $r_P = \Delta V_P/\Delta I_P$, and transconductance $g_m = \Delta I_P/\Delta V_g$) are interrelated as shown.

$$\mu = \frac{\Delta V_P}{\Delta V_g} = \frac{\Delta I_P}{\Delta V_g} \times \frac{\Delta V_P}{\Delta I_P}$$

or

$$\mu = g_m r_P$$

Thus, if two of the triode constants are known, the third may be determined without reference to a graph by using the above equation. For example, we found from the plate characteristic graph that $g_m = 0.002$ mho and $r_P = 10{,}000$ ohms. We can find the amplification factor by substituting in the equation $\mu = g_m \times r_P$. Thus, $\mu = 0.002$ mho $\times$ 10,000 ohms = 20. This is, of course, the same value obtained earlier from the graph. Note that the amplification factor ($\mu = \Delta V_P/\Delta V_g$) does not have any units, because it is volts divided by volts (or mhos times ohms).

13.6 LOAD LINE

Returning to the triode, we will show that only specific values of plate current and voltage are possible for a particular circuit. (See Figure

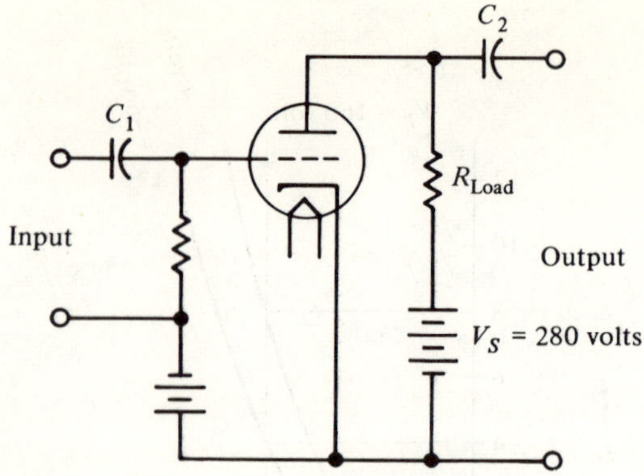

FIGURE 13.12

13.12.) We already know that as the grid is made more negative with respect to the cathode, eventually the cutoff voltage is reached, and no current flows in the triode. If we reduce the plate current to zero ($I_P = 0$) by adjusting the grid voltage, then the plate voltage will equal the supply voltage ($V_P = E = 280$ volts) at this point. This is because there is no current flowing in the circuit and hence no voltage is dropped across the load resistor (the entire potential is applied across the tube). We will plot this point ($I_P = 0$, $V_P = 280$) later. But first let us consider what happens when we connect a wire from cathode to plate in the triode. The wire short-circuits the triode, so there is no voltage drop across the tube ($V_P = 0$). We next measure plate current with an ammeter and find $I_P = 10$ mA. I_P can also be found using Ohm's Law, $I_P = E/R_{\text{Load}}$. This gives us a second point ($I_P = 10$, $V_P = 0$).

Now we are ready to plot both of the above points on the plate characteristic graph and connect them with a straight line. (See Figure 13.13.) This line is called the load line for the amplifier. Suppose we decide on an operating voltage for the grid of $V_g = -4$ volts. The point at which the curve, representing a grid bias of -4 volts, intersects the load line is called the operating point. For our example, the values of plate current and plate voltage corresponding to this operating point are $I_P = 5$ mA and $V_P = 130$ volts. Note that I_P was found by projecting the operating point onto the current axis (drawing a horizontal line from the operating point to the axis); V_P was found by projecting the operating point onto the voltage axis (drawing a vertical line from the operating point to the axis).

If an ac signal, which varies 4 volts from peak to peak, is placed on the grid of this triode amplifier biased at $V_g = -4$ volts, the corresponding variations in plate current and plate voltage can be found using

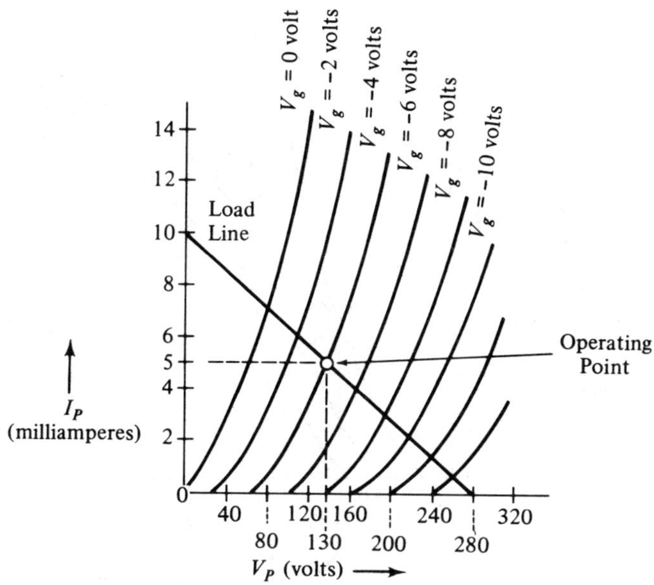

FIGURE 13.13

the load line as shown in Figure 13.14. First, since the total change in V_g is 4 volts and the grid is biased at -4 volts, the grid signal must vary from -2 volts to -6 volts. Finding the points at which plate characteristic curves with $V_g = -2$ volts and $V_g = -6$ volts intersect the load line, we draw lines from these points and from the operating point perpendicular to the load line. The waveform for V_g is shown on these lines. Next, projecting these three points onto the current axis, we find that plate current varies from 4 mA to 6 mA, and the waveform for I_P is shown on these horizontal lines of projection. Projecting the same three points onto the voltage axis, we see that plate voltage varies from 100 volts to 160 volts. Notice that V_P, shown on the vertical lines, is out of phase with I_P and V_g. To restate the explanation of this inversion, remember that plate current (I_P) increases because of a decrease in plate resistance (r_P); even though I_P increases, voltage ($V_P = I_P r_P$) decreases because of the decrease in r_P. The waveforms for input, V_g, I_P, V_P, and output of this amplifier are shown in Figure 13.15.

The peak-to-peak value of the input signal to the grid was 4 volts, and the peak-to-peak value of the output plate voltage was 60 volts. Applying our equation for the amplification factor, $\mu_{\text{Stage}} = \Delta V_P/\Delta V_g =$ 60 volts/4 volts $= 15$. This is the amplification of the triode amplifier or of one stage of amplification. Recall that the amplification of this triode with the same plate characteristic graph was earlier found ideally

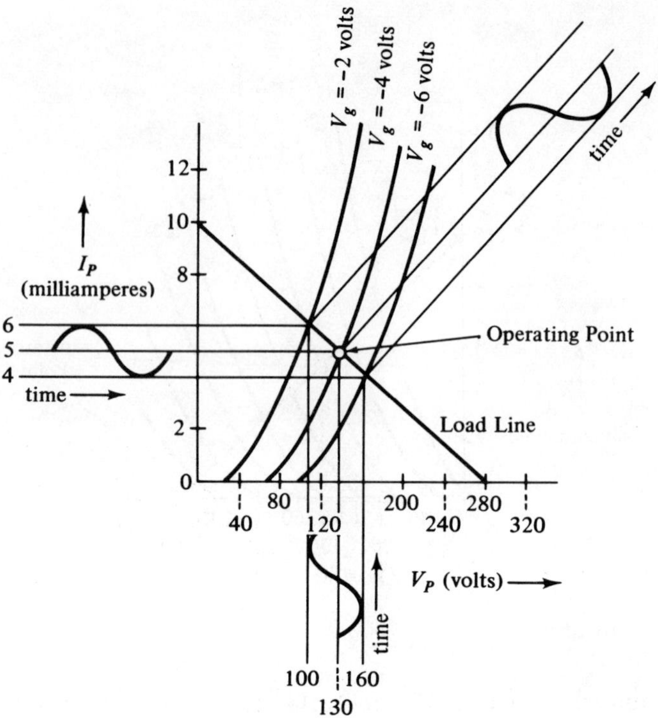

FIGURE 13.14

to be $\mu = 20$. The reason the voltage amplification is reduced in our triode amplifier is that some of the voltage is dropped across the load resistance; this voltage then is not part of the output, which is taken across the tube.

13.7 INTERELECTRODE CAPACITANCES

Before turning our attention to a more careful examination of grid voltages, we should observe that the triode, with its three electrodes (cathode, plate, and grid) separated by a nonconductor, will display three interelectrode capacitances. (See Figure 13.16.) The largest interelectrode capacitance is between the plate and cathode (C_{pk}); C_{pk} is the largest in that the two surfaces (plate and cathode) have the largest surface area and are farthest apart. Other capacitances exist between plate and grid (C_{pg}) and between grid and cathode (C_{gk}). Since these interelectrode capacitances are extremely small, in the neighborhood of a picofarad (0.000000000001 farad), the capacitive reactances ($X_C =$

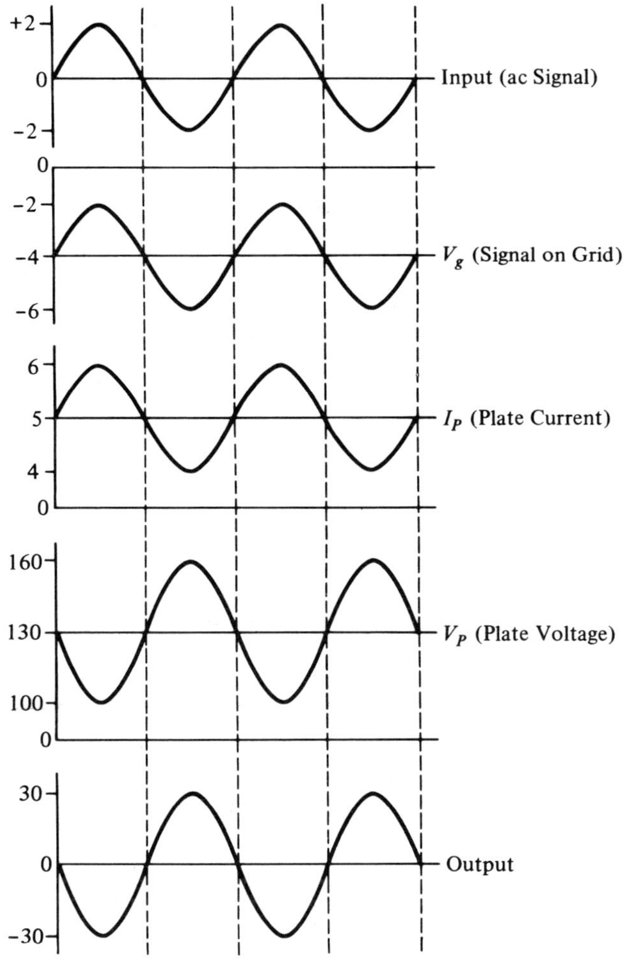

FIGURE 13.15

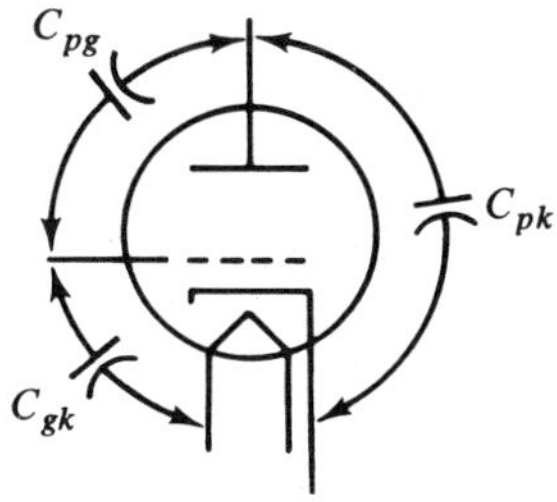

FIGURE 13.16

$1/2\pi fC$) are normally quite large. However, these capacitances do couple the electrodes of the triode, and, as the frequency of the input signal becomes larger, the capacitive reactances become smaller, limiting the application of the triode as an amplifier.

13.8 FIXED BIAS

Normally, a triode has a voltage applied between grid and cathode which controls the flow of electrons from cathode to plate. If the grid is positive with respect to the cathode, it has no control over the flow of electrons; in fact, it will attract some of the electrons to itself, and there will be a small grid current. This undesirable grid current is small, because the cross-sectional area of the grid is small; hence most of the electrons will miss the target and pass on through the grid to the plate. With the grid negative, no grid current will flow, and the grid can control the flow of electrons.

Since the grid actually opposes the flow of electrons, it is biased against them. The voltage between the grid and cathode may be referred to as reverse bias, as grid bias, or simply as bias. The plate of the triode, which attracts electrons, may be thought of as forward biased. Actually, the terms forward and reverse bias are usually reserved for transistors, and the grid-to-cathode voltage of a triode is simply referred to as bias.

This bias, or negative grid-to-cathode voltage, may be obtained in several ways. Batteries can be used to obtain the dc voltages required in the ac amplifier. (See Figure 13.17.) The "A" battery in the triode

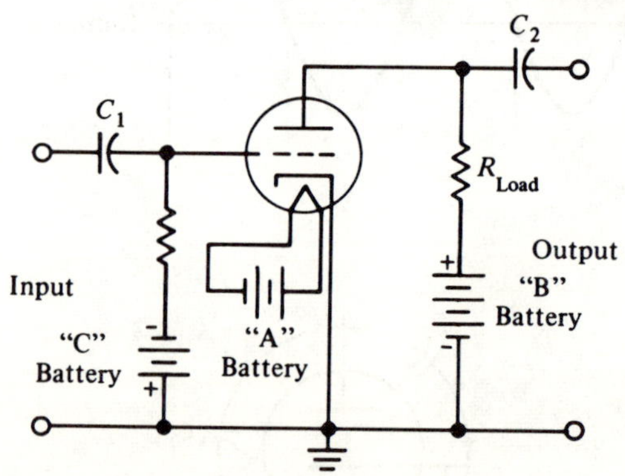

FIGURE 13.17 *Triode amplifier, grounded cathode.*

amplifier supplies the filament, or heater, voltage. The "B" battery supplies voltage for the plate. The "C" battery supplies the grid with its negative voltage necessary to control the flow of electrons from cathode to plate. Actually, the steady dc applied to the grid can come from a dc power supply as well as a battery. Because of the short life of batteries, when a triode amplifier is to be used in a fixed location rather than being portable, it is desirable to replace the "A," "B," and "C" batteries with a dc power supply. Since the external source of voltage, whether battery or power supply, is not affected by the ac signal, it is referred to as fixed bias. Notice that the cathode of the fixed bias triode is common to both the plate voltage supply and the fixed grid bias supply. Thus, the voltage at this common point, or ground, is the same for the plate circuit and the grid circuit. With fixed bias, we have a common, or grounded, cathode.

If the voltage divider of the dc power supply does not supply a negative bias voltage for the grid, an external voltage divider in the amplifier may be used as shown in Figure 13.18. Resistors R_2 and R_3

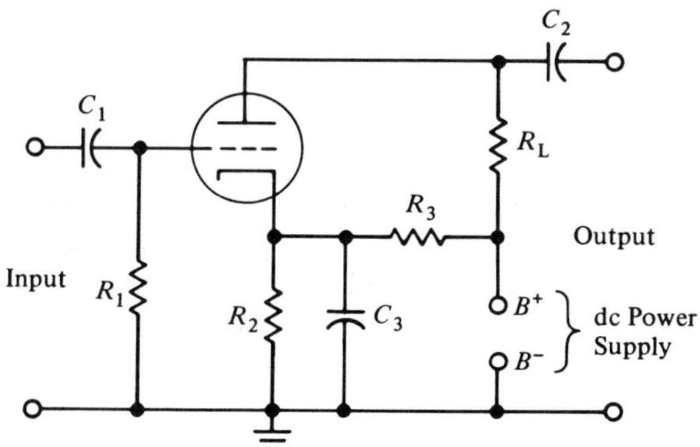

FIGURE 13.18 *Triode amplifier, voltage divider.*

are in series with the dc power supply (indicated as B^+ or B^-). This series circuit forms a voltage divider with part of the dc voltage supplied being dropped across R_2 and the rest across R_3. Since electrons flow from the bottom of R_2 to the top, the top of R_2 must be more positive than the bottom. Thus, the cathode is positive with respect to the bottom of R_2. The bottom of R_2 is connected through R_1 to the grid. Since there is no current flow from cathode to grid, there is no current in R_1 and hence no voltage drop across R_1. This means that the grid is negative with respect to the cathode (the bottom of R_2 is negative with respect to the cathode) and, in addition, the grid bias voltage is equal to the voltage drop across R_2 (the grid is at the same potential as the bottom of R_2).

Returning to our dc or B^+ supply voltage, we find that the current about the plate circuit causes voltage drops by the load resistance (V_L), the tube (V_P), and the cathode resistor (V_2). $V_{\text{dc supply}} = V_L + V_P + V_2$. There are voltage drops across R_2 due to: (1) the voltage divider network and (2) the plate current. (See Figure 13.19.) These two voltage drops are in the same direction and thus add to give the bias voltage (the total voltage drop across R_2).

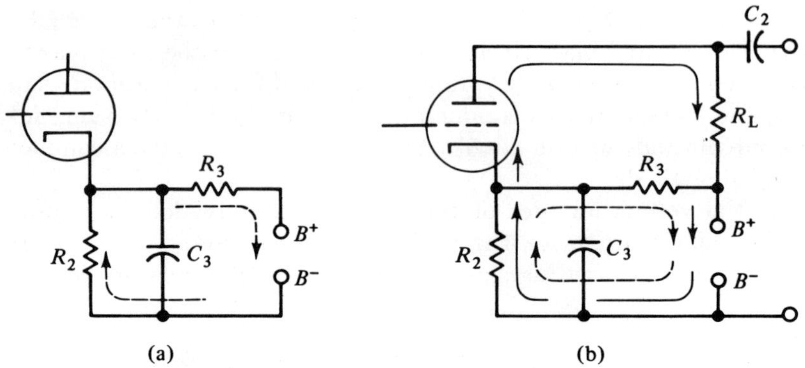

FIGURE 13.19 *(a) Voltage divider network; (b) plate circuit.*

The capacitor C_3 is called a bypass capacitor. Its purpose is to help maintain a steady voltage drop across R_2. Any ac variation in plate current, due to the input signal, would normally change the voltage drop across the cathode resistor R_2. The bypass capacitor offers a low reactance to ac variations, allowing the ac to go around, or bypass, the cathode resistor and thus preventing changes in the voltage drop across R_2. The purpose of input and output coupling capacitors C_1 and C_2, which is to remove dc, has already been discussed.

13.9 CATHODE BIAS

A simplification of the voltage divider as a source of grid bias gives still another method of obtaining the necessary grid voltage: cathode bias or self-bias. The self-bias circuit differs from that of the voltage divider bias only in that resistor R_3 is absent. (See Figure 13.20.) Again, plate current flows from B^- through R_2 to the cathode, from cathode to plate, across the load resistor to B^+. There are voltage drops across the cathode resistor ($V_2 = I_P R_2$), the tube (V_P), and the load (V_L). The top of R_2 is positive with respect to the bottom, and, since there is no current in R_1, the grid is at the same potential as the bottom of R_2 (negative with respect to the cathode). Notice that if no current flows in the plate

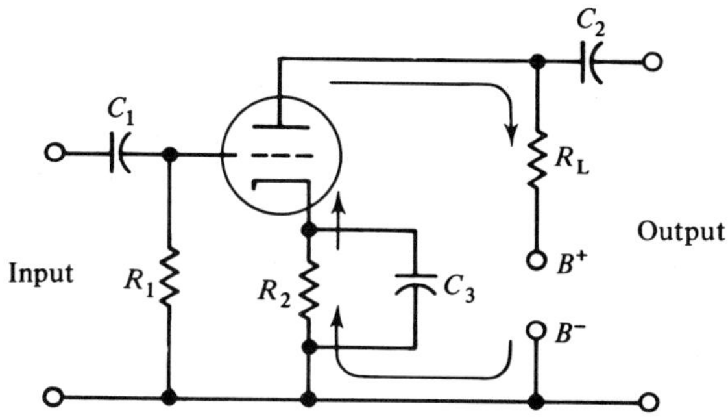

FIGURE 13.20 *Self-bias or cathode bias.*

circuit, then there is no voltage drop across R_2 and hence no grid potential. In the voltage divider bias circuit, there was always some grid voltage since current could flow in the voltage divider network (R_2 and R_3) even when the tube did not conduct.

The capacitor C_3 is a bypass capacitor. As described with the voltage divider bias, it maintains a steady dc voltage drop across the cathode resistor R_2 by passing any ac signal through itself rather than R_2.

Let us now consider the consequences of omitting the bypass capacitor. When the incoming signal goes positive, more electrons reach the plate, and plate current increases. With an increase in plate current ($I_P\uparrow$), the voltage drop across the cathode resistor will increase ($V_2\uparrow = I_P\uparrow R_2$). Increasing the voltage drop across R_2 means an increase in grid bias, that is, the grid is made more negative. With the grid more negative (or less positive), the incoming positive ac signal is at a lower dc potential (that is, the bias voltage is more negative), and the plate current decreases, counteracting the initial increase in plate current.

When the incoming signal goes negative and reduces the plate current, then the voltage drop across the cathode resistor and the grid bias both decrease. As the grid becomes less negative (more positive), the incoming ac signal rides at a higher dc potential, and plate current increases, counteracting its original decrease.

Therefore, the absence of the bypass capacitor in the cathode bias circuit causes the current to be self-regulating, that is, increases and decreases in current caused by changes in the input signal are opposed by adjustments within the circuit. The self-regulating feature of cathode bias without bypass capacitor does protect the plate from excessive current, but it also reduces the response, or amplification, of the triode. This reduction of changes in plate current is called degeneration or negative feedback. The desired changes in plate current are reduced, or degenerated, by opposite changes in grid bias.

13.10 GRID LEAK BIAS

Suppose no bias voltage is externally supplied to the grid of a triode. Then the grid is initially at zero dc potential with respect to the cathode. When an ac input signal on the grid goes positive, the grid becomes positive with respect to the cathode and attracts some of the electrons coming from the cathode. These negative electrons on the grid will charge capacitor C_1, as shown in Figure 13.21. When the incoming signal begins

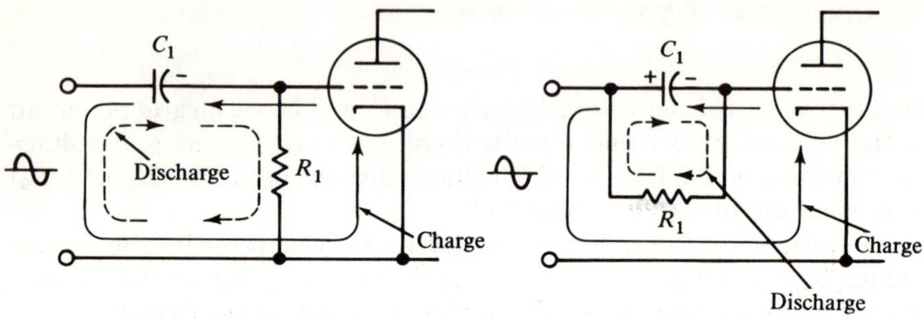

FIGURE 13.21 *Grid leak bias.*

to go negative, capacitor C_1 begins discharging through R_1. The time constant of circuit R_1C_1 is large in comparison to the period of the incoming signal, and hence C_1 discharges only slightly. This maintains a negative voltage, or bias voltage, on the grid. The electrons, which leave the tube via the grid, give rise to the name grid leak bias. The size of the grid leak bias depends on the amplitude of the input signal (the more positive it becomes, the more electrons it can attract to charge the capacitor) and on the area of the grid (the smaller the target, the fewer electrons will hit it).

13.11 COMBINATION BIAS

If there is no input signal, the triode may be damaged by excessive plate current. In order to protect the triode from excessive current, the grid leak bias is often combined with cathode bias. This combination bias circuit looks the same as the cathode bias circuit, but it is such that capacitor C_1 can serve both as a coupling capacitor and to maintain a grid leak bias. (See Figure 13.22.) The cathode bias gives some steady dc potential on the grid which is negative with respect to the cathode; however, if the ac input signal on this dc is such that it will cause the

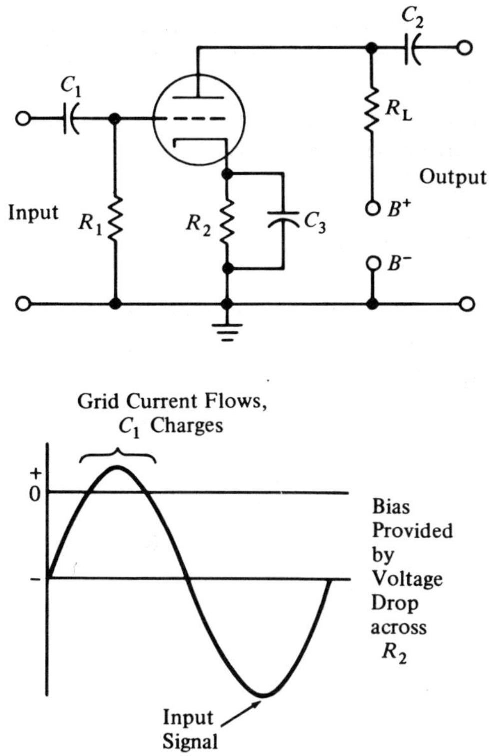

FIGURE 13.22

grid to go positive during some fraction of each cycle, then a grid current will flow, capacitor C_1 will charge, and thus provide a grid leak bias. The two bias voltages add to give the grid bias and thus the voltage drop across the cathode resistor need not be as large as when there is no grid leak bias. The combination bias thus provides a stable negative grid bias made up of cathode bias and grid leak bias, with the cathode bias acting as a protective bias to prevent large current flow when there is no input signal.

Questions

(1) Why is it that the grid of a triode can stop, or cut off, the flow of electrons from cathode to plate when the grid is just a few volts negative with respect to the cathode and the plate is several hundred volts positive with respect to the cathode?

(2) What is the phase relation of (a) plate current, (b) plate voltage, and (c) voltage across the load resistor to the signal voltage on the grid of a triode amplifier?

(3) Bias is the negative dc voltage on the grid of a triode. How is this bias voltage obtained in the grid leak bias circuit?

Suggested Experiments

(1) Triode Characteristics. The variation in plate current with changes in grid voltage when the plate voltage is kept constant may be recorded and graphed. (See Figure 13.23.) From the graph of this tube

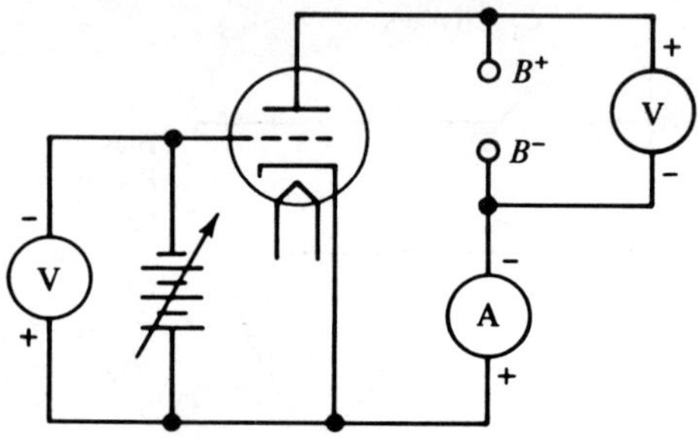

FIGURE 13.23

characteristic, you may calculate amplification factor, transconductance, and plate resistance of the triode.

CHAPTER 14

Transistors

14.1 JUNCTION TRANSISTOR

The vacuum tube triode was a three-electrode device consisting of cathode, plate, and grid. We will now consider a three-element semiconductor device commonly called a transistor.

We will first look at the junction transistor since it represents a common type of transistor construction and since we have already studied the PN junction diode. Recall from our study of the PN junction diode that the doping of germanium with atoms containing five valence electrons produces N-type semiconductor. Doping germanium with atoms containing three valence electrons produces P-type semiconductor. (See Figure 14.1.) Fusion of the two types of semiconductors produces a

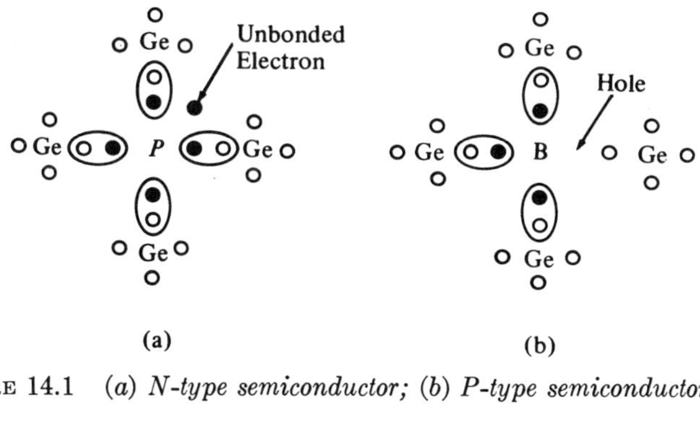

FIGURE 14.1 (*a*) *N-type semiconductor;* (*b*) *P-type semiconductor.*

diode which, when forward biased, permits the flow of electrons from N to P. (See Figure 14.2.)

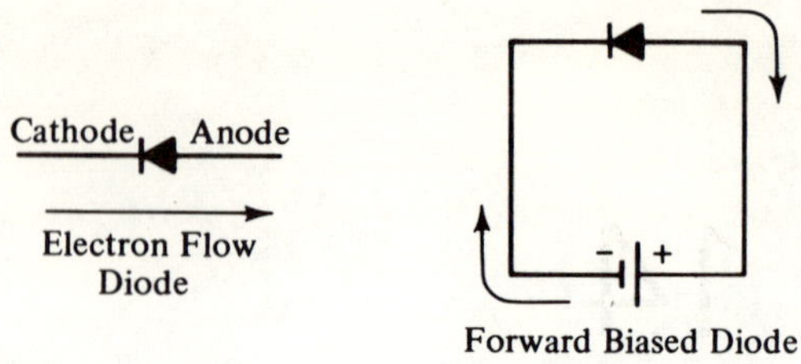

FIGURE 14.2

If we make a sandwich of semiconductors, we have a transistor. The two possible types of junction transistors are shown with their schematic diagrams in Figure 14.3.

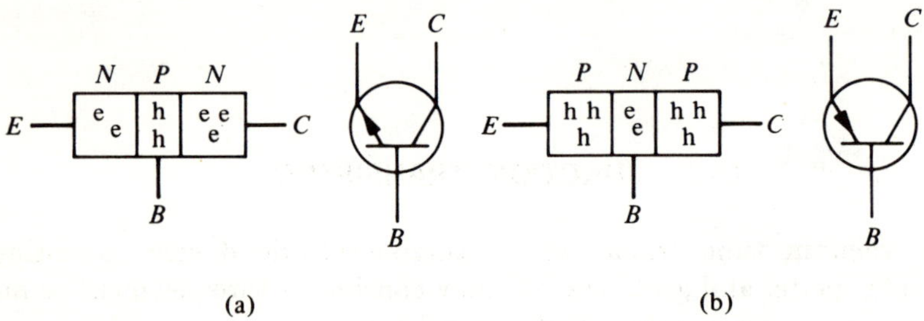

FIGURE 14.3 *(a) NPN transistor; (b) PNP transistor.*

14.2 *NPN* TRANSISTOR

The *NPN* transistor may be thought of as two diodes. (See Figure 14.4.) The first (*EB*) is forward biased. Electrons entering the *N*-type semiconductor repel other electrons toward the *EB* junction. Holes in the *P*-type semiconductor, which occur when electrons are removed from the *P*-type by the positive terminal of the battery, are attracted to the *EB* junction where they combine with the free electrons from the *N*-type.

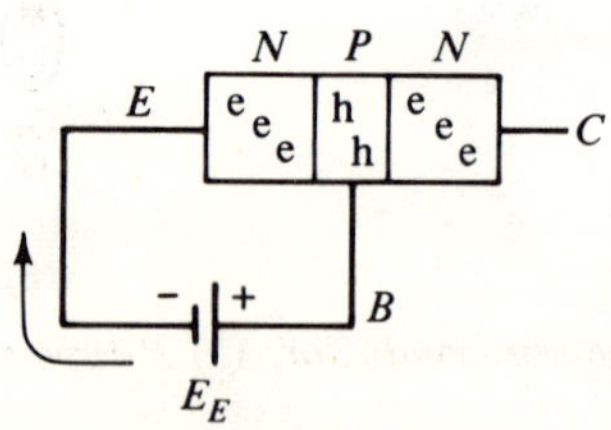

FIGURE 14.4

Thus, electrons enter the *N*-type, other electrons are removed from the *P*-type, and current flows.

The second diode *BC* is reverse biased and thus has a high resistance to current flow. (See Figure 14.5.) However, as discussed earlier, there will be a very small flow of current due to minority carriers in both *P*- and *N*-type semiconductors.

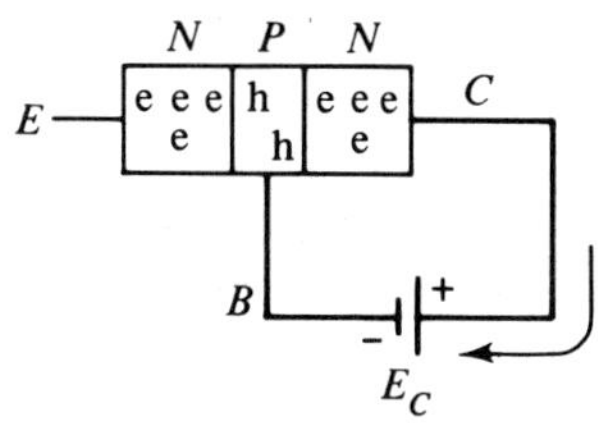

FIGURE 14.5

The transistor differs from the two diodes in both its construction and operation. (See Figure 14.6.) The middle semiconductor material

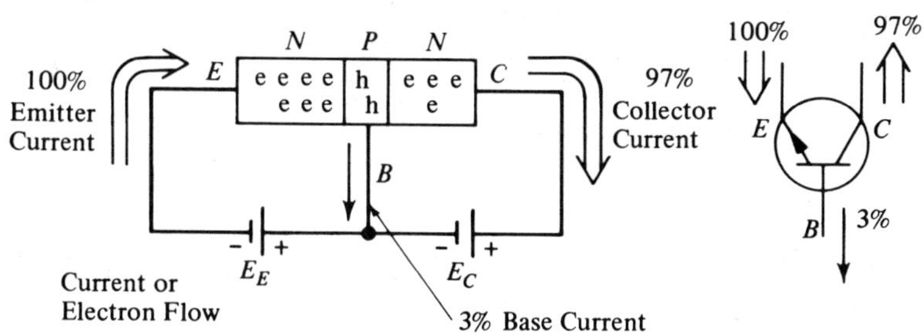

FIGURE 14.6

in a transistor is very thin in comparison to the outer layers. Electrons from battery E_E enter the *N*-type semiconductor on the left (the emitter *E*) and repel other electrons toward the *EB* junction; holes in the *P*-type semiconductor, caused by the removal of electrons from the *P*-type, are repelled toward the *EB* junction just as in the diode. However, because there are comparatively few holes in the *P*-type semiconductor available to combine with electrons from the *N*-type semiconductor and because the *P*-type semiconductor layer is so thin, only a few electrons will leave the *P*-type semiconductor for the E_E battery. Most of the electrons from the emitter *E* will pass through the thin *P*-type semiconductor (called the base *B*) without combining with any holes and will go into the *N*-type semiconductor on the right. Since this *N*-type semiconductor collects most of the electrons, it is called the collector *C*. The electrostatic repulsion between these electrons and the free electrons already present in the N-type material causes electrons to be repelled to the battery E_C. Notice that electrons entering base B from the emitter E can take either of two paths: the path produced by battery E_E, which attracts just a few electrons, and the path produced by both batteries E_E and E_C, which attract the majority (approximately 97 percent) of the electrons.

14.3 *PNP* TRANSISTORS

The *PNP* junction transistor is the reverse of the *NPN*. The batteries are also reversed in order to maintain a forward bias on the emitter and a reverse bias on the collector. (See Figure 14.7.) The explanation of

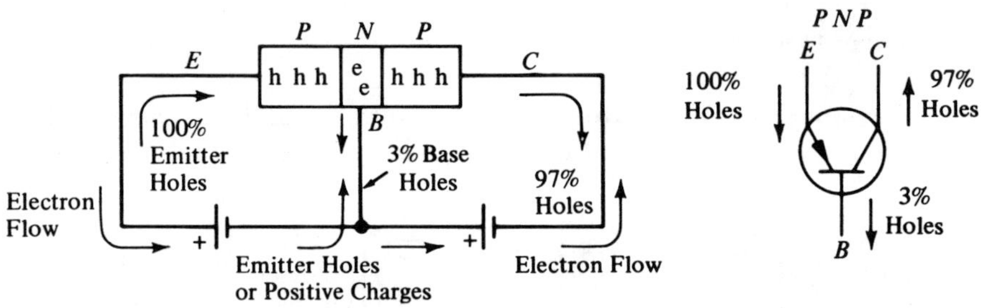

FIGURE 14.7

how the *PNP* junction transistor works is identical to that of the *NPN*, except we now consider that positive charges, or holes, are being emitted and collected. Any confusion should be eliminated by recalling that the flow of positive charges in one direction is equivalent to the flow of electrons in the opposite direction. Remember then that in the *PNP* transistor, the emitter emits positive charges, whereas in the *NPN* transistor, the emitter emits electrons.

14.4 TRANSISTOR VERSUS TRIODE

The transistor and triode are similar in certain respects: Each has a source of electrons (emitter or cathode) and each has a receiver of electrons (collector or plate). The flow of electrons from source to receiver is controlled in the transistor by the base and in the triode by the grid. (See Figure 14.8.) However, the control action differs in the two devices. In the triode circuit, the grid is negative with respect to the cathode, and no electrons flow from cathode to grid. In contrast, in the transistor circuit, the base is positive with respect to the emitter, and there is a

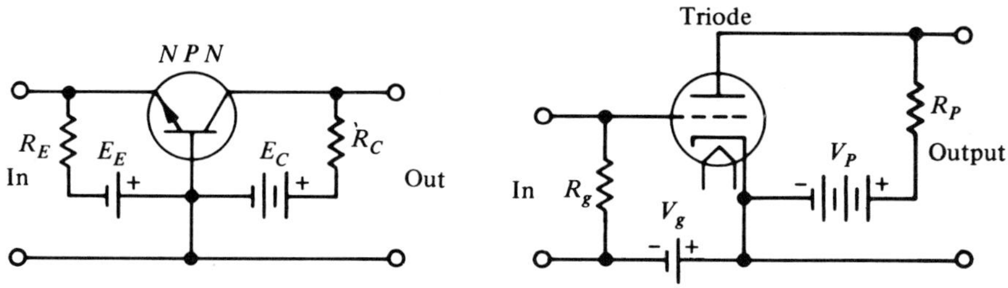

FIGURE 14.8

small flow of current from emitter to base. This current flow is limited basically by the thinness of the base.

14.5 COMMON-BASE TRANSISTOR AMPLIFIER

Since the emitter-base voltage can control the flow of current in the transistor, amplification of this voltage, or of power, is the principal application of the transistor. We will next discuss three circuits which use transistors for amplification. These three variations are called common base, common emitter, and common collector. The term common refers to a point in the circuit to which several different voltages are connected. You may recall the same term was used for ground potential (common ground) when we studied voltage dividers.

The common-base transistor amplifier, as the name implies, has its base at ground potential, that is, several different voltages are connected to the base element of the transistor. (See Figure 14.9.) We will consider

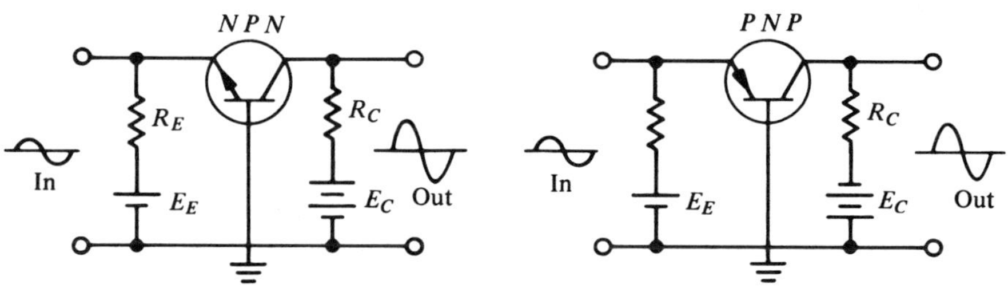

FIGURE 14.9 *Common-base transistor amplifier.*

both *NPN* and *PNP* transistors. Since electron flow is a more familiar concept, we will follow the *NPN* circuit more carefully.

In the *NPN* circuit, the forward bias voltage is negative with respect to ground, whereas the reverse bias voltage is positive with respect to ground. When the input signal goes positive, the forward bias becomes less negative, which causes a reduction in emitter current. There will then necessarily be a reduction in collector current. With this decreased collector current, less current flows in R_C, and the voltage drop across R_C decreases. If we consider the closed loop containing the voltage supply (E_C) and two components which offer opposition to current flow (the load resistance R_C and the collector base of the transistor), we recognize that each of these oppositions to current flow causes a voltage drop and that the sum of the voltage drops about the closed loop must equal the supply voltage. $E_C = V_{CB} + V_{\text{Load}}$ (all voltages positive). Since the supply voltage (E_C) remains constant, when the voltage drop across the load decreases ($V_{\text{Load}}\downarrow$), the voltage drop across the collector base of the transistor must increase ($V_{CB}\uparrow$). An increase in a positive voltage indicates a positive going signal. The output voltage is taken across the collector base (output voltage $= V_{CB}$), so the output voltage increases (has its positive half cycle) when the input voltage increases. Thus, the voltage output of the *NPN* common-base transistor amplifier, unlike that of the triode amplifier, is in phase with the input.

To follow the effects of the input signal in the *PNP* circuit, we first note that the forward bias is positive and the reverse bias is negative with respect to ground. When the input signal goes positive, the forward bias increases (becomes more positive), causing an increase in emitter current and collector current. The increased collector current will cause an increase in the voltage drop across the load R_C. Using the same closed loop as in the *NPN* circuit, we still have the relation $E_C = V_{CB} + V_{\text{Load}}$. But this time all the voltages are negative. Thus, an increased voltage drop across the load means that V_{Load} becomes more negative. Since E_C is constant, V_{CB} must become less negative to maintain the equality. V_{CB} becoming less negative indicates a positive going output voltage. Thus, the input and output voltages are in phase in the *PNP* common-base transistor amplifier.

The current gain (given the symbol α, alpha) of a common-base transistor amplifier is the ratio of the collector current to the emitter current. $\alpha = I_C/I_E$. Since there will always be some current flow in the base, the collector current will be less than the emitter current, and the current gain in the common-base transistor amplifier will be less than 1.0. Usually, α is in a range from 0.92 to 0.98.

If we compare the low resistance of the forward biased emitter to the extremely high resistance of the reverse biased collector, we find the resistance gain of the junction transistor is quite large. Typical values of emitter-base resistance are from 50 ohms to 1000 ohms, while the collector-base resistance is in a neighborhood of from 100,000 ohms to 2,000,000 ohms. Thus, the gain in resistance ($R_{\text{Gain}} = R_{CB}/R_{EB}$) might be as high as 2000.

Recalling Ohm's Law, we can find the ideal voltage gain of the common-base transistor amplifier: $V_{\text{Gain(ideal)}} = V_{\text{out}}/V_{\text{in}} = V_{CB}/V_{EB} = (I_C \times R_{CB})/(I_E \times R_{EB}) = \alpha R_{\text{Gain}}$. Since α is nearly 1, the voltage gain ideally is approximately equal to R_{Gain}.

Knowing the current gain, voltage gain, and resistance gain, we can calculate power gain. Recall that power is found with the equation $P = VI = I^2R$. Thus, $P_{\text{Gain(ideal)}} = V_{\text{Gain}} \times I_{\text{Gain}} = I_{\text{Gain}}{}^2 R_{\text{Gain}} = \alpha^2 R_{\text{Gain}}$. These values for V_{Gain} and P_{Gain} are ideal values which assume no energy loss through heating.

A form of cathode bias may be applied to the common-base transistor amplifier to eliminate the forward bias battery (E_E). Consider the *NPN* circuit in Figure 14.10. Following base current through the closed

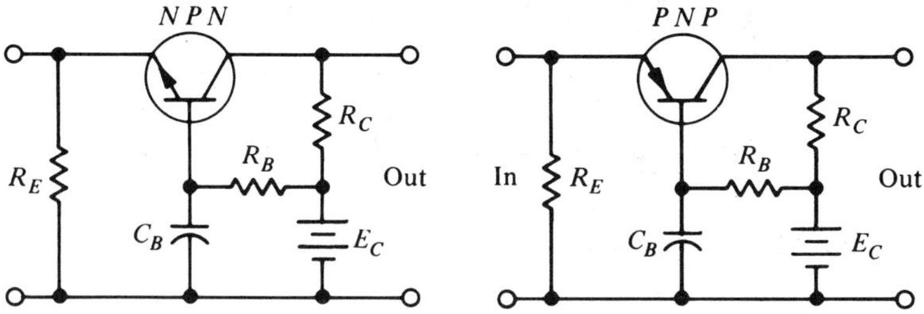

FIGURE 14.10 *One-battery common-base transistor amplifier.*

loop containing E_C with voltage drops across R_E, the emitter-base junction, and R_B, we see that the base is positive with respect to the emitter (forward bias). Following collector current through the loop containing E_C and voltage drops across R_E, the emitter-base-collector junctions, and R_C, we recognize that the collector is more positive than the base which gives the necessary reverse bias. In addition, the voltage drop across R_C should be less than that across R_B in order to make the reverse bias voltage greater than the forward bias voltage. The capacitor C_B is a bypass capacitor which passes the ac output signal and thus prevents it from being felt across the emitter base, so that variations in emitter-base voltage are caused only by the input signal. Notice that although the base is not at dc ground potential, ac can pass through C_B to ground, and so the base is at ac ground potential.

14.6 COMMON-EMITTER TRANSISTOR AMPLIFIER

The common-emitter transistor amplifier has its emitter at ground potential. (See Figure 14.11.) In the common-emitter amplifier, the output is out of phase with the input. We will explain this phase inversion

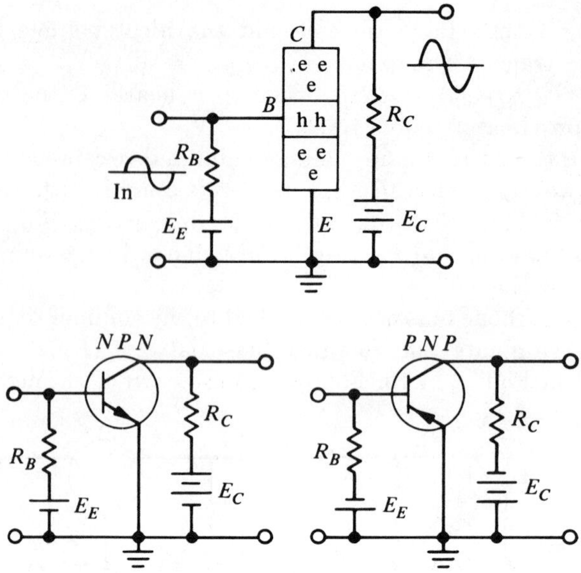

FIGURE 14.11 *Common-emitter transistor amplifier.*

in the *NPN* circuit; a similar explanation holds for the *PNP* circuit. In the *NPN* common-emitter amplifier, both the forward and reverse bias voltages are positive with respect to ground. When the incoming signal goes positive, the base-to-emitter voltage becomes more positive and the forward bias increases. This causes an increase in base and collector current. An increased collector current means more voltage is dropped across the load resistance R_C. Since V_{Load} increases and since $E_E = V_{EC} + V_{\text{Load}}$ with E_E constant, an increase in V_{Load} implies a decrease in V_{EC}. The output is taken across the emitter collector (output $= V_{EC}$), so output voltage decreases when input voltage increases —a phase inversion.

You should have noticed that the polarities of the batteries are the same in the common emitter as in the common base. For example, with *NPN* transistors in both common-emitter and common-base circuits, the emitter is negative with respect to the base and the collector is positive with respect to the emitter. (See Figure 14.12.) These polarities would both be reversed for *PNP* junction transistors.

In the common base, the input signal is applied to the emitter and affects emitter current; in the common emitter, the input signal is applied to the base and directly affects base current. Thus, the current gain of the common emitter (called β, beta) is the ratio of collector current (I_C) to base current (I_B). This ratio will be large, since base current is rather small. $\beta = I_C/I_B$ and usually ranges from 25 to 50.

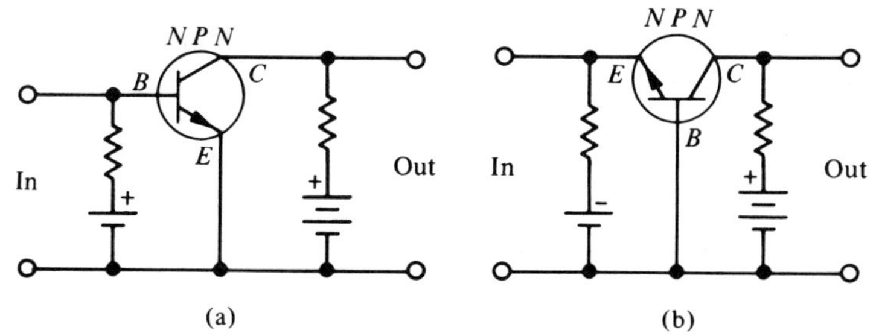

FIGURE 14.12 *(a) Common emitter; (b) common base.*

The input resistance of the common emitter is a bit larger than the input resistance of the common base, and the output resistance is considerably smaller. Typical values of input emitter-base resistance range from 500 ohms to 1500 ohms and of output collector-emitter resistance from 30,000 ohms to 50,000 ohms. Thus, the resistance gain of the common emitter ($R_{\text{Gain}} = R_{CE}/R_{BE}$) can vary from 20 to 100.

By applying Ohm's Law, the voltage gain for the common-emitter amplifier can be written in terms of current gain (β) and resistance gain. Since voltage gain in an amplifier is the ratio of output voltage to input voltage, $V_{\text{Gain(ideal)}} = V_{\text{out}}/V_{\text{in}} = V_{CE}/V_{BE} = I_C R_{CE}/I_B R_{BE} = \beta \times R_{\text{Gain}}$. Typical values for the common-emitter amplifier would be between 300 and 1000.

Recalling our expression for power, we can write $P_{\text{Gain(ideal)}} = V_{\text{Gain}} I_{\text{Gain}} = I_{\text{Gain}}{}^2 R_{\text{Gain}} = \beta^2 R_{\text{Gain}}$, which can range from 300 to 10,000. The power gain in the common emitter is extremely large, and it is thus well suited for the output stage of amplification when power is needed to operate devices such as loudspeakers.

Eliminating the forward bias battery and leaving just one battery in the common-emitter circuit are quite easily accomplished as shown in Figure 14.13. The voltage drop across R_C is less than that across R_B;

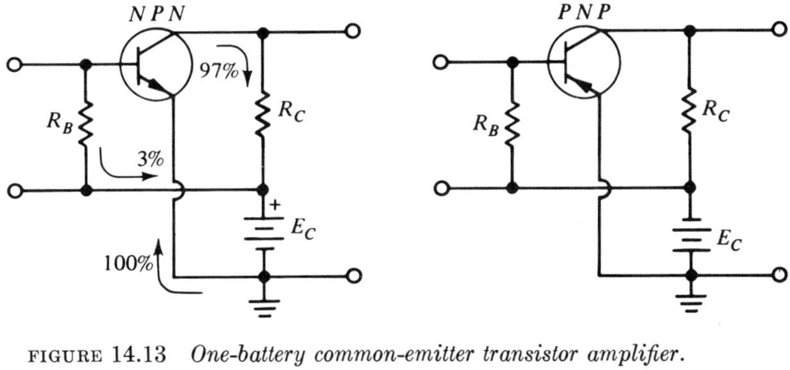

FIGURE 14.13 *One-battery common-emitter transistor amplifier.*

hence the collector is more positive than the base in the *NPN* circuit which gives the necessary reverse bias. Also, since there is a voltage drop across R_B, the base is positive with respect to the emitter (forward biased) in the *NPN* circuit.

14.7 COMMON-COLLECTOR TRANSISTOR AMPLIFIER

The third variation of the transistor amplifier is the common collector, sometimes referred to as the emitter follower. Unlike the other two transistor amplifiers we have studied, the common collector's output resistance is lower than its input resistance. Figure 14.14 shows schematic

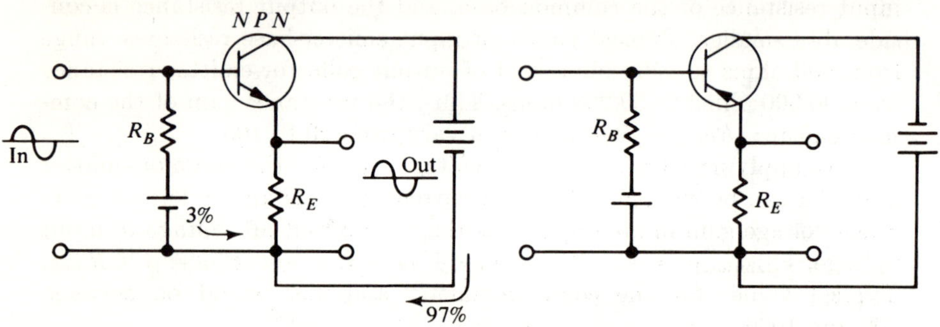

FIGURE 14.14 *Common-collector transistor amplifier.*

diagrams of the common-collector amplifier. Notice that the output is taken across the emitter resistor (R_E) rather than across the transistor. In the common-collector amplifier, the output across the emitter resistance will be in phase with the input signal. Consider the *NPN* transistor circuit. When the input signal goes positive, this voltage and the positive forward bias add to increase the positive base-to-emitter voltage. This increase in forward bias causes increases in base current and collector current and a corresponding increase in emitter current. An increased emitter current means that the voltage drop across the emitter resistance, which is the output voltage, must increase. Therefore, when the input signal of a common-collector amplifier goes positive, the output voltage signal also goes positive; that is, the input and output signals are in phase.

If we consider the resistance of the emitter-base junction to be in series with the emitter resistance (R_E), we see that a small fraction of the input voltage will be dropped across the junction and the remainder across R_E. (See Figure 14.15.) Thus, the output will always be less than the input voltage, and the voltage gain will be less than one. $V_{\text{Gain}} = V_{\text{out}}/V_{\text{in}}$ is less than 1.

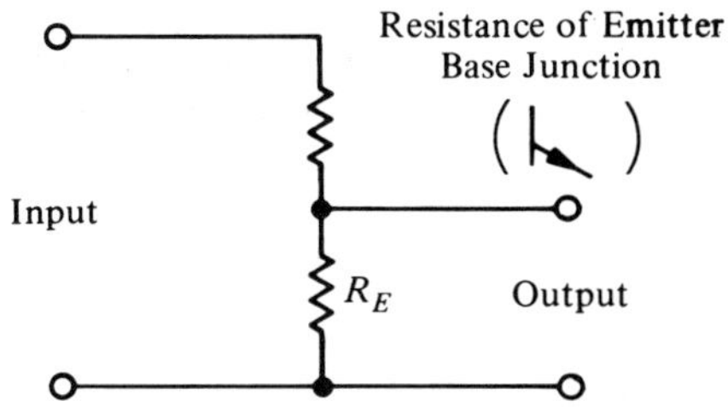

FIGURE 14.15

The current gain of the common-collector transistor amplifier is the ratio of the emitter current (output) to the base current (input). $I_{\text{Gain}} = I_E/I_B$ is usually between 25 and 50. This is about the same current gain as the common emitter.

The resistance of the common-collector circuit steps down. The input resistance varies from 20,000 ohms to 500,000 ohms, whereas the output resistance is usually from 500 ohms to 2000 ohms.

The power gain of the common collector is calculated in the same way as power gain for the other circuits considered. $P_{\text{Gain}} = V_{\text{Gain}} I_{\text{Gain}} = I_{\text{Gain}}{}^2 R_{\text{Gain}}$ varies from 10 to 100. This power gain for the common-collector amplifier is smaller than that of the other two transistor amplifiers. The principal use of the common collector is impedance matching. High impedance devices, such as a crystal phonograph pickup, may be coupled to a common-emitter amplifier via a common-collector amplifier, which can have a high input impedance.

Eliminating the forward bias battery in the common-collector circuit is accomplished in the following way (Figure 14.16). In the *NPN* circuit, the voltage drop across R_B makes the base less positive than the collector, so the collector base is reverse biased. The forward biasing of

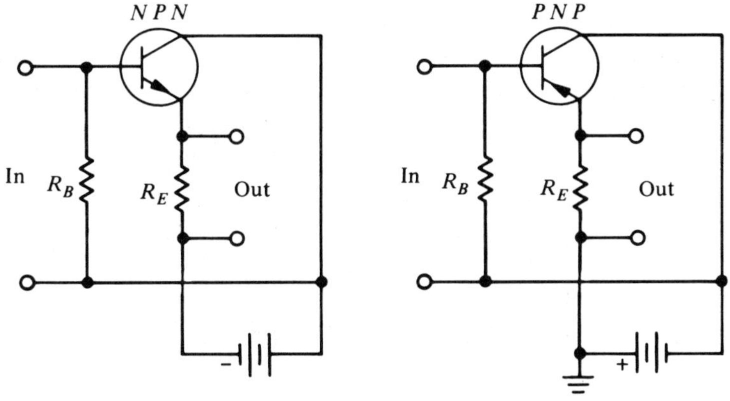

FIGURE 14.16 *One-battery common-collector transistor amplifier.*

the emitter base is obvious if we follow the loop containing the supply voltage and the three resistances (R_E, the small resistance of the emitter-base junction, and R_B).

Questions

(1) How may one distinguish a *PNP* from an *NPN* transistor in a circuit, assuming that the emitter, base, and collector leads are properly identified?

Suggested Experiments

(1) Transistor Characteristics. Collector current may be observed for different values of base-emitter forward bias voltage and collector-emitter reverse bias voltage. This basic experiment may be repeated for a *PNP* transistor. (See Figure 14.17.)

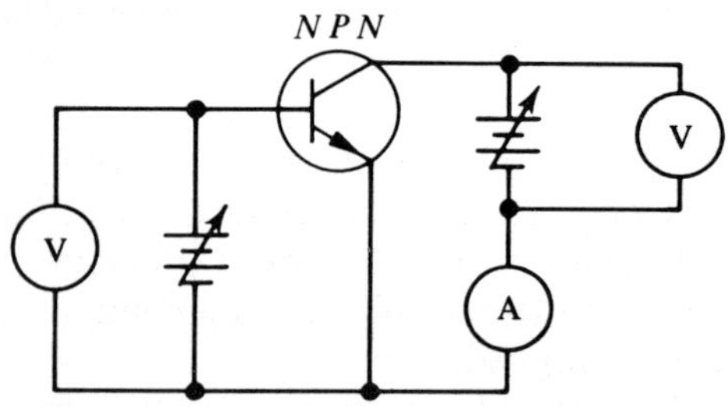

FIGURE 14.17

(2) Phase Relation and Amplification of Transistor Amplifiers. The amplification of a transistor amplifier, either common emitter, common base, or common collector, may be observed by comparing the height, or amplitude, of the input signal from the audio oscillator to that of the output signal from the amplifier. At the same time, any phase shift between input signal and output signal may be observed. (See Figure 14.18.)

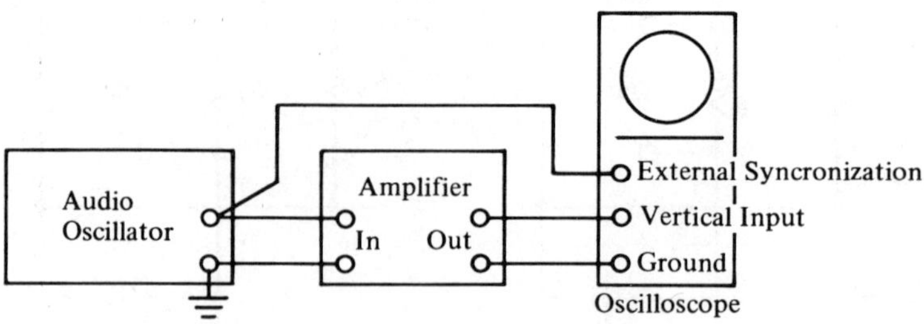

FIGURE 14.18

CHAPTER 15

Power, Current, and Voltage Ratios

15.1 LOGARITHMS

Since many amplifiers are designed to operate loud speakers, a knowledge of the response of the human ear to sound intensity is important. The human ear does not usually perceive small changes in sound intensity; instead, the ear's ability to distinguish differences in sound intensity is said to be logarithmic, or proportional. For example, the human ear can normally recognize a sound whose intensity has been doubled as being twice as loud.

Current, voltage, and power gains or losses can also be described logarithmically. Logarithms are based on a table which shows the values obtained by raising some number to higher and higher powers. In Table 15.1, we use the number 10, which is always used for common logarithms. Notice that the values obtained increase quite rapidly (10, 100, 1000, and so on).

TABLE 15.1

$$\begin{aligned} 10^x &= N \\ 10 = 10^1 &= 10 \\ 10 \times 10 = 10^2 &= 100 \\ 10 \times 10 \times 10 = 10^3 &= 1000 \\ 10 \times 10 \times 10 \times 10 = 10^4 &= 10{,}000 \\ 10 \times 10 \times 10 \times 10 \times 10 = 10^5 &= 100{,}000 \end{aligned}$$

But this rapid increase in these values is due to just a unit increase in the power or exponent (1, 2, 3, and so on). If we let the number used (in this case 10) be represented by the letter b, the exponents by x, and the values obtained (in this case 10, 100, 1000, and so on) by N, we can express the procedure used to find N with the equation $b^x = N$. (For example, the next line of the above table would be $10^6 = 1{,}000{,}000$). The b is said to be the base, and the x, besides being the exponent, is also called the logarithm of the number N.

Notice that, when the base is changed, the same exponent gives different values of N. Table 15.2 shows two different bases: a base of 2 and a base of 3.

TABLE 15.2

Base 2	Base 3
$2 = 2^1 = 2$	$3 = 3^1 = 3$
$2 \times 2 = 2^2 = 4$	$3 \times 3 = 3^2 = 9$
$2 \times 2 \times 2 = 2^3 = 8$	$3 \times 3 \times 3 = 3^3 = 27$

We will use a base of 10 for our logarithmic expression of voltage, current, and power gains. Thus, the equation above can be written $10^x = N$; since x was said to be the logarithm of the number N when the base 10 is used, we rewrite this equation as $x = \log_{10} N$, which is read x is equal to the log to the base 10 of the number N. You will avoid confusion in working with logarithms by simply remembering that these last two equations have the same meaning: x is the log of a number N just means that x is the power to which 10 must be raised to give N. For example, what is x if $x = \log_{10} 1000$? Looking at the table, which shows powers of 10, we see that when the exponent is 3 ($x = 3$), the number will be 1000 ($N = 1000$). $3 = \log_{10} 1000$ or $10^3 = 1000$.

The number N in the equation $x = \log_{10} N$ can have almost any value. For example, N could be the number obtained by taking the ratio of power out to power in for an amplifier or some other electrical device. Our equation would then be written, $x = \log_{10} (P_{\text{out}}/P_{\text{in}})$. If the power output of an amplifier is 1000 watts when its input power is 10 watts, the equation would read $x = \log_{10}$ (1000 watts/10 watts) or $x = \log_{10} 100$. From the base 10 table, we know that when $N = 100$, the exponent is 2.0. Therefore, $2.0 = \log_{10} 100$, and this power ratio is expressed logarithmically as 2.0.

15.2 THE BEL

As we have mentioned, the reason for expressing power ratios logarithmically comes from the fact that the human ear hears logarithmically. This connection with the science of sound in seen in the name given to the logarithm or exponent (x). Dr. Alexander Graham Bell, who invented the telephone, is remembered in the area of electronics by referring to the x in the equation $x = \log_{10} (P_{out}/P_{in})$ as a bel. Thus, in our problem, the power increase was 2.0 bels.

Since the bel represents such a large increase in power, most changes will be described in terms of some fraction of a bel. The prefix deci, meaning one-tenth ($\frac{1}{10}$), is used to provide smaller units. One decibel (written 1 dB) is one-tenth of a bel (1 dB $= \frac{1}{10}$ bel), or one bel equals ten decibels (1 bel $=$ 10 dB). Thus our equation, which gave the power ratio in bels, can be changed to express the power ratio in decibels. Power ratio in decibels $=$ 10 $\times$ power ratio in bels $= 10 \log_{10} (P_{out}/P_{in})$. For convenience, in the discussion that follows, we will use Table 15.3 which

TABLE 15.3

Ratio	dB	Ratio	dB	Ratio	dB	Ratio	dB
1.0	0.0000	3.3	5.1851	5.6	7.4819	7.9	8.9763
1.1	0.4139	3.4	5.3148	5.7	7.5587	8.0	9.0309
1.2	0.7918	3.5	5.4407	5.8	7.6343	8.1	9.0849
1.3	1.1394	3.6	5.5630	5.9	7.7085	8.2	9.1381
1.4	1.4613	3.7	5.6820	6.0	7.7815	8.3	9.1908
1.5	1.7609	3.8	5.7978	6.1	7.8533	8.4	9.2428
1.6	2.0412	3.9	5.9106	6.2	7.9239	8.5	9.2942
1.7	2.3045	4.0	6.0206	6.3	7.9934	8.6	9.3450
1.8	2.5527	4.1	6.1278	6.4	8.0618	8.7	9.3952
1.9	2.7875	4.2	6.2325	6.5	8.1291	8.8	9.4448
2.0	3.0103	4.3	6.3347	6.6	8.1954	8.9	9.4939
2.1	3.2222	4.4	6.4345	6.7	8.2607	9.0	9.5424
2.2	3.4242	4.5	6.5321	6.8	8.3251	9.1	9.5904
2.3	3.6173	4.6	6.6276	6.9	8.3885	9.2	9.6379
2.4	3.8021	4.7	6.7210	7.0	8.4510	9.3	9.6848
2.5	3.9794	4.8	6.8124	7.1	8.5126	9.4	9.7313
2.6	4.1497	4.9	6.9020	7.2	8.5733	9.5	9.7772
2.7	4.3136	5.0	6.9897	7.3	8.6332	9.6	9.8237
2.8	4.4716	5.1	7.0757	7.4	8.6923	9.7	9.8677
2.9	4.6240	5.2	7.1600	7.5	8.7506	9.8	9.9123
3.0	4.7712	5.3	7.2428	7.6	8.8081	9.9	9.9564
3.1	4.9136	5.4	7.3239	7.7	8.8649	10.0	10.0000
3.2	5.0515	5.5	7.4036	7.8	8.9201		

Power ratio in dB $= 10 \log_{10} (P_2/P_1)$
If input and output resistances are equal, then
Voltage ratio in dB $= 10 \log_{10} (V_2/V_1)^2$
Current ratio in dB $= 10 \log_{10} (I_2/I_1)^2$

gives power, voltage, and current ratios in decibels. Let us consider an example in which the power gain of an amplifier is from 5 watts to 10 watts; the power is doubled. Although we do not need the equation to use Table 15.3, we will substitute in the equation anyway for review. dB $= 10 \log_{10} (P_{out}/P_{in}) = 10 \log_{10}$ (10 watts/5 watts) $= 10 \log_{10} 2.0$.

The actual power ratio is 10 watts/5 watts = 2.0; the power is doubled. To find the equivalent power ratio expressed in decibels, we look at the conversion table and read the dB value as 3.01 dB. Thus, 3.01 dB = $10 \log_{10}$ (10 watts/5 watts). When the power is doubled, the decibel increase is 3.01 dB. Correspondingly, if the power were halved, the power ratio in decibels would be −3.01 dB. A power loss is indicated by a negative sign.

15.3 ZERO POWER INPUT

If a device has zero power input and a large power output, there is a problem in expressing the power ratio (P_{out}/P_{in}) because division by zero ($P_{in} = 0$) is not defined in our number system. To overcome this difficulty, we often use an extremely small power input of 6 mW (milliwatts) = 0.006 watt in place of zero power input.

Suppose an inexpensive hi-fi amplifier generates 3 watts of power. What is its power ratio in decibels, if no input power is listed? Utilizing the above convention (P_{in} = 0.006 watt), we have dB = $10 \log_{10} (P_{out}/P_{in})$ = $10 \log_{10}$ (3 watts/0.006 watts) = $10 \log_{10} 500$. The values for power ratio in Table 15.3 only go from 1.0 to 10.0. In order to use the table to find this power ratio in decibels, we have to reduce the power ratio of 500 to a value between 1 and 10; this is done by dividing 500 by 10 as many times as necessary. 500/10 = 50 (1st division); 50/10 = 5.0 (2nd division). The power ratio 5.0 appears in the table, so it is only necessary to divide twice. For each division required we will have to add 10 dB to the value from the table. From the table, a power ratio of 5.0 is equivalent to 6.99 dB. Therefore, a power ratio of 500 is equivalent to 6.99 dB + 10 dB + 10 dB = 26.99 dB. The decibel output of the hi-fi amplifier is 26.99 dB = $10 \log_{10}$ (3 watts/0.006 watt).

Quite often, electronic devices are rated in decibels. For example, a microphone may be listed as −50 dB. This means that its power output may be as low as the equivalent of 50 dB below "zero" power (6 mW). By applying our equation, we can observe how little power this microphone actually puts out and hence what a big job the amplifier will have to do. Substituting in the equation dB = $10 \log_{10} (P_{out}/P_{in})$, we have −50 dB = $10 \log_{10}$ (P_{out}/0.006 watt). Dividing both sides of the equation by 10 gives −5 dB = $\log_{10}$ (P_{out}/0.006 watt). Recalling that $x = \log_{10} N$ and $N = 10^x$ are equivalent expressions, we can rewrite the above equation as P_{out}/0.006 watt = 10^{-5} (where $N = P_{out}$/0.006 watt and $x = -5$). To solve this equation for output power, we must multiply both sides of the equation by 0.006 watt. We have then (P_{out}/0.006 watt) × 0.006 watt = 10^{-5} × 0.006 watt or P_{out} = 0.00001 × 0.006 watt; P_{out} = 0.00000006 watt = 0.06 μW (1 μW = 1 microwatt =

0.000001 watt). This is 6 one hundred millionth's of a watt, an extremely small output.

15.4 VOLTAGE AND CURRENT RATIOS

We can use the power equations, $P = V \times I = V^2/R$ and $P = I^2R$, to write voltage and current ratios in terms of decibels. Substituting I^2R for power in our decibel expression, dB $= 10 \log_{10} (P_{out}/P_{in}) = 10 \log_{10} (I_{out}^2 R_{out}/I_{in}^2 R_{in})$. If input and output resistances are equal, then they cancel and the expression becomes dB $= 10 \log_{10} (I_{out}^2/I_{in}^2)$. Since output and input currents are both squared, we rewrite the equation as dB $= 10 \log_{10} (I_{out}/I_{in})^2$.

We can obtain another form of this same ratio in terms of voltage by substituting V^2/R for power. dB $= 10 \log_{10} (P_{out}/P_{in}) = 10 \log_{10} (V_{out}^2/R_{out})/(V_{in}^2/R_{in})$. Again, if input and output resistances are equal, this becomes dB $= 10 \log_{10} (V_{out}^2/V_{in}^2)$ or dB $= 10 \log_{10} (V_{out}/V_{in})^2$. Note that if input and output resistances are equal, all three of these ratios give the same decibel value, which can represent power, voltage, or current ratios.

Suppose the input voltage of some device is 4 volts and the output voltage of the same device is 8 volts. We apply our equation to find the voltage ratio in decibels as follows. dB $= 10 \log_{10}$ (8 volts/4 volts)^2 $= 10 \log_{10} 2^2 = 10 \log_{10} 4.0$. From our table, a ratio of 4.0 has a decibel value of 6.021 dB. Thus, voltage or power ratio in decibels = 6.021 dB for this device.

15.5 THE VOLUME UNIT

The volume level meter, often seen on tape recorders and broadcast booth consoles, is an ammeter, whose scale is read neither in amperes nor in decibels but in volume units. Remember that to express a power ratio in decibels, a power input of zero was taken to be 0.006 watt. The value 0.006 watt served as the zero reference level. The volume unit (vu) has its zero reference level at 0.001 watt. Thus, the equation for power ratio expressed in volume units is similar to that for power ratio expressed in decibels; only the value of P_{in} at zero input is different. vu $= 10 \log_{10} (P_{out}/0.001$ watt).

Suppose the volume level meter on a hi-fi amplifier reads 40 vu. We can easily calculate the power output at this level. 40 vu $= 10 \log_{10} (P_{out}/0.001$ watt). Dividing both sides of the equation by 10, 4 vu $= \log_{10} (P_{out}/0.001$ watt). Since this is a logarithm expression with $N = P_{out}/$ 0.001 watt and $x = 4$, we can rewrite it as $P_{out}/0.001$ watt $= 10^4$. Multiplying both sides of the equation by 0.001 watt, we have $P_{out} = 10^4 \times 0.001$ watt $= 10{,}000 \times 0.001$ watt $= 10$ watts.

We said that these volume level meters are actually ammeters but that the scale or the meter is in volume units, and the reading can be used to determine power output. It may seem inconsistent that an ammeter, which actually measures current, can be used to find power output. This is possible only because the power ratio in volume units has been calculated for different output currents [vu $= 10 \log_{10} (P_{out}/P_{in}) = 10 \log_{10} (I_{out}/I_{in})^2$], and these volume unit values placed on the same scale of the ammeter in place of the corresponding currents. Thus, although the ammeter measures the output current, the scale reading is in volume units.

15.6 POWER RATIOS MEASURED WITH A VOLTMETER

Many ac voltmeters have decibel scales although they actually measure a voltage drop across some known resistor. In audio frequency (20–20,000 Hz), the voltage drop is measured across a 500-ohm resistor; in telephone work, a 600-ohm resistor is used.

Consider an ac voltmeter, calibrated in decibels, which uses a 500-ohm resistor. What voltage drop across the resistor will have a meter reading of zero decibel? At zero decibel, the power is 6 mW (0.006 watt), and the voltage drop across the 500-ohm resistor can be found using the equation $P = V^2/R$. Substituting, 0.006 watt $= V^2/R = V^2/500$ ohms. Multiplying both sides of the equation by 500 ohms, $V^2 = 0.006$ watt $\times$ 500 ohms $= 3$ watt-ohms. We are looking for a number which, when multiplied by itself, gives 3. $V^2 = 1.732 \times 1.732 = 3$. Hence $V = 1.732$ volts. That is, when the voltmeter measures a voltage drop of 1.732 volts across the 500-ohm resistor, the meter reading is at a power ratio of 0 dB.

This information is useful if we wish to measure power ratios in decibels with a voltmeter calibrated in volts. Suppose an ac voltmeter is across a 500-ohm resistor which is attached to the output of an amplifier whose power input is zero and whose input and output resistances are equal. (See Figure 15.1.) What is the equivalent decibel reading if the

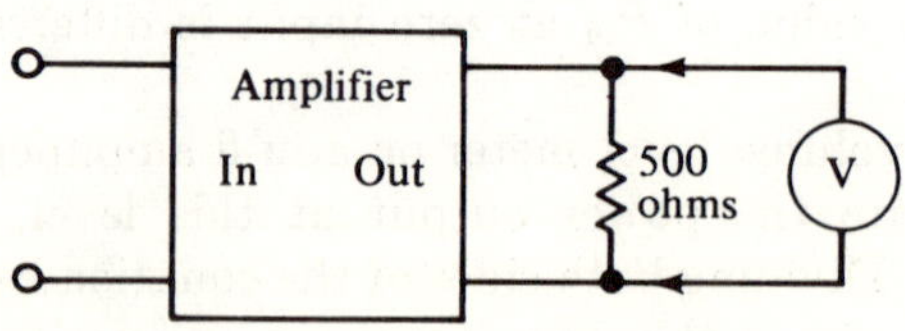

FIGURE 15.1

voltage drop across the resistor is 5.2 volts? We know dB = 10 $\log_{10}$ $(V_{out}/V_{in})^2$. Since the input power is "zero" (0.006 watt), by our previous work the input voltage must be 1.732 volts. Thus, dB = 10 $\log_{10}$ (5.2 volts/1.732 volts)^2 = 10 $\log_{10} 3^2$ = 10 $\log_{10}$ 9.0. Referring to our table, the decibel value corresponding to an actual ratio of 9.0 is 9.542 dB. Thus, the amplifier has an output of 9.542 dB above a 0-dB level.

Questions

(1) What determines how large the power gain in decibels is when the input power is zero?

(2) If 0-dB power is 0.006 watt and 0-vu power is 0.001 watt, what is the difference in these zero levels in decibels?

CHAPTER 16

Classification of Amplifiers

16.1 FREQUENCY RESPONSE

Amplifiers are devices to increase the voltage or power of an ac signal. Classification of the various types of amplifiers gives an indication not only of the wide variety of amplifiers but also of some of their limitations. We had already begun to make a rough division of amplifiers into types when we observed that some have low current outputs and are classified as voltage amplifiers, while others have large current responses and are designated as power amplifiers. This latter type is often used to operate loudspeakers.

The range of frequency (number of cycles each second, in this book referred to as hertz) to which the amplifier will respond supplies another method of classification. The human ear will respond to frequencies ranging approximately from 20 Hz to 20,000 Hz. This range is called the audible, or audio, frequency range. Audio frequency amplifiers, such as high fidelity amplifiers for records and tape recordings, are designed to have their best response within this relatively narrow range of audio frequencies (AF). Radio frequencies (RF) cover such a broad range from 20,000 Hz to upwards of 30,000,000,000 Hz (20 kHz to 30,000 MHz) that any amplifier which responds to even a small portion of this range is called a radio frequency (RF) amplifier. Video amplifiers, used in television and radar, have a very broad frequency response from approximately 20 Hz to 6 MHz (6,000,000 Hz). Later, we will explain how such a broad response can be obtained.

16.2 AMPLIFIER COUPLING

The method by which one stage of amplification is connected to another gives a way of identifying different types of amplifiers. We have, for example, resistance–capacitance (RC) coupling, transformer coupling, direct coupling, and impedance coupling.

16.3 CLASS A AMPLIFIERS

The dc level of the incoming signal with respect to the cutoff point (in a triode the point where the plate current is zero due to the grid being too negative to allow electrons to pass) is also used to classify amplifiers. This level of bias roughly divides amplifiers into three classes. Returning to the transfer characteristic curve of a triode, we recall that the cutoff point occurs when the grid voltage (V_g) is negative and the plate current (I_p) is zero. If the amplifier is biased far enough above cutoff that the dc level of the grid does not drop below cutoff even when the signal is negative going, the amplifier is called a class A amplifier. (See Figure 16.1.)

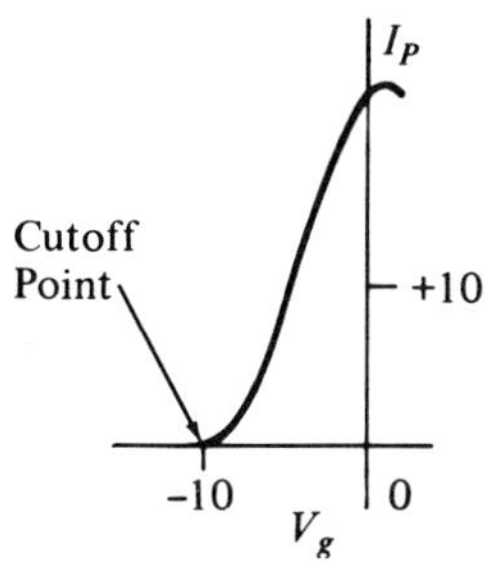

FIGURE 16.1 (*Note:* V_P *is constant transfer characteristic.*)

Another consideration must be whether the signal will cause the grid to go positive, allowing grid current to flow. If the positive half cycle of the input signal does not cause the grid to go positive, the subscript 1 will be associated with the class of amplifier. On the other hand, if the grid does draw current due to its going positive, the subscript 2 will be used. Figure 16.2 illustrates class A_1 and class A_2 amplifiers. With a class A_1 amplifier, if the signal is placed on the linear portion of the curve as shown rather than near the upper or lower bend, the plate current and consequently the plate voltage waves will have little or no distortion.

Ideally, the output power of an amplifier should be equal to the input power. Unfortunately, a great deal of energy is lost by fast electrons hitting the plate and heating it; this loss of energy naturally decreases the

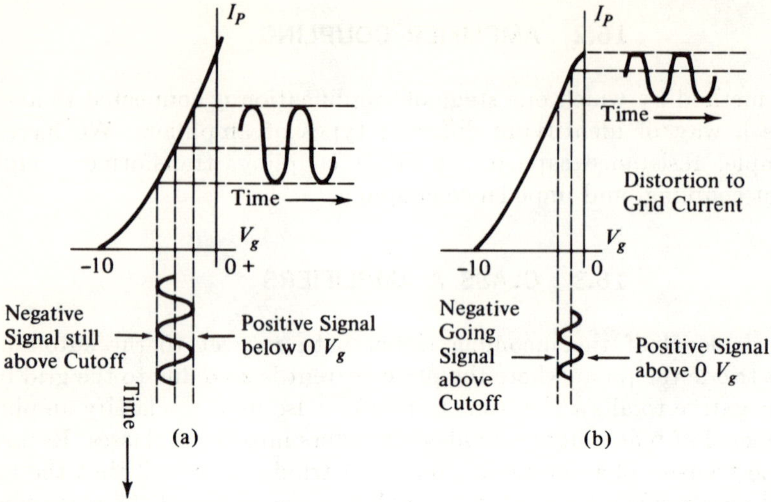

FIGURE 16.2 *(a) Signal on grid class A_1; (b) signal on grid class A_2.*

power output. Thus, the output power of the vacuum tube amplifier depends on how little energy is dissipated in heating the plate or on how efficient the plate is. The plate efficiency, represented by η(eta), is the ratio of the output power of the ac output signal to the input power furnished by the dc supply in the plate circuit. To express this ratio as a percent we multiply by 100.

$$\eta = \frac{\text{ac power out}}{\text{dc power in}} \times 100$$

Let us now consider a class A_1 amplifier and determine its approximate plate efficiency. The triode is cathode biased at $V_g = -5$ volts and an input signal of 6 volts peak to peak is shown on the plate characteristic curve. (See Figure 16.3.) With the bias at -5 volts and the plate current at 50 mA, the plate voltage is 200 volts, as seen on the graph. Thus, the dc input power can be calculated. $P_{\text{in}} = V_P I_P = 200$ volts $\times$ 0.05 A $= 10$ watts. Note that the grid voltage was constant (no input signal) and so this value does represent dc input power.

To determine the ac power output of this class A_1 amplifier, we must substitute the effective values of ac current and voltage in the power equation. Referring to Figure 16.3, we see that the peak value of the ac plate voltage is 350 volts $-$ 200 volts $=$ 150 volts; the effective value of ac plate voltage may be calculated from the expression $V_{\text{Effective ac}} =$ 0.707 $V_{\text{Peak ac}} = 0.707 \times 150$ volts $= 106$ volts. Returning to the graph, we can also find the peak value of ac plate current ($I_{\text{Peak ac}} = 80$ mA $-$

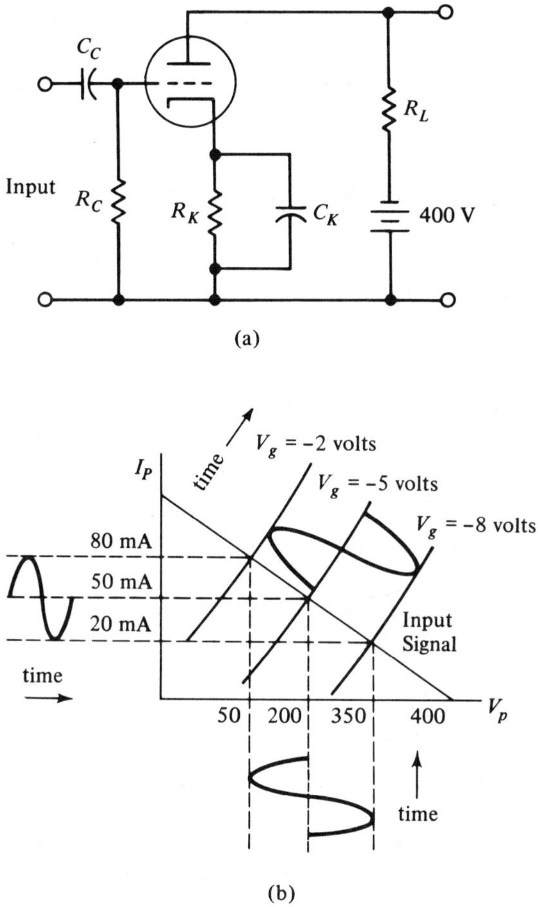

FIGURE 16.3 *(a) Triode class A_1 amplifier; (b) plate characteristic.*

50 mA = 30 mA) and calculate the effective value ($I_{\text{Effective ac}} = 0.707 \times I_{\text{Peak ac}} = 0.707 \times 30$ mA = 21.2 mA). Now applying the power equation, we have $P_{\text{out}} = V_{\text{Effective ac}} \times I_{\text{Effective ac}} = 106$ volts $\times$ 0.0212 A = 2.226 watts, which represents the power of the ac output signal taken across the triode.

Knowing both dc power in and ac power out for this amplifier, we can find plate efficiency. η = (ac power out/dc power in) $\times$ 100 = (2.22 watts/10 watts) $\times$ 100 = 22.2 percent. The efficiency of the class A_1 amplifier is rather low (in our example only 22.2 percent).

The plate loss can easily be calculated by subtracting the output power from the input power. In this case, plate loss = dc input power − ac output power = 10 watts − 2.22 watts = 7.78 watts. This 7.78 watts of power is lost to the output because of the heating of the plate mentioned earlier. Both low plate efficiency and low power output are characteristic of the class A amplifier, and thus this type of amplifier is principally used as a voltage amplifier.

16.4 CLASS B AMPLIFIERS

If the bias voltage is established at cutoff, the amplifier is a class B amplifier. (See Figure 16.4.) When the input signal on the grid goes negative,

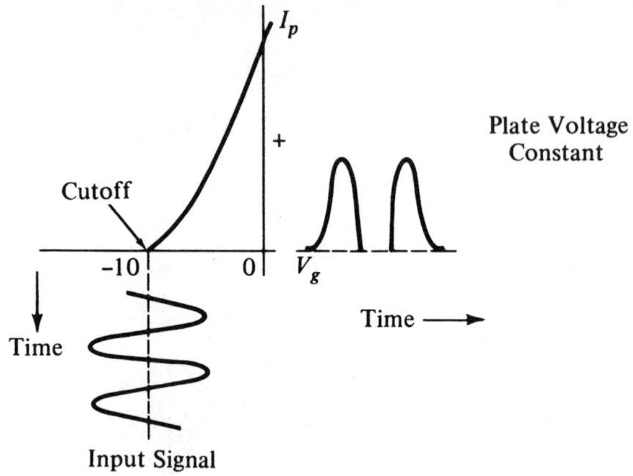

FIGURE 16.4 *Transfer characteristic class B_1 amplifier.*

it falls below the cutoff point and no plate current flows (there is no output). Thus, the only time there will be any output current or voltage is when the input signal is positive. This means that the output of a class B amplifier looks like the output of a half-wave rectifier.

The output wave is distorted in two ways. First, as we just mentioned, there is no negative half cycle. Second, the shape of the positive waveform will be somewhat distorted. (See Figure 16.5.) This second

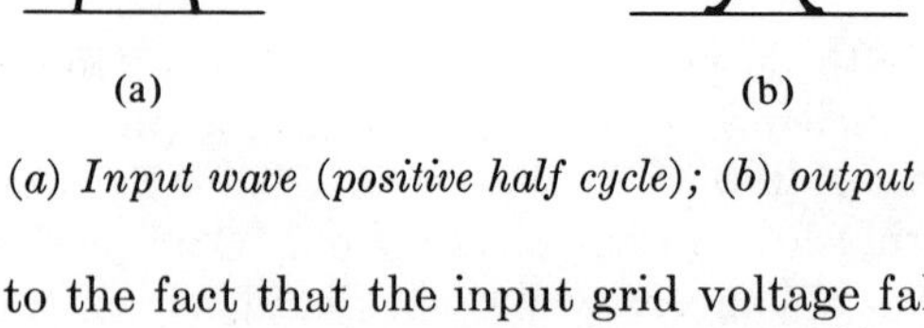

FIGURE 16.5 *(a) Input wave (positive half cycle); (b) output wave.*

distortion is due to the fact that the input grid voltage falls on the nonlinear portion of the transfer characteristic curve; this portion of the graph does not reflect changes in grid voltage to changes in plate current in a linear manner; instead, the same change in grid voltage will produce different changes in plate current when taken on different sections of the nonlinear graph. This causes some distortion in transferring from input to output waves.

The problem of plate heating losses is not as great in the class B amplifier as it was in the class A. In fact, the plate efficiency of the class B amplifier ranges from 40 percent to 60 percent, so this type of amplifier has a much higher power output. To utilize this high power output, yet overcome the output distortion, two class B amplifiers are employed in a push–pull arrangement (Figure 16.6).

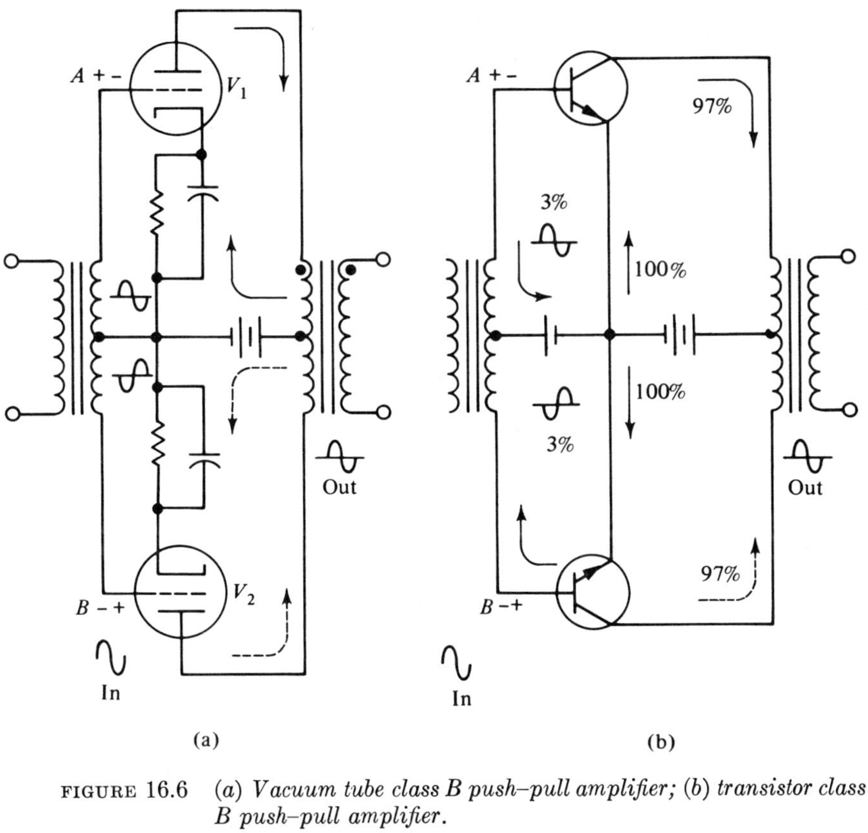

FIGURE 16.6 *(a) Vacuum tube class B push–pull amplifier; (b) transistor class B push–pull amplifier.*

Looking at the vacuum tube amplifier, when the incoming signal goes positive, point *A* is more positive than point *B*, and tube V_1 conducts. Remember that a class B amplifier is biased at cutoff, so tube V_2 will not conduct when point *B* is negative. Current flows through V_1, through the center tapped primary transformer winding (the load), and back to the supply battery. When the signal goes negative, point B becomes positive with respect to point A, so tube V_2 conducts with current flowing through the load and back to the supply. Note that current is moving in the opposite direction to that which flowed in the load during the positive half cycle. In the past, we have taken the output across the tube which made the output out of phase with the input. However, since the output of the push–pull amplifier is across the center tapped primary transformer winding, which serves as the load, and since the input is also transformer coupled, the input and output will be in phase. The transformer input coupling limits the frequencies for which this push–pull amplifier will be effective; later, we will see how this frequency range can be increased by using a paraphase amplifier in place of the input transformer. The waveforms for the push–pull amplifier are shown in Figure 16.7.

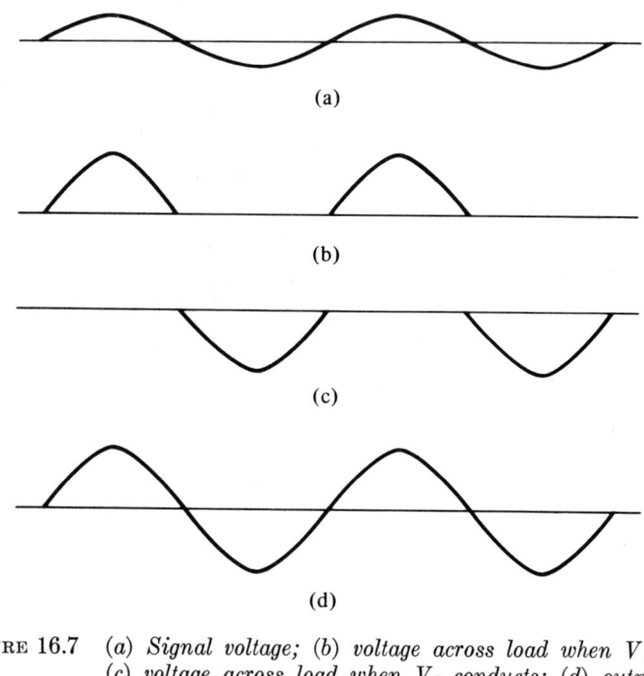

FIGURE 16.7 *(a) Signal voltage; (b) voltage across load when V_1 conducts; (c) voltage across load when V_2 conducts; (d) output class B push–pull amplifier.*

16.5 CLASS C AMPLIFIERS

Class C operation occurs when the grid is biased below cutoff. Thus, the tube conducts only for a short time during the positive half cycle of the input signal. The transfer characteristic curve in Figure 16.8 shows

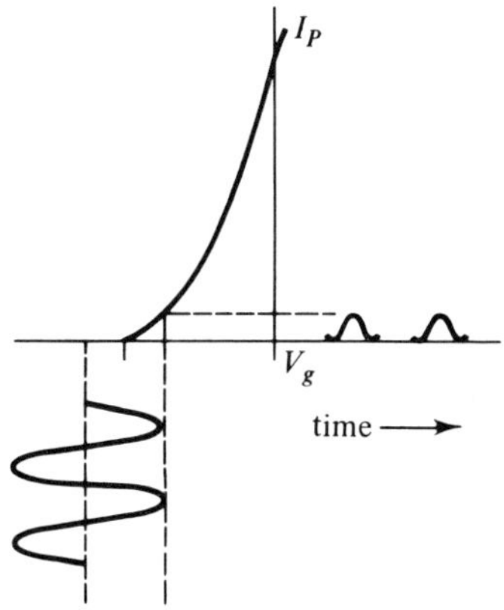

FIGURE 16.8 *Transfer characteristic class C amplifier.*

this bias point with the resulting high distortion of the plate current. Because of this extremely high distortion, class C amplifiers lack the output fidelity to be used to amplify audio frequencies. However, when we are concerned with radio transmitters, the high plate efficiency (60 percent to 80 percent) of the class C amplifier provides the largest radio frequency power output of any of the three classes of amplifiers.

Questions

(1) What determines how much of the input signal to an amplifier is reproduced at the output?

CHAPTER 17

Amplifier Coupling

17.1 RC COUPLING

Since the amplification of one stage is normally not sufficient to bring weak incoming signals up to workable output strength, several stages of amplification are usually joined in cascade fashion so that the output of one stage becomes the input to the next. The most common type of coupling of one stage to another is resistance–capacitance coupling, which we studied when we talked about time constants. In order to pass the signal on to the next stage without introducing any distortion, the time constant (1 TC = RC) must be long in comparison to the period (T = $1/f$). The dc component of the plate voltage, if allowed to pass to the next stage, would affect its level of bias. Thus, the coupling capacitor serves the purpose of blocking this dc while passing the ac with minimum distortion. (See Figure 17.1.) RC coupling is noted for its good frequency response over the audible range. An increased range of response to include the video range can be obtained by a reduction in overall gain in signal strength.

RC coupling is normally associated with junction transistor amplifiers. (See Figure 17.2.) Capacitor C_1 blocks the flow of dc from the first stage to the second, but it does not interfere with the forward bias voltage of the second stage. RC coupled transistor amplifiers are used in audio amplifiers.

17.2 INDUCTIVE COUPLING

A variation on the RC filter exchanges the load resistor for a coil. (See Figure 17.3.) The inductive reactance (the load) is directly proportional

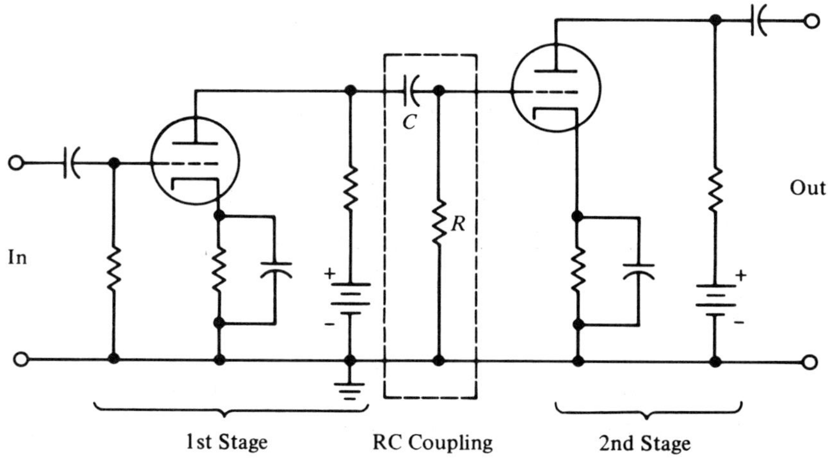

FIGURE 17.1

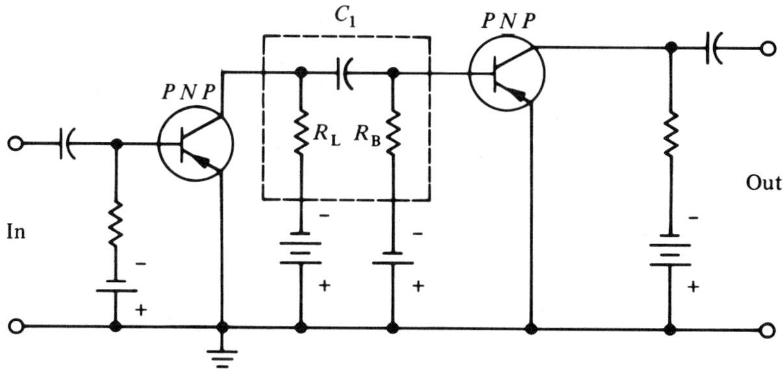

FIGURE 17.2 *RC coupled common-emitter transistor amplifier.*

to frequency ($X_L = 2\pi fL$). Thus, at higher frequencies, the inductive reactance increases, and more voltage is dropped across the load, giving greater amplification. Since amplification varies with frequency, the inductively coupled amplifier is used only over a small range of rather high frequencies. The coil has a low resistance to dc, so only a small portion of the dc voltage in the plate circuit is dropped across the load; this allows for a larger plate voltage.

Now consider a transistor circuit with inductance coupling. (See Figure 17.4.) Since the resistance of the coil is small, the dc power loss in the collector circuit is negligible, and the amplifier is more efficient than it would be with a collector resistor. Again, the inductive reactance, and hence the amplification, vary with frequency, just as in the inductance coupled triode amplifier above.

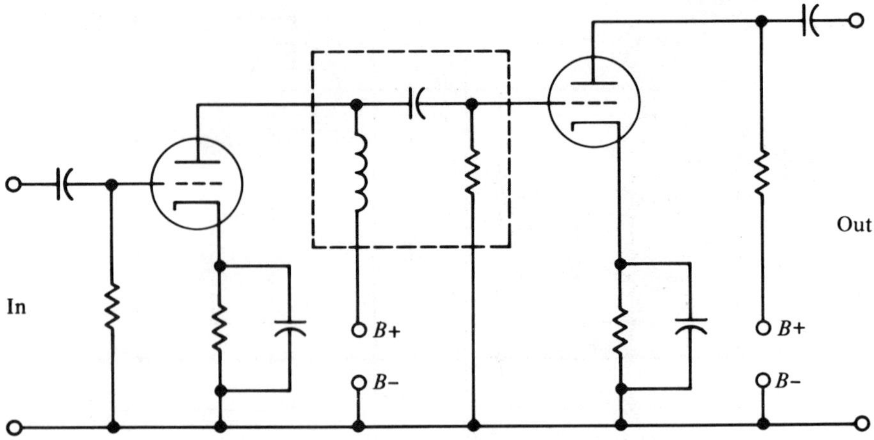

FIGURE 17.3 *Inductance coupled amplifier.*

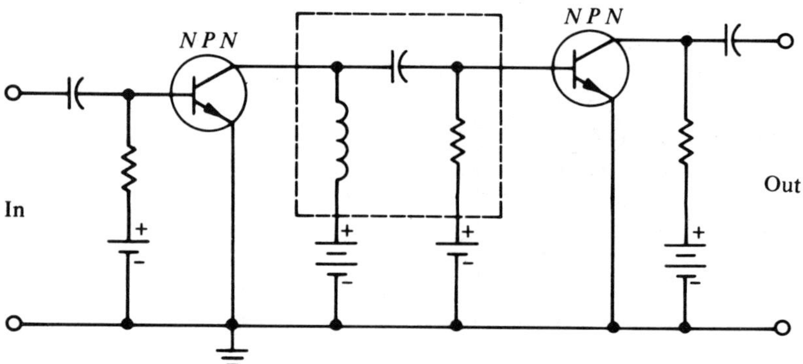

FIGURE 17.4 *Inductance coupled common-emitter transistor amplifier.*

17.3 TRANSFORMER COUPLING

We are aware that the transformer, like the capacitor, does not pass dc, so it can also be used as a coupling device. In addition, we have seen that the transformer has excellent impedance matching ability, which makes possible maximum power transfer from one stage to the next. Comparing the transformer coupled amplifier to one with RC coupling, we see that the transformer has replaced the plate load resistor, the coupling capacitor, and the grid resistor. (See Figure 17.5.) Because the primary coil in the transformer offers very little resistance to dc, the plate voltage of triode V_1 is nearly at B^+ potential; since no dc can be induced from the primary into the secondary coil, a coupling capacitor is not

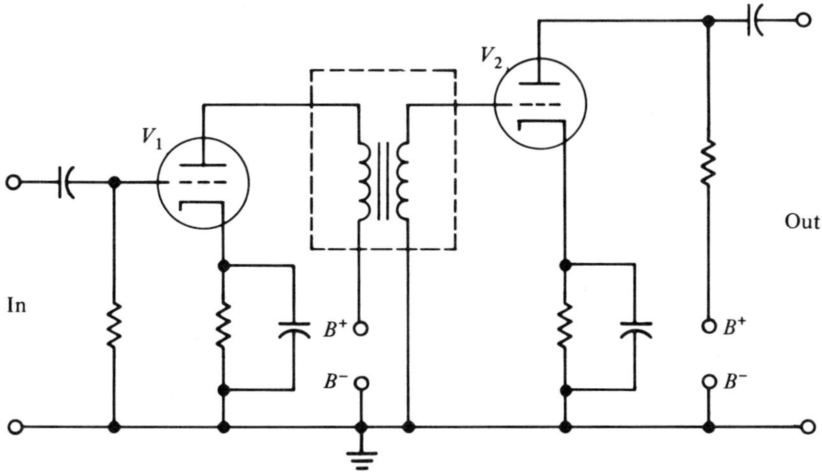

FIGURE 17.5 *Transformer coupled amplifier.*

necessary. The resistance of the secondary transformer winding serves as the grid resistor for V_2.

The transformer coupled transistor amplifier also utilizes the transformer to block dc and to match the output impedance of one stage to the input impedance of the next. (See Figure 17.6.) Capacitor C_1 blocks

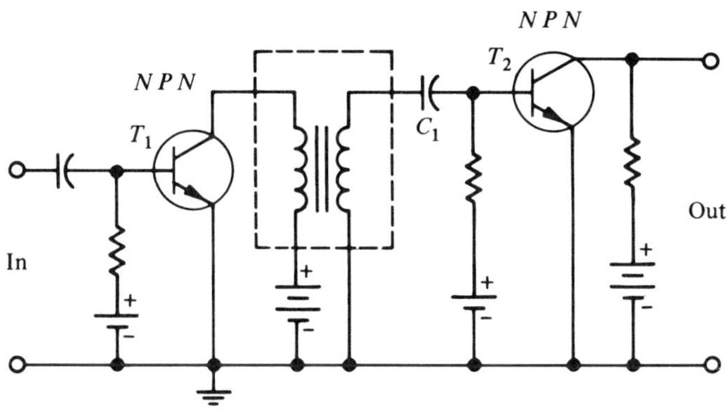

FIGURE 17.6 *Transformer coupled transistor amplifier.*

additional dc bias, which would otherwise be placed on the base of transistor T_2 by the voltage drop across the resistance of the secondary coil (this voltage drop would occur if dc current were allowed to flow in the secondary); C_1, of course, passes the ac input signal to the base of T_2.

As with inductance coupling, transformer coupling has the disadvantage of quite limited frequency response; in addition, transformers are more expensive than capacitors and resistors. The principal advantage of transformer coupling is its excellent power output made possible by the low resistance of the coils.

17.4 TUNED CIRCUIT COUPLING

Another form of transformer coupling, the tuned circuit, consists of two parallel LC circuits both with the same resonant frequency. (See Figure 17.7.) Remember that at resonance, the impedance of a coil and

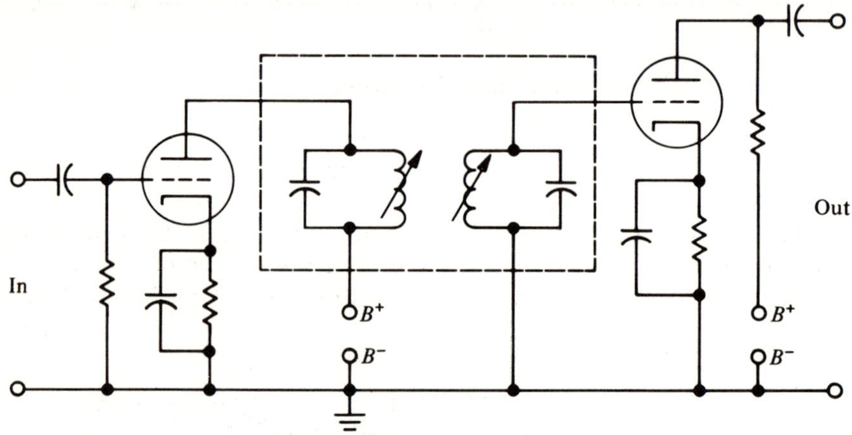

FIGURE 17.7 *Tuned circuit coupling.*

capacitor in parallel is a maximum. The output of the first stage is taken across the parallel LC circuit, which displays high impedance at resonance and thus gives good amplification of signals at the resonant frequency. Signals not at resonance find a small load impedance and are not effectively amplified. The magnetic coupling to the second stage eliminates the dc. The resonant frequency of the parallel LC circuit in the second stage is the same as that in the first, so amplified signals at resonant frequency meet high impedance and are placed on the grid of the second triode. Any signals not at resonant frequency meet a much lower impedance and are shunted to ground. Tuned circuit coupling is often used in radio frequency amplifiers, since it only amplifies signals which are within the bandpass at resonant frequency. The inductance of the coils can be varied by moving a metal core in and out, and thus the resonant frequency of the LC circuits can be changed, that is, the circuit can be tuned to a desired resonant frequency.

17.5 DIRECT COUPLING

We will consider one more method of coupling the output of one stage of amplification to the input of the next. This next type of coupling is called direct coupling in that the output of the first stage is applied directly to the input of the second. The implication of this direct coupling is that each successive stage must operate at a higher voltage than the preceding one.

In the directly coupled amplifier of Figure 17.8, notice that the plate of V_1 is connected directly to the grid of the next tube, V_2. Since the plate

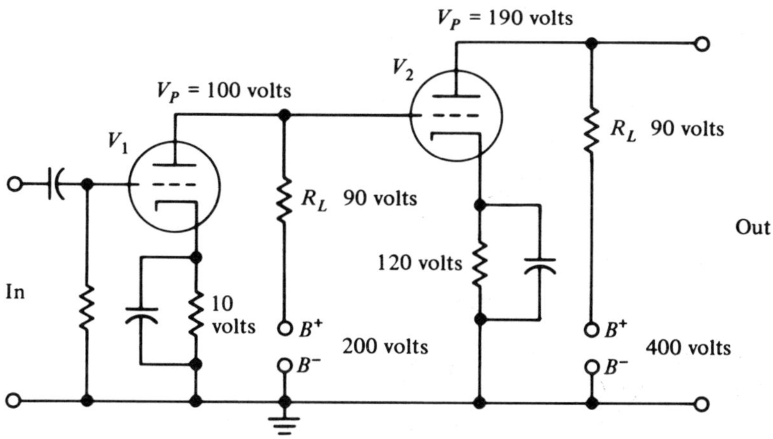

FIGURE 17.8 *Directly coupled amplifier.*

of V_1 is positive, the grid of V_2 has a large positive dc voltage impressed on it in addition to the ac signal from V_1. But the grid of a triode must be negatively biased with respect to the cathode; to accomplish this, the cathode of V_2 is made even more positive than the grid of V_2 (or the plate of V_1). In our example, the voltage drop across V_1 (the plate voltage) is 100 volts, and the voltage drop across the cathode resistor of the first stage is 10 volts. The dc voltage applied to the grid of V_2 is the potential between ground and the plate of V_1 (110 volts). In order to make the cathode of V_2 more positive than the grid of V_2, the voltage drop across the cathode resistor of the second stage is 120 volts. This larger voltage drop naturally requires a larger supply voltage for the second stage. The dc power supply for a directly coupled amplifier must be well regulated so as to maintain the proper B^+ supplies for each stage of amplification.

A directly coupled transistor amplifier utilizes both *NPN* and *PNP* transistors. (See Figure 17.9.) Notice that the collector of the first tran-

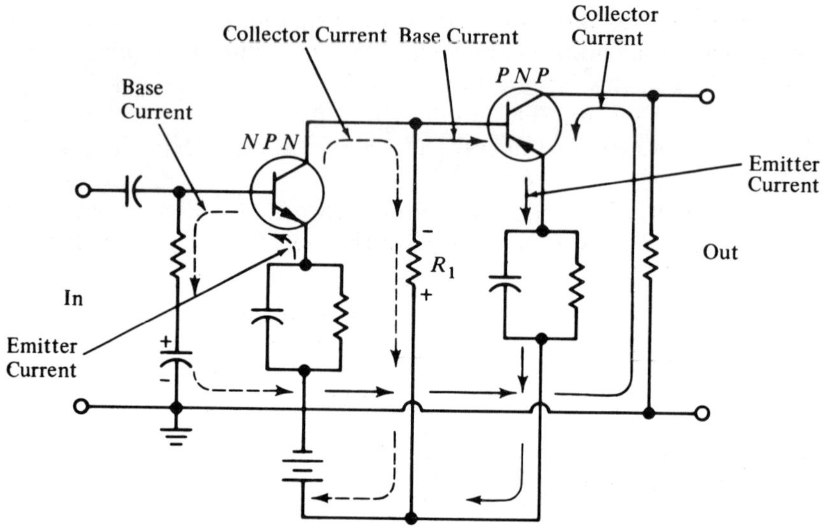

FIGURE 17.9 *Directly coupled transistor amplifier.*

sistor is directly connected to the base of the second. The collector current in the first stage (*NPN* transistor) produces a voltage drop across R_1, which provides the forward bias for the *PNP* transistor in the second stage. As the *NPN* collector current increases due to an increase in the incoming signal, the forward bias voltage of the *PNP* transistor increases, and an increased current flows in the second stage. Thus, the output of the *NPN* transistor in the first stage becomes the input for the second stage and is further amplified by the *PNP* transistor.

17.6 DIRECTLY COUPLED VTVM

The VTVM (vacuum tube voltmeter) is one application of the directly coupled triode amplifier. Since the directly coupled amplifier is limited to low-frequency ac signal or to dc voltage, we will use the VTVM as a very sensitive dc voltmeter. (See Figure 17.10.)

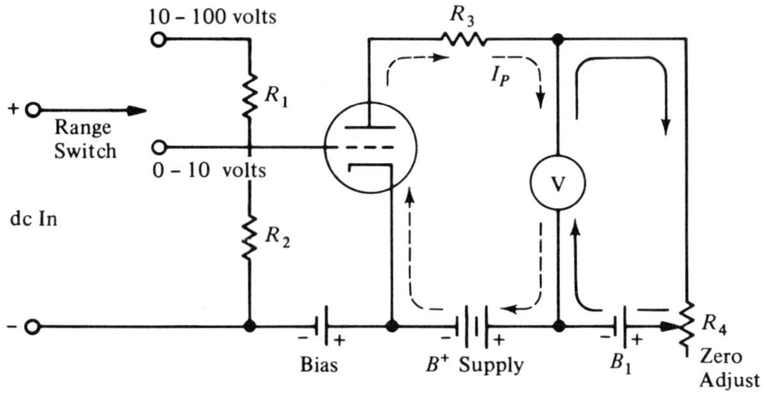

FIGURE 17.10 *VTVM (vacuum tube voltmeter).*

The triode is biased near the beginning of the linear portion of the transfer characteristic curve. (See Figure 17.11.) Thus, even before a dc

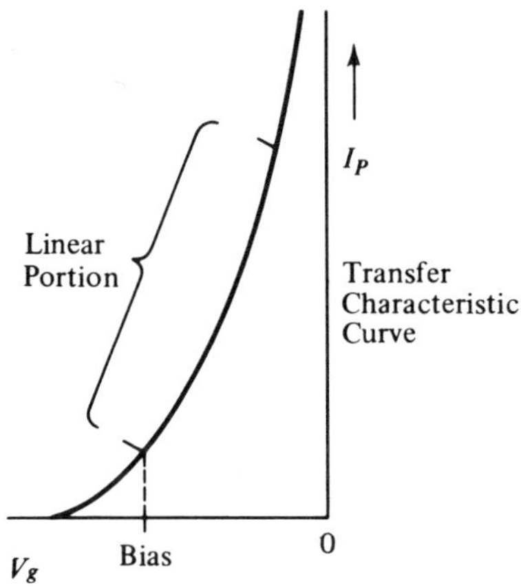

FIGURE 17.11

input signal is introduced to the VTVM, a small plate current (I_P) will flow through the tube from the B^+ supply battery, through resistor R_3, through voltmeter V (deflecting its pointer), and back to the battery. Battery B_1 also causes a current to flow in the voltmeter in the opposite direction to I_P. Variable resistor R_4 (zero adjust) can be adjusted so that this current from B_1 is equal to I_P; since the two currents are then equal but in the opposite direction in V, they cancel each other, and no current flows in the voltmeter (its pointer is at the zero position). The dc input voltage is placed across the range switch, which is simply a voltage divider consisting of resistors R_1 and R_2. If the input voltage is small enough (in our example, 0–10 volts), it can be placed directly on the grid. If the input voltage is larger, the input is connected to the upper terminal of the switch, and part of the voltage is dropped across R_1. The remainder, dropped across R_2, is placed on the grid. This prevents the grid from becoming too positive. A different scale on the meter face is used for each range. (See Figure 17.12.) When the input is applied to the grid of the

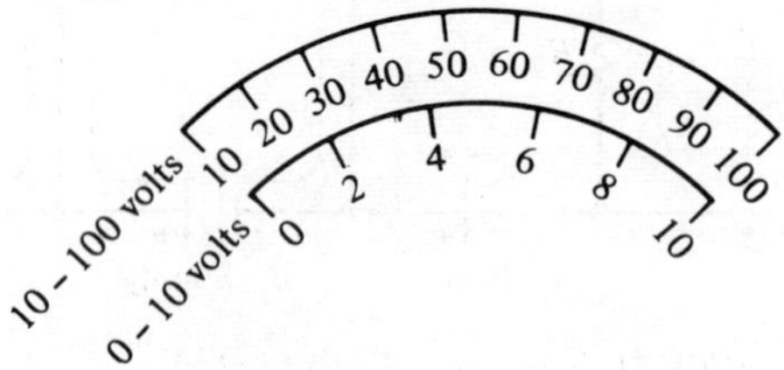

FIGURE 17.12

tube, the grid becomes more positive, and the plate current increases. This causes the dc voltmeter to deflect proportionately. Since the input signal is amplified by the triode, the VTVM can measure small dc voltages to which the basic voltmeter V would not have been sensitive due to the high resistance of the meter movement.

Questions

(1) Consider some analogies between amplifier coupling and the game called "Pass It" in which a story is passed from one person to another and the final story is compared to the original.

(2) Distinguish between the gain of an amplifier and the amplification factor μ of a tube.

CHAPTER 18

Broad Frequency Response in an Amplifier

18.1 UNIFORM GAIN

A uniform gain in output voltage over a wide range of frequencies is best obtained with an RC coupled amplifier. Because of certain limitations of the triode and the circuit wiring, some modifications of the basic RC amplifier are necessary to reach the goal of linear output (uniform gain). In the audio frequency range (20–20,000 Hz), the problems of frequency response are not as great as with video amplifiers (video range:

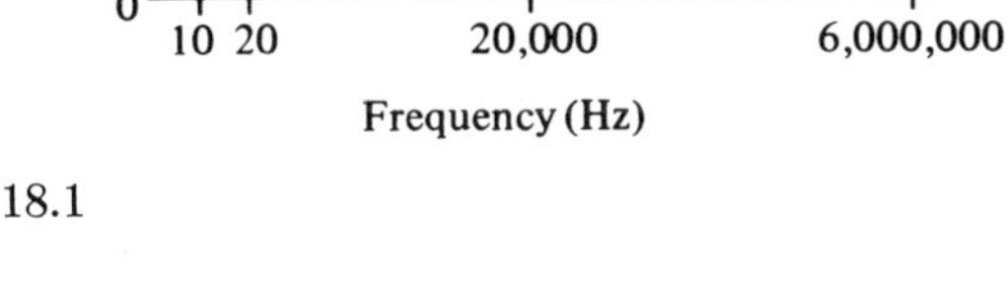

FIGURE 18.1

10–6,000,000 Hz). (See Figure 18.1.) Therefore, in discussing audio amplifiers, we will concentrate on obtaining maximum gain (Gain $= V_{out}/V_{in}$).

To obtain maximum gain, a large plate load resistance is used. (See Figure 18.2.) Remember that $V_{Supply} = V_{Load} + V_{Plate}$ where the output (V_{Plate}) is taken between plate and ground. When the signal goes positive, causing even a small plate current (I_P) to pass through the large load

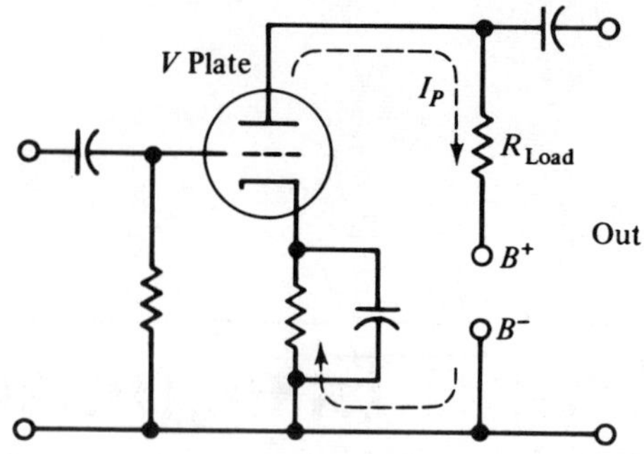

FIGURE 18.2

resistance (R_{Load}), it will produce a large voltage drop (V_{Load}). Because of the phase inversion already discussed, an increased V_{Load} drives the output (V_{Plate}) more negative. A larger load thus increases the amplitude of the output and hence gives a greater voltage gain.

18.2 HIDDEN CAPACITANCES

Broadening the frequency response of the RC coupled amplifier to the video range requires attention to frequencies both above and below the audible range. For improved high-frequency response, we will have to overcome certain hidden capacitances, whose capacitive reactances ($X_C = 1/2\pi fC$) are so small at high frequencies that they tend to shunt the signal to ground. These hidden capacitances, which include some of the interelectrode capacitances that we have already discussed, are: (1) the output capacitance (C_o) between plate and cathode of the triode in the first stage, (2) the input capacitance (C_i) between grid and cathode in the second stage, and (3) the distributive capacitance (C_d) caused by the proximity of components and connecting wires. (See Figure 18.3.)

18.3 SHUNT COMPENSATOR

Before turning to other types of vacuum tubes, such as the tetrode and pentode, to overcome some of these tube capacitances, we will discuss several methods of improving high-frequency response in amplifiers which use the more familiar triode. The first technique, which is simply a form of inductance coupling called a shunt compensator, places a coil in series with the load resistor. (See Figure 18.4.) This coil can be thought of as being in parallel with the unwanted hidden capacitances. If the

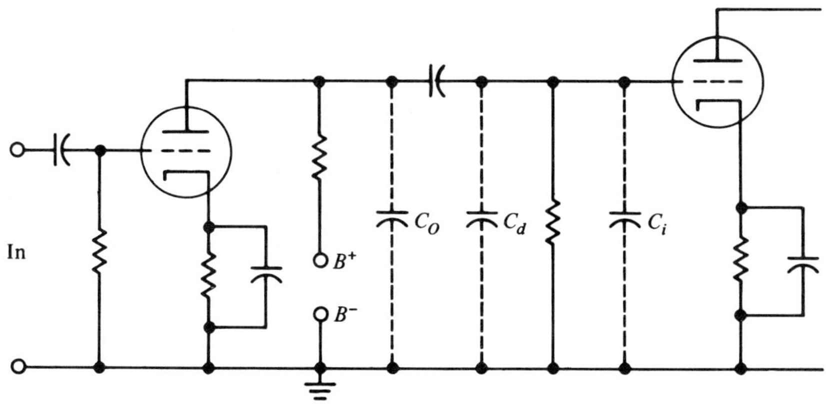

FIGURE 18.3 *Hidden capacitances.*

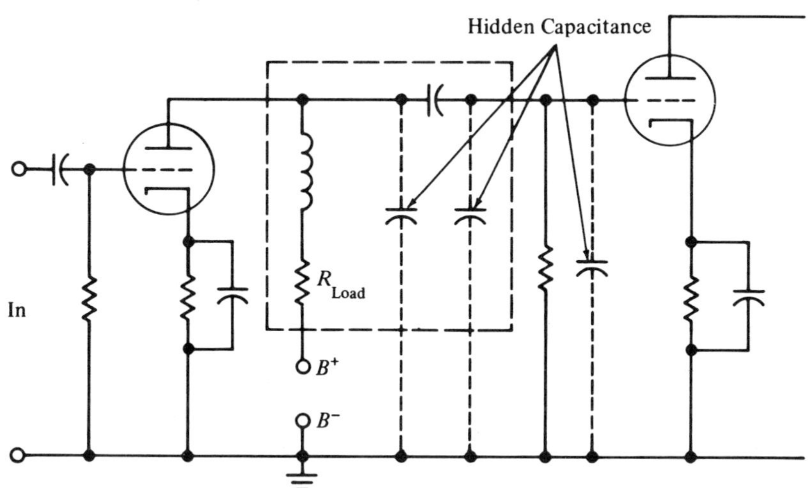

FIGURE 18.4 *Shunt compensator.*

inductance of the coil is chosen so that it resonates with these hidden capacitances at the frequency where the gain would normally begin to fall off, then the parallel LC combination will offer a high load impedance at the resonant frequency and thus improve the gain and extend the frequency range. Since the inductive reactance increases and the capacitive reactance decreases as the frequency increases beyond the resonant point, the parallel circuit then takes on the characteristic of X_C, the smaller reactance. You may wish to review the section on parallel resonance if the above statement seems foreign to you. Thus, the shunt compensator can only slightly extend the frequency response before the hidden capacitances again start to ground the output signal and reduce the gain. (See Figure 18.5.)

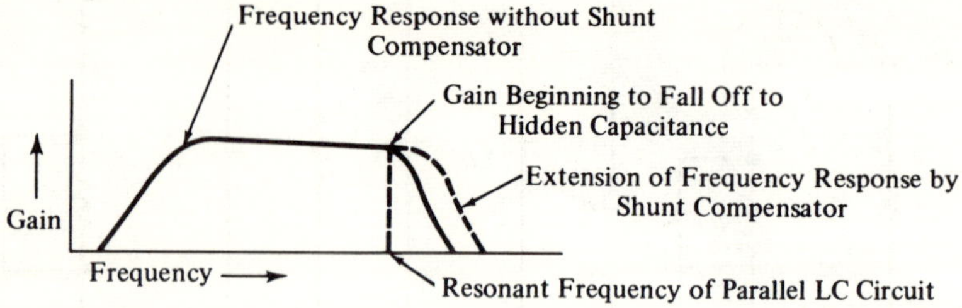

FIGURE 18.5

18.4 SERIES COMPENSATOR

Another way to avoid losing part of the high-frequency output from the first stage before it can be amplified by the second stage is to place a coil in series with the input capacitor to the second stage (series compensator). (See Figure 18.6.) Remember in our study of resonant circuits, we

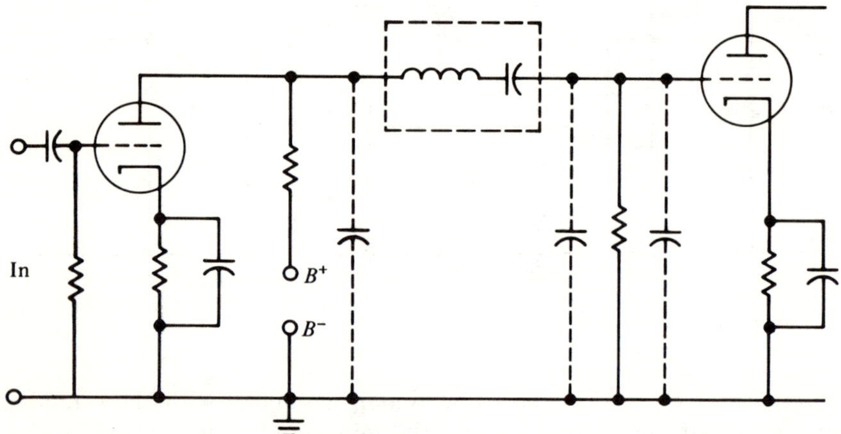

FIGURE 18.6 *Series compensator.*

observed that the impedance of a coil and capacitor in series was at a minimum at the resonant frequency. The series compensator has its resonant frequency at the point where gain would normally begin to fall off; at this point, it offers a low impedance path and thus allows more current to reach the grid of the second stage.

18.5 SERIES–SHUNT COMPENSATOR

The series and shunt compensator networks may be combined for increased gain and better response at high frequencies. (See Figure 18.7.)

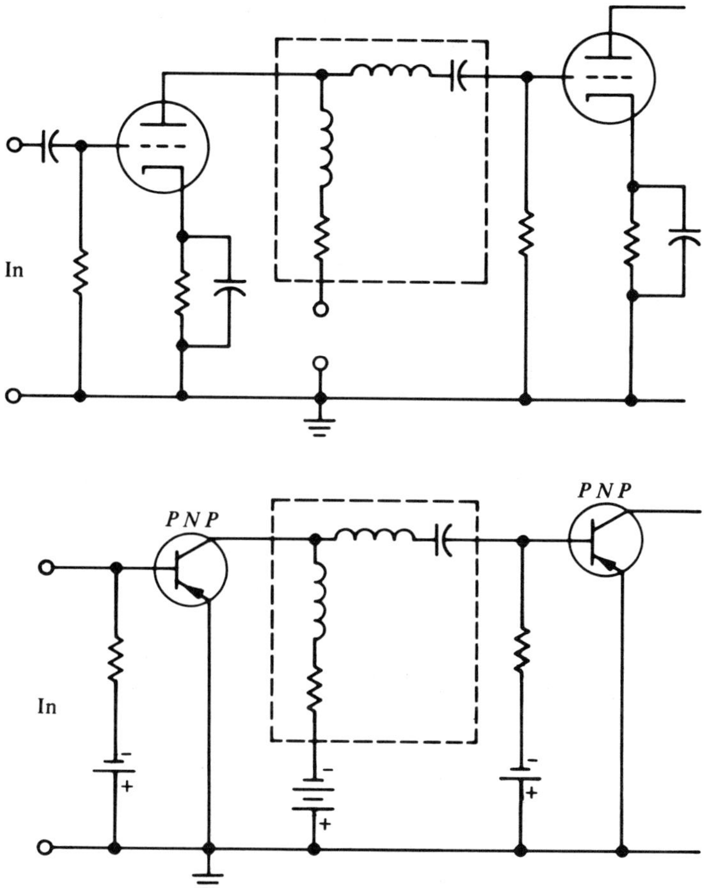

FIGURE 18.7 *Series–shunt high-frequency compensator.*

Notice that they may also be applied to transistor circuits to overcome input and output capacitances.

18.6 LOW-FREQUENCY COMPENSATOR

Since capacitive reactance is inversely proportional to frequency, at low frequencies the coupling capacitor has a high reactance, which reduces the signal to the second stage. Trying to overcome this high reactance by simply increasing the size of the coupling capacitor is unsatisfactory because of the stray interstage capacitances thus introduced. In order to compensate for this high reactance at low frequencies, we have to improve low-frequency signal strength without affecting the higher-frequency signals. The low-frequency compensating network consists of a capacitor (C_f) and a resistor (R_f) placed in the load circuit. (See Figure 18.8.) At

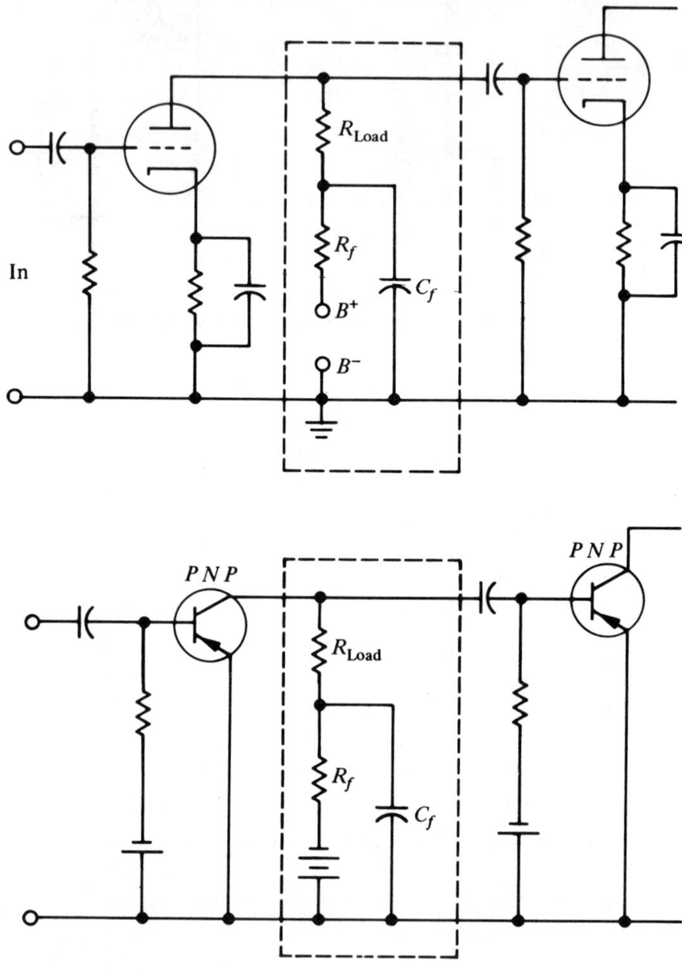

FIGURE 18.8 *Low-frequency compensator network.*

high frequencies, capacitor C_f has an extremely small capacitive reactance and passes the signal around R_f; thus, the voltage drop is just across R_{Load}. At low frequencies, however, the capacitive reactance increases, and the total impedance of the parallel circuit containing R_f and C_f also increases. This causes an increase in total load resistance (voltage is dropped across both R_{Load} and the $R_f\ C_f$ combination) which, as we have seen, increases gain.

18.7 NEGATIVE FEEDBACK TO IMPROVE HIGH-FREQUENCY RESPONSE

In placing these additional components in the circuit to act as high- and low-frequency compensators, we sacrifice overall gain for the broader-frequency response needed in video amplifiers. By utilizing negative feedback, we can further improve frequency response again at the expense of overall gain. Remember when we talked about the role of the bypass capacitor in cathode biasing, we recognized that its absence resulted in negative feedback. That is, without the bypass cathode capacitor, an incoming positive signal, which increases the plate current, will also increase the voltage drop across the cathode resistor. This larger voltage drop places a more negative bias on the grid. But making the grid more negative lowers the dc potential of the incoming positive signal, thus decreasing the plate current. Negative, or degenerative, feedback causes any increase in plate current to be countered with a corresponding decrease, and vice versa. A similar process makes it possible to extend the high-frequency range of a cathode biased amplifier by eliminating its bypass cathode capacitor. When the incoming high-frequency signals are weak, there is only a small increase in plate current. Thus, the voltage drop across the cathode resistor is much smaller at higher frequencies. This smaller voltage drop means a decreased grid bias (grid is more positive) at higher frequencies. With the grid more positive, the dc level of the weak incoming high-frequency signals is higher than that of other signals, and their effect is increased by the resulting higher plate current.

Another method of obtaining negative feedback is to connect a capacitor between the plate and grid of the triode. In a sense, this capacitor already exists in the triode due to the interelectrode capacitance between plate and grid (C_{pg}); such a capacitor does not actually have to be placed in the triode to produce negative feedback, because it is already there. Although just a fraction of the output from the plate is fed back through this capacitance to the grid, the relative size of the feedback regulates grid voltage. When the plate voltage is large, the feedback increases. Since there is a polarity inversion between plate voltage and the input signal, an increase in plate voltage will cause a decrease in grid voltage. This makes the grid more negative and causes less current to flow with a corresponding decrease in plate voltage. The weak high-frequency signals, which produce small plate voltages, cannot cause such a large degenerative feedback. Thus, they do not drive the grid as negative as the other stronger signals, and the plate current for high-frequency signals is increased.

The schematic in Figure 18.9 shows high- and low-frequency compensators combined with degenerative feedback to produce a video amplifier.

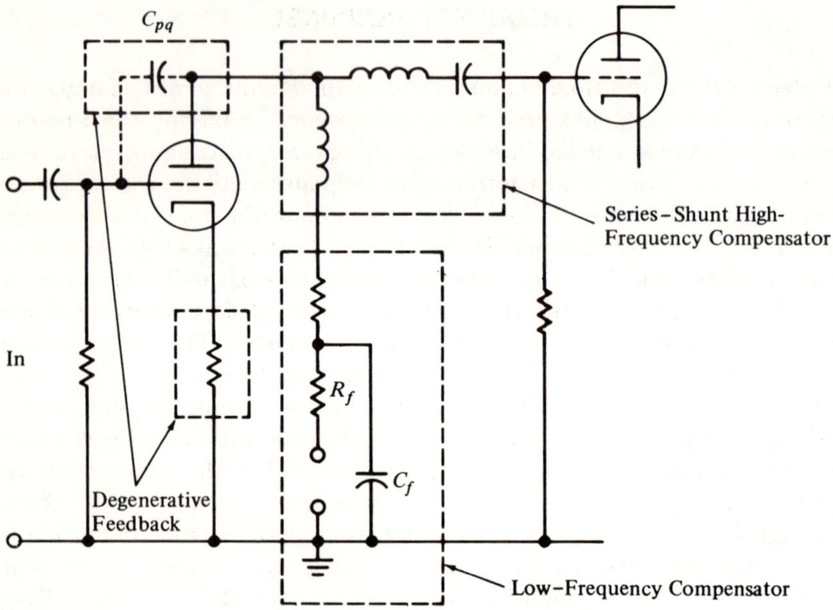

FIGURE 18.9 *Compensated video amplifier.*

Figure 18.10 indicates the range of frequency response and the relative gain in signal strength for both an uncompensated RC coupled amplifier and a video amplifier. As can be seen, gain is sacrificed for increased frequency response.

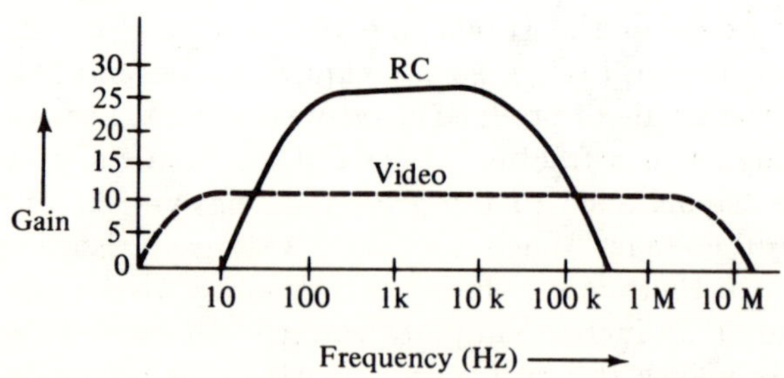

FIGURE 18.10

Questions

(1) What type relation, if any, exists between gain and frequency response in a video amplifier?

Suggested Experiments

(1) Frequency Response of an RC Coupled Amplifier. The gain of an RC amplifier at different frequencies can be determined by comparing voltage output to voltage input. Gain = V_{out}/V_{in}. First calibrate the oscilloscope to measure voltage as explained in the section on oscilloscopes. Adjust the audio or RF oscillator to a specific frequency and voltage output, with the vertical input of the oscilloscope attached to the output of the oscillator. Now move the vertical input lead to the output of the RC amplifier and observe the output voltage on the scope. Compute the gain at that frequency. Then change the frequency and repeat the above process. (See Figure 18.11.)

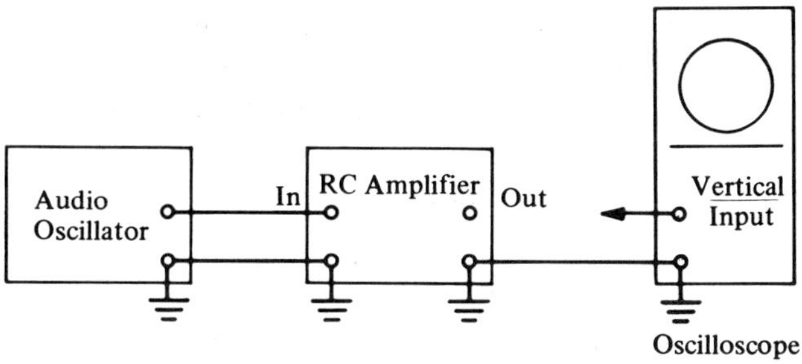

FIGURE 18.11

(2) Frequency Response of an Amplifier. The gain of any amplifier may be calculated and graphed as described in the above experiment.

CHAPTER 19

Multigrid Vacuum Tubes

19.1 THE TETRODE

The pentode vacuum tube, a five-element three-grid tube, is used more frequently than the triode in wide-frequency response amplifiers because of its low interelectrode capacitances. Before studying the pentode, we should briefly mention the four-element vacuum tube called the tetrode. (See Figure 19.1.) The tetrode has a second grid, the screen grid, which

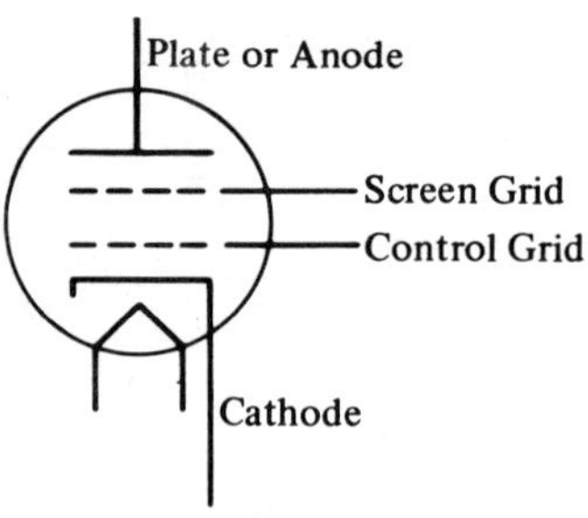

FIGURE 19.1 *Tetrode.*

is placed between the control grid and the plate to reduce their interelectrode capacitance (C_{pg}). Without the screen grid some of the electrons that get past the control grid do not have enough energy to reach the plate; they build up a negative space charge between plate and control grid which creates an interelectrode capacitance. When the screen grid is placed close to the control grid and a positive potential is placed on the screen grid, the electrons that pass the control grid are accelerated toward the plate. This reduces any negative space charge and the resulting interelectrode capacitance. The relative size of R_{SC} and R_{Load} must be such that the plate is more positive than the screen grid. A voltage drop across R_{SC} is produced by the small current, which flows when electrons strike the positive screen grid. Any ac signal that might be attracted to the screen grid, due to its being positive, is shunted to ground via capacitor C_{SC}; thus, the screen grid is at ac ground potential but at a positive dc potential. (See Figure 19.2.)

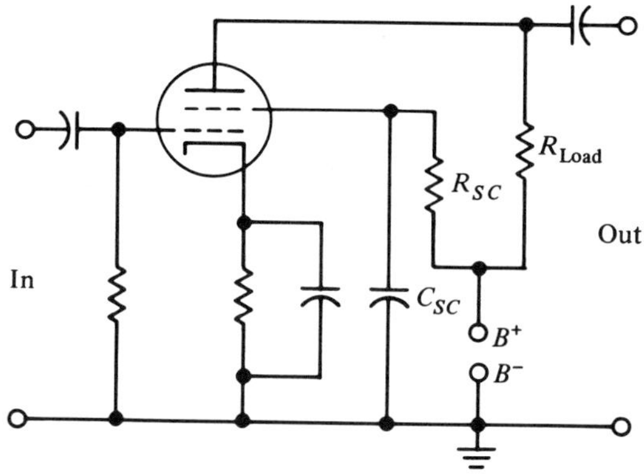

FIGURE 19.2 *Tetrode amplifier.*

19.2 THE PENTODE

The screen grid accelerates electrons toward the plate, and the plate, being positive, attracts the electrons to itself. These electrons may be traveling at such high speeds when they strike the plate that they remove other surface electrons from the plate. The phenomenon is much like throwing a stone into a lake and observing the splash. Electron "splash" from the plate is called secondary emission. Those electrons emitted from the plate build up a space charge between plate and screen grid. To reduce secondary emission and to prevent any electrons, which are emitted, from flowing to the screen grid, a third grid called the suppressor grid is introduced between screen grid and plate. (See Figure 19.3.) The suppressor grid is kept at ground or cathode (near ground) potential. Electrons after passing the screen grid see a negative charge, which slows them down before they strike the plate and thus suppresses secondary emission. In addition, any secondary electrons, which are emitted from the plate, are reflected back toward the plate by the negative suppressor grid. In this way, screen grid current due to secondary electrons is suppressed.

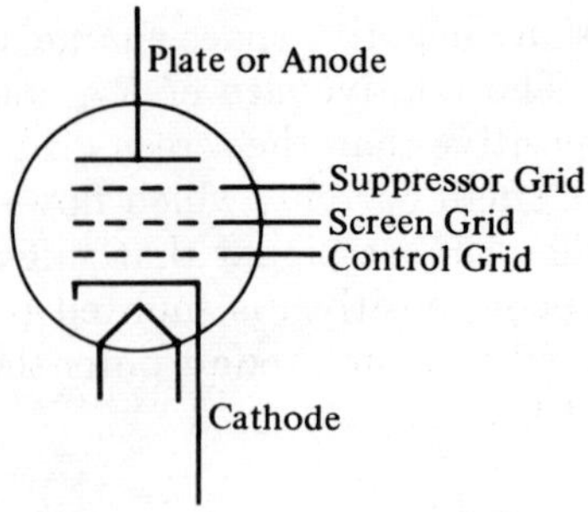

FIGURE 19.3 *Pentode.*

We have seen that the reduction of distance between electrodes by the insertion of additional grids reduces interelectrode capacitances. This reduction of interelectrode capacitances enables an amplifier to increase its frequency range. The schematic in Figure 19.4 shows a pentode amplifier with compensating networks for high and low frequencies.

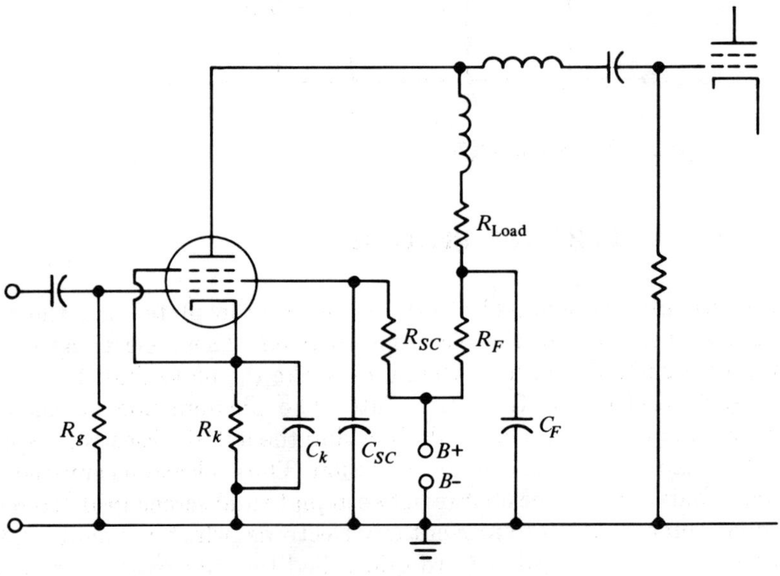

FIGURE 19.4 *Pentode video amplifier.*

CHAPTER 20

Amplifier Modifications

20.1 GROUNDED GRID AMPLIFIER

Radio frequency amplifiers, which amplify some small portion of the radio frequency range (20,000 to 30,000,000,000 Hz), utilize resonant principles to increase the gain of these high-frequency signals. The RF input signal is placed on the cathode rather than on the grid, which is grounded. (See Figure 20.1.) This causes the cathode voltage to vary with respect

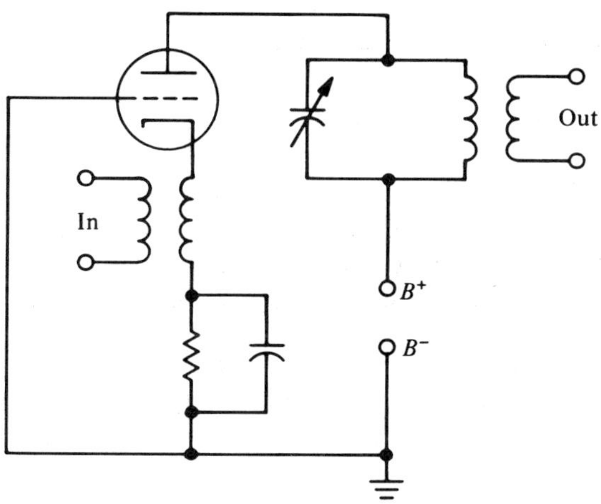

FIGURE 20.1 *Grounded grid RF amplifier.*

to the grid, plate current is affected, and the voltage is amplified. The output voltage is taken across a parallel LC circuit, which serves as the plate load resistor. The circuit is adjusted so that its resonant frequency is the same as the frequency of the signal to be amplified. This gives high impedance and good gain at resonance. The reason for introducing the input signal on the cathode instead of the grid is to reduce the plate-to-cathode capacitance by reducing the difference between cathode and plate potential. In addition, the plate-to-grid capacitance (C_{pg}) forms a negative feedback path, of the type already discussed, which at high frequencies can result in an output when there is no input. With the grounded grid any feedback is grounded, and this possibility is eliminated.

20.2 PENTODE AMPLIFIER

This plate-to-grid capacitance can be reduced enough to make the grounded grid unnecessary by using a pentode instead of a triode. (See Figure 20.2.) In this circuit, the grid resistor has been replaced by the

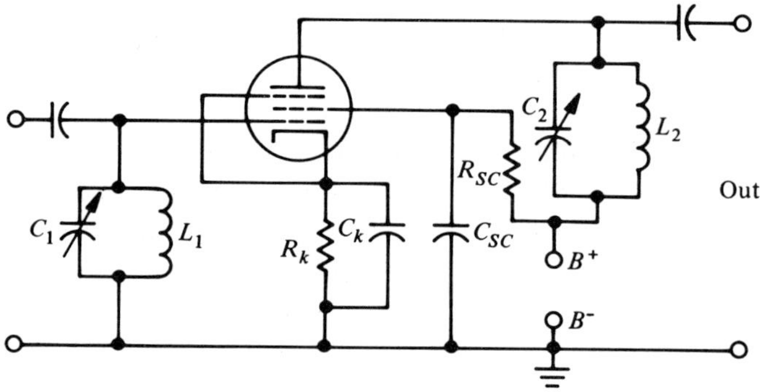

FIGURE 20.2 *Pentode RF amplifier.*

parallel circuit L_1C_1, which offers some selectivity for the incoming signal in that any signal outside the desired frequency range finds low impedance and is shunted to ground, whereas any signal near the resonant frequency of L_1C_1 meets high impedance and is placed on the grid. The plate load is the parallel circuit L_2C_2, which, of course, offers high impedance at resonance. In both LC circuits, the variable capacitance is adjusted to make the circuit resonate in the middle of the desired radio frequency range. Cathode bias (R_kC_k) is used when the input signal is small and a bias of only a few volts is required. If the amplifier were class B or class C, cathode bias would probably be ineffective and a fixed bias would be required.

Recall that in class B and class C operation, plate current only flows during all or some of the input signal's positive half cycle. Thus, when the plate load was just a resistor, class B and class C amplifiers did not have a continuous voltage output. However, when the plate load is an LC circuit as shown above, capacitor C_2 charges while plate current flows. After plate current has reached its peak, capacitor C_2 starts discharging through L_2. The discharge current flowing through L_2 builds a magnetic field about the coil. After discharge, this magnetic field collapses and induces a current in the same direction as the discharging current, which charges C_2 oppositely to its original charge. While no current flows, C_2 is discharged and charged in the opposite direction; this means that when no current flows, there is an output voltage, which is in the opposite direction to that produced when current does flow. Thus fluctuations, or oscillations, in the parallel LC circuit produce a continuous voltage

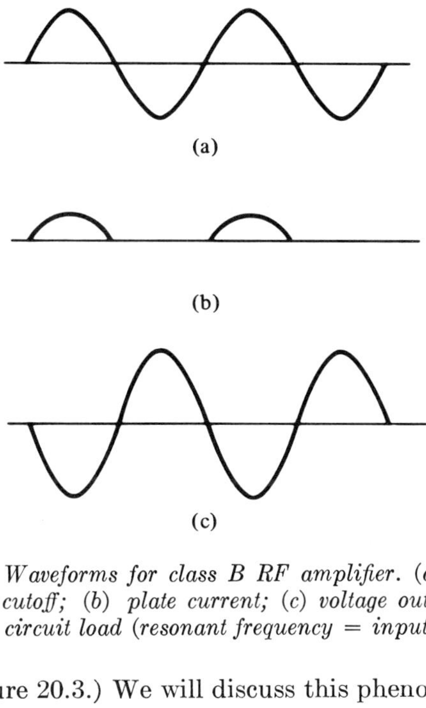

(a) (b) (c)

FIGURE 20.3 *Waveforms for class B RF amplifier. (a) Grid voltage, bias at cutoff; (b) plate current; (c) voltage output across parallel LC circuit load (resonant frequency = input frequency).*

output. (See Figure 20.3.) We will discuss this phenomenon of oscillation in an LC circuit more thoroughly when we study oscillators.

20.3 HARMONIC GENERATOR

If the L_2C_2 parallel resonant circuit were tuned to some multiple (or harmonic) of the incoming signal frequency, the RF amplifier would act as a frequency multiplier (sometimes called a harmonic generator). The output would be at the higher harmonic frequency because the frequency of oscillation of L_2C_2 is this higher frequency. Figure 20.4 shows the waveforms for a class C RF amplifier whose resonant frequency of parallel circuit L_2C_2 is twice that of the input frequency.

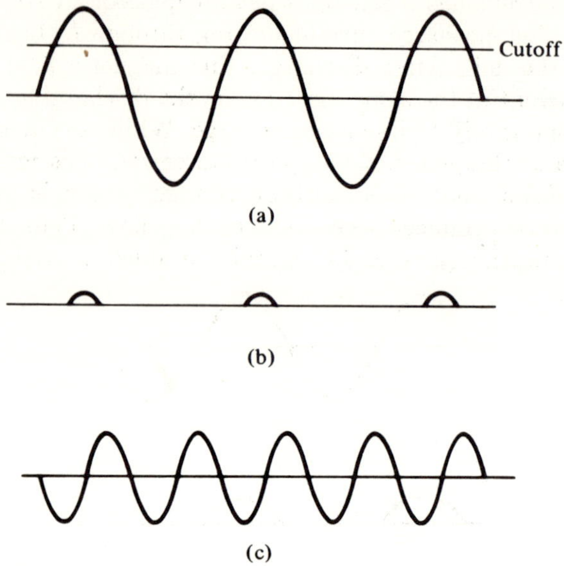

FIGURE 20.4 *Waveforms for class C RF harmonic amplifier. (a) Grid voltage, bias below cutoff; (b) plate current; (c) voltage output across parallel LC circuit load (resonant frequency equals twice input frequency).*

If the "pulse" of plate current were not repeated periodically, the oscillations in L_2C_2 would quickly die out. Later we will consider other forms of oscillators, which can be tuned to provide almost any desired frequency.

20.4 DECOUPLING

Many of the amplifier circuits we have studied show the plate current flowing through the plate voltage supply (B^+). This naturally causes plate supply voltage to vary with the signal. In order to eliminate this ac superimposed on the output of the dc power supply, the amplifier stage must be isolated, or decoupled, from the power supply. The ac must be able to return to the cathode without passing through the dc power supply. The application of an RC network (R_DC_D) whose time constant is large in comparison to the period, allows the ac to bypass the dc supply with minimum distortion. R_D makes the path through C_D a lower impedance path for ac than the path through the power supply. (See Fig-

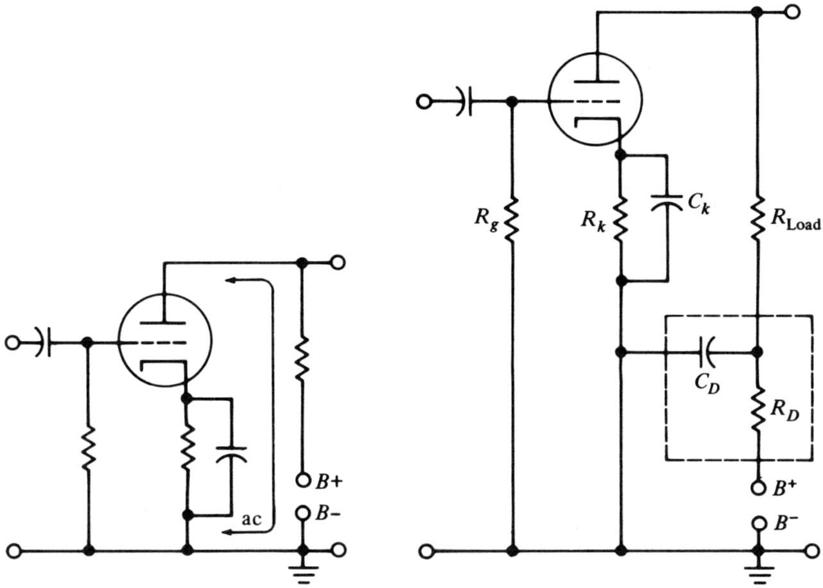

FIGURE 20.5 *(a) ac signal across dc power supply; (b) decoupling dc supply from ac signal.*

ure 20.5.) For simplicity, we will not continue to show the decoupling networks in the circuit diagrams, but we do want to emphasize the importance of decoupling in maintaining a steady dc plate supply voltage.

20.5 FILAMENT CIRCUIT

We have neglected to discuss the filament voltages in our vacuum tube circuits. These voltages are normally obtained from a secondary tap on the dc voltage supply. Unless isolation is required, as in a full-wave rectifier, all of the filaments are connected in series to an ac secondary tap on the transformer. (See Figure 20.6.) The filament circuit contains, in addition, a switch, a variable resistor which is adjusted to apply the rated value of voltage to the filaments, and a thermistor (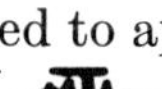). The thermistor is a safety device which, unlike resistors, has a negative temperature coefficient of resistance. Most resistors experience an increase in resistance when they are heated; thermistors, which are made of manganese, cobalt, copper, and nickel, experience a decrease in resistance when heated. When the filament circuit is first turned on, the resistance of the filaments is low, and the resistance of the thermistor is high. Once the filaments reach incandescence, their resistance has increased, which would tend to

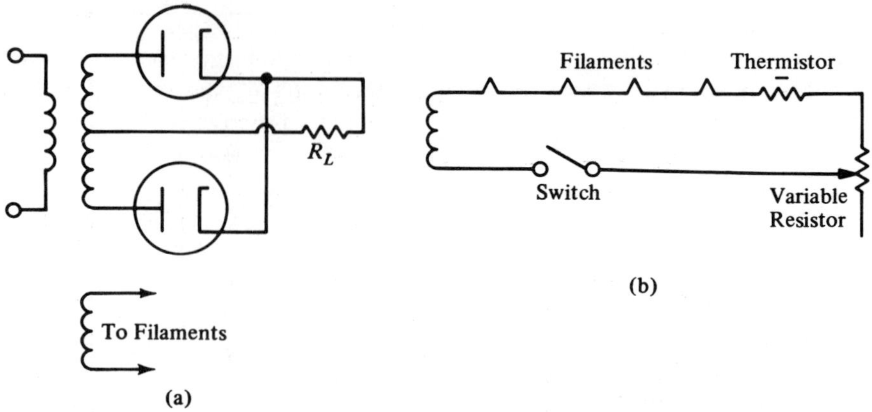

FIGURE 20.6 *(a) dc voltage supply; (b) filament circuit.*

decrease the current. But the thermistor, being physically close to the tubes, has also been heated, and its resistance has decreased, which would allow current to increase. These two effects counteract each other, the current remains constant, and the delicate filaments are thus protected from extreme variations in current.

20.6 VOLUME CONTROL

The control of gain, or volume control, in an RC amplifier is quite easily accomplished by making the grid resistor variable. The variable grid resistor is actually a voltage divider. (See Figure 20.7.) Moving the arm

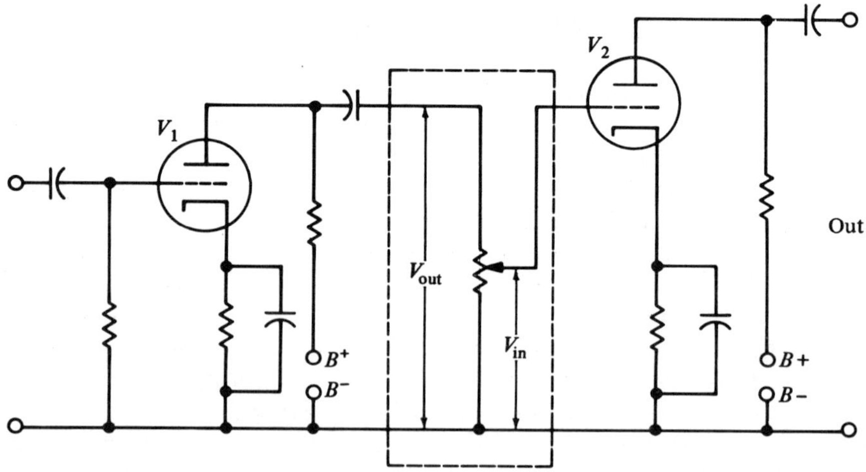

FIGURE 20.7 *Volume or gain control on an RC coupled amplifier.*

of the variable resistor to the top makes the input voltage to the grid of V_2 a maximum (it is equal to the entire output voltage from V_1). When the arm is moved farther down, only that part of the output voltage dropped between the arm and ground is felt on the grid of V_2. When the arm is at the bottom of the grid resistor, the input voltage is zero.

In the RC coupled transistor amplifier, the input signal causes changes in base-emitter voltage, and hence the gain is controlled by a variable base resistor. (See Figure 20.8.) As the arm of the variable resis-

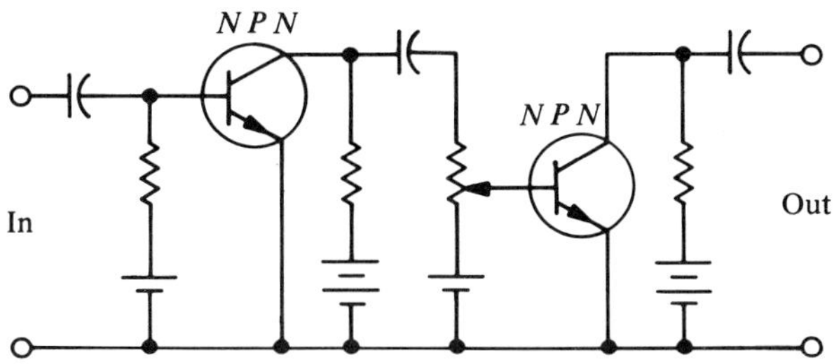

FIGURE 20.8 *Volume or gain control on an RC coupled transistor amplifier.*

tor is moved farther down, less of the output voltage from the first stage is applied to the base of the second stage, and voltage and volume decrease.

This same type of variable resistor volume control can be used on transformer coupled amplifiers. (See Figure 20.9.)

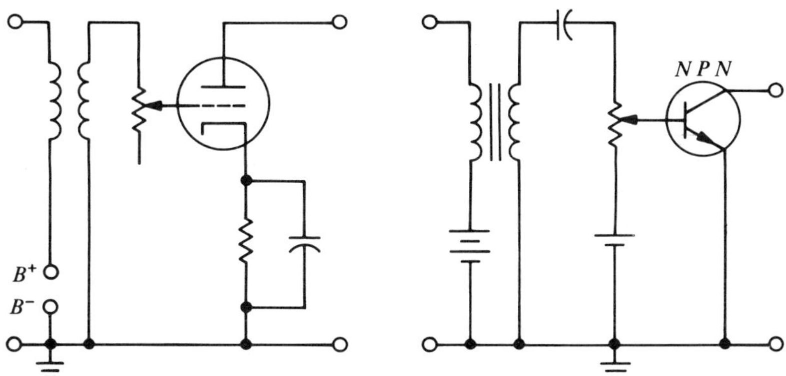

FIGURE 20.9 *Volume control on transformer coupled amplifiers.*

20.7 BASS AND TREBLE BOOST

In high-fidelity amplifiers, not only is volume control provided, but bass and treble controls are also added to increase listening pleasure by accentuating these areas of frequency response. To boost gain for bass frequencies, which range from 20 to 440 Hz (up to about middle A on the piano), we can use a volume control with the addition of capacitor C_B and resistor R_B. (See Figure 20.10.) The variable resistor acts as a volume

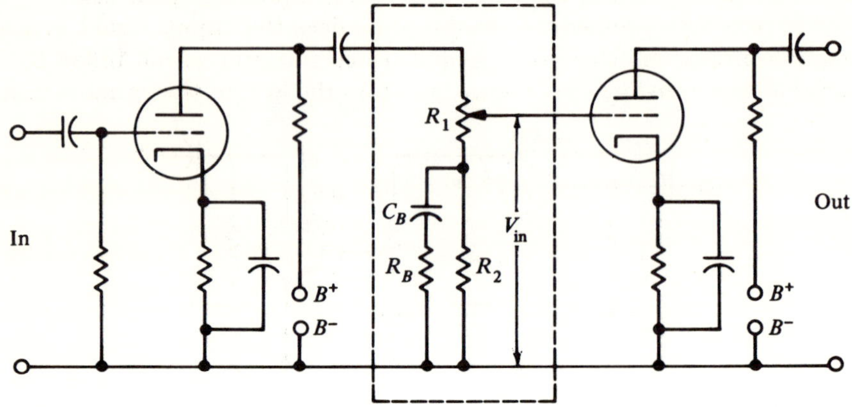

FIGURE 20.10 *Bass boost.*

control. But the gain at lower frequencies is improved by the presence of C_B. This capacitor has a high capacitive reactance at low frequencies ($X_C = 1/2\pi fC$). Since it is in parallel with R_2, the total impedance is larger at lower frequencies; this causes a stronger signal to be placed on the grid of the next stage at low frequencies. Resistor R_B is necessary to maintain some impedance at high frequencies, when the capacitive reactance is quite small.

Treble response in high-fidelity amplifiers can be improved using a similar circuit to that for bass boost. (See Figure 20.11.) Capacitor C_T bypasses high-frequency signals around resistor R_T. The high-frequency output is thus dropped entirely across R_3, whereas lower-frequency output signals are dropped across both R_3 and the parallel combination R_TC_T. This means there is less loss due to resistance in the high-frequency signals than in the lower-frequency signals, and a boost in treble range signals results.

20.8 CATHODE FOLLOWER

The cathode follower should remind you of the common-collector transistor amplifier in that it has a high input impedance and a low output impedance. Previously, the output of a vacuum tube amplifier has been taken across the tube; the cathode follower has its output across the cathode resistor, which thus replaces the load resistor. (See Figure 20.12.)

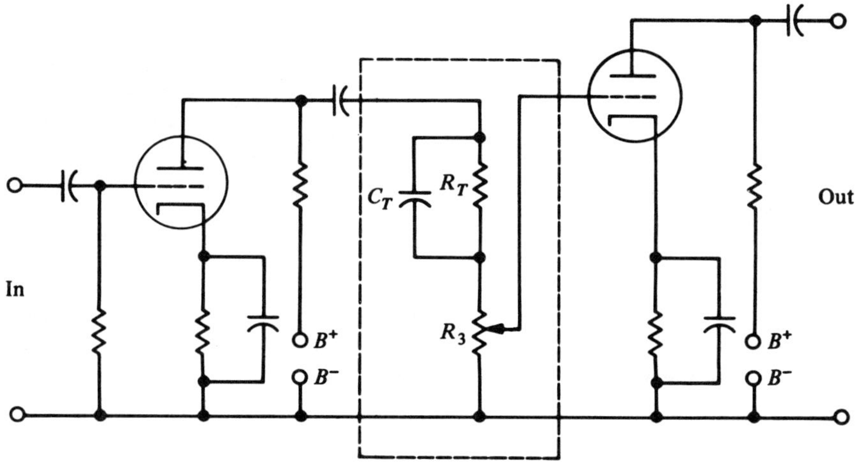

FIGURE 20.11 *Treble boost*

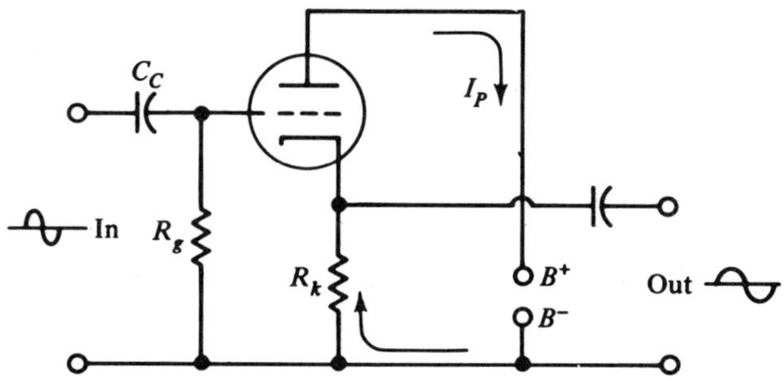

FIGURE 20.12 *Cathode follower.*

When an incoming signal on the grid goes positive, the plate current increases. An increased plate current causes an increase in output voltage across the load (cathode resistor R_k). Thus, we see that the output is in phase with the input.

Note that to obtain any output, the input signal must be felt across R_k; so there cannot be a bypass capacitor around R_k. With no bypass capacitor, negative feedback will occur. When a positive going signal is applied to the grid, the plate current increases, causing an increased voltage drop across R_k. This makes the cathode more positive and increases the bias (the grid appears to be more negative with respect to the cathode). A more negative bias allows less plate current to flow and thus reduces the output voltage. When the input signal on the grid goes negative, less plate current flows, causing a decreased voltage drop across R_k (output). But this also makes the cathode less positive, decreasing the bias, allowing more plate current to flow, and increasing the output voltage. Whether the incoming signal goes positive or negative, degenerative feedback reduces the amplitude of the output voltage. Thus, the voltage gain of the cathode follower ($V_{Gain} = V_{out}/V_{in}$) is less than one.

Since negative feedback opposes both increases and decreases in output, the input appears to meet a very high impedance. Input impedance of the cathode follower is in the neighborhood of 100 kiloohms. Since the output is taken across the cathode resistor, the output impedance is relatively small. An output impedance of 500 ohms would be representative.

If phase inversion is desired and gain is not needed, the triode amplifier or the common-emitter transistor amplifier may be used with either a voltage divider on the input (like that used for volume control) or degenerative feedback controlling gain.

20.9 PARAPHASE AMPLIFIERS

The limited frequency response of the transformer, not to mention its weight and expense, led us to suggest the use of the cathode follower across the output of a power amplifier to match the load impedance. The input of a push–pull class B or C amplifier was, you recall, a center tapped transformer coupling. With its secondary coil being center tapped, the transformer offered two outputs when provided with a single input. One output went to each of the two tubes or transistors in the push–pull amplifier. In order to eliminate the input transformer for the push–pull amplifier, we introduce the paraphase amplifier. Like the center tapped transformer, the paraphase amplifier takes one input and provides two outputs, which are 180° out of phase.

We will discuss several types of paraphase amplifiers and indicate some of their good and bad points. As with all the other circuits mentioned in this text, only a few of a rather long list of circuits, which perform similar tasks, are brought to your attention. This is not to ignore other circuit variations on a theme, but rather to expedite the description of a typical circuit. As you extend your reading and work with more electronic circuits, you will come across hundreds of types of dc power supplies, amplifiers, and oscillators. But we hope that our study of basic circuit types will be helpful in enabling you to recognize their purpose and follow their method of obtaining a desired circuit response.

A single tube (single transistor) paraphase amplifier with outputs taken across cathode and plate (emitter and collector) is shown in Figure 20.13. Looking at the vacuum tube paraphase amplifier, when the input signal goes positive, the plate current increases, causing the voltage across the cathode to increase and the plate voltage to decrease. Thus,

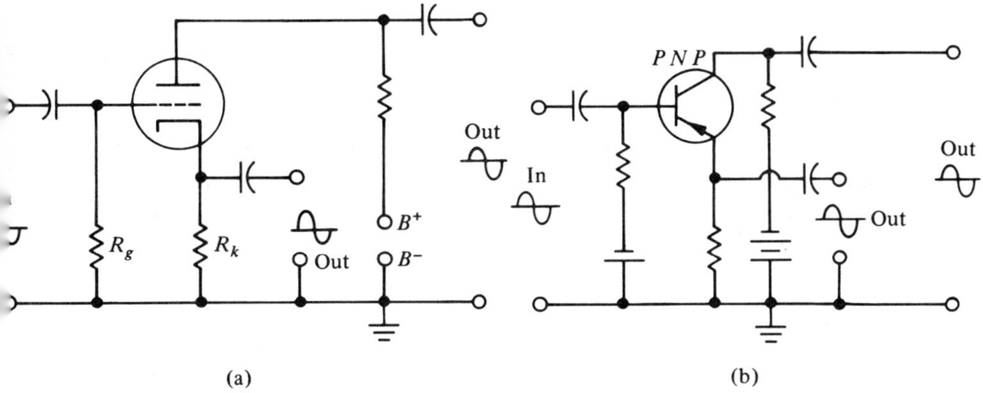

FIGURE 20.13 *(a) Vacuum tube paraphase amplifier; (b) transistor paraphase amplifier.*

the two outputs, across the cathode resistor and across the tube, will be out of phase. If R_{Load} and R_k are equal, the output signals will be equal in magnitude. Since there is no bypass capacitor around R_k, negative feedback exists, and the gain will thus be less than one. Therefore, the amplitude of the output signal will be less than that of the input; thus, the single tube paraphase amplifier is similar to the cathode follower.

The second type of paraphase amplifier employs two tubes, and its gain is greater than one. One problem, which occurs with two-tube paraphase amplifiers, is that the aging of two vacuum tubes is usually unequal. Therefore, in order to maintain equal amplitude outputs, it is necessary to provide some means of adjustment either in the input or output of one of the tubes. (See Figure 20.14.) When an incoming signal goes positive, plate current in V_1 increases, and plate voltage, which is the first output, decreases. A small part of the first output is placed on the grid of V_2 via variable resistor R_{g2}. This signal is amplified by V_2, and the second output, which is taken across V_2, will be out of phase with this signal. The second output is thus in phase with the original input to V_1. If tube V_2 becomes weak with age, then R_{g2} is adjusted to put a larger signal on the grid of V_2 and thus strengthen its output. If, on the other hand, V_1 becomes weaker, R_{g2} is adjusted to decrease the output of V_2 to the lower level of V_1. In this second case, the output of both tubes is lessened so they can remain equal in magnitude.

An inexpensive two-tube paraphase amplifier utilizes an unbypassed cathode resistor, which is common to both tubes. (See Figure 20.15.) When the plate current of V_1 changes, the common cathode resistor reflects this change onto the grids of both tubes, producing negative feedback for V_1 but placing a signal on the cathode of V_2. When a positive

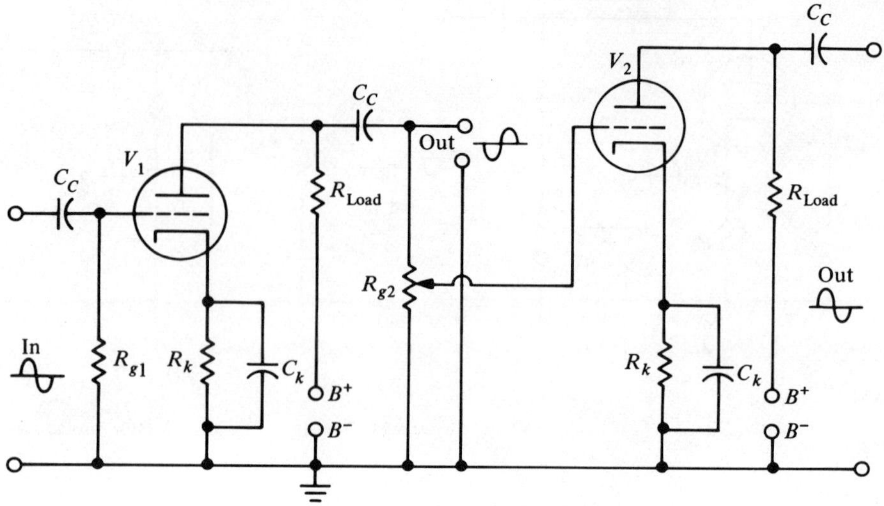

FIGURE 20.14 *Two-tube paraphase amplifier.*

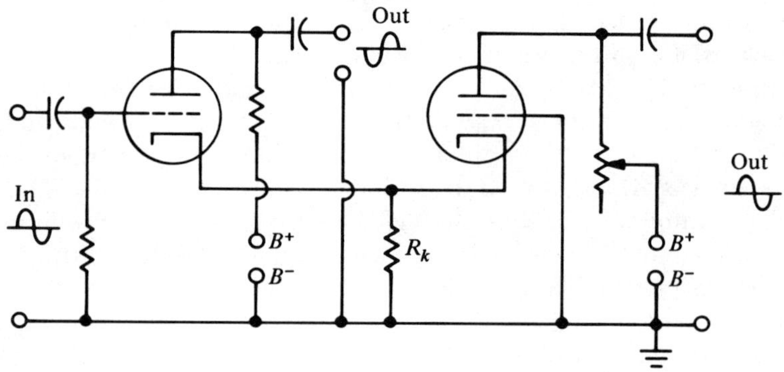

FIGURE 20.15 *Cathode coupled paraphase amplifier.*

going input causes an increase in plate current in V_1, the plate voltage of V_1 decreases. Therefore, the first output is inverted. The second output is taken across V_2. The voltage which is dropped across cathode resistor R_k appears on the cathode of V_2. This voltage across R_k is in phase with the input signal, so the signal applied to the cathode of V_2 is in phase with the input. A positive going signal on the cathode makes it more positive with respect to the grid, or the grid is made more negative with respect to the cathode. Thus, a positive going signal on the cathode appears to be the same as a negative going signal on the grid, and this apparent signal on the grid is out of phase with the original input. The output across V_2 will be out of phase with the grid signal and hence in phase with the original input. The adjustment of the variable load resistor or V_2 will equalize the outputs of V_1 and V_2.

Using a modified block diagram form, we will indicate the probable insertions of both paraphase amplifiers and cathode followers to replace transformers in an amplifier system. (See Figure 20.16.) These amplifier

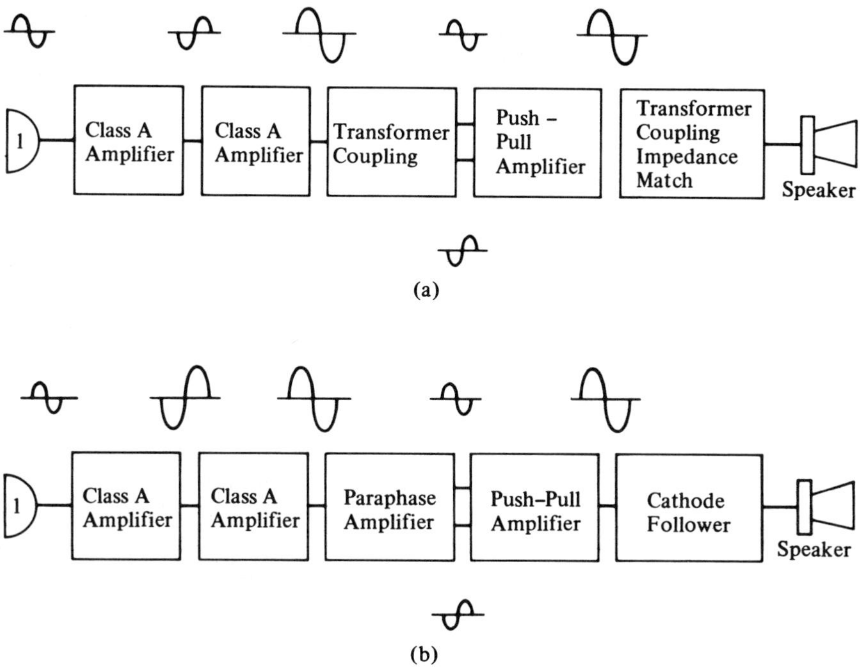

FIGURE 20.16 *(a) Amplifier system utilizing transformer coupling for impedance matching; (b) amplifier system utilizing paraphase coupling and cathode follower for impedance matching.*

systems should also give some indication of the applications of the amplifier circuits we have studied.

Questions

(1) To protect the filament circuit of a vacuum tube instrument, we suggested the use of a thermistor. We could have used a fuse instead of a thermistor. How do these two protective devices differ (with respect to resistance)?

(2) In the cathode follower, the cathode bias resistor is also the load resistor. Why is the gain of the cathode follower not increased when the load resistance is increased?

Suggested Experiments

(1) The Cathode Follower. The gain of a cathode follower may be calculated with the aid of an audio oscillator and a calibrated oscilloscope. Phase shifts may also be observed. The vertical input lead of the oscilloscope is first placed at the input of the cathode follower, and the waveform observed. Then it is moved to the output of the cathode follower. (See Figure 20.17.)

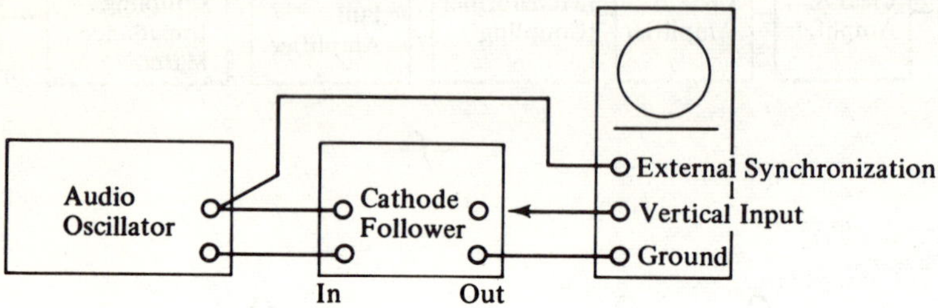

FIGURE 20.17

(2) The Paraphase Amplifier. The amplitude and phase relation of the two outputs and the single input of the paraphase amplifier may be observed by using the audio oscillator and calibrated oscilloscope. The waveform at the input of the amplifier is observed by placing the vertical input lead at the input of the amplifier. Each of the outputs may be observed by placing the vertical input lead of the oscilliscope at output 1, then output 2 of the paraphase amplifier. (See Figure 20.18.)

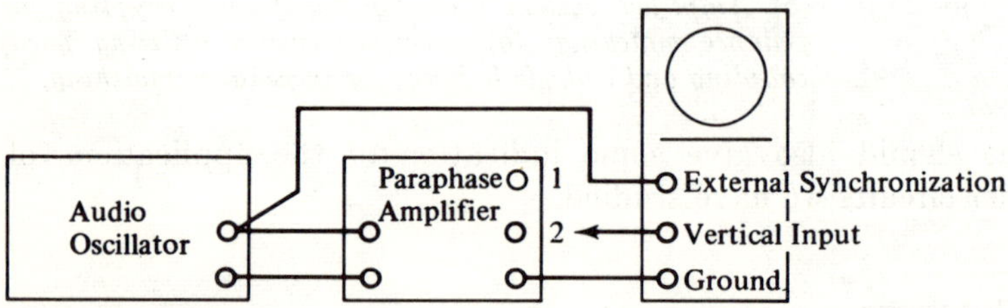

FIGURE 20.18

CHAPTER 21

Oscillators

21.1 AN IDEAL OSCILLATOR

Earlier we explained how ac voltage and current could be generated by rotating a coil of wire in a magnetic field. When this mechanical method is used, the frequency (number of cycles, or oscillations, each second) is limited by how fast the coil can be rotated. In order to generate a broader spectrum of frequencies, we will use an electronic oscillator.

Let us first consider an ideal oscillator, consisting of a capacitor and a pure inductor in parallel. Ideally, this LC circuit contains no resistance; only capacitive and inductive reactances are present. We begin with a charge on the capacitor. The capacitor discharges through the coil, the flow of electrons building a magnetic field about the coil. When the capacitor is completely discharged, the flow of electrons stops, and the magnetic field collapses. As the magnetic field about the coil begins to collapse, it induces a flow of electrons in the coil. The direction of flow must be such as to oppose the collapse of the magnetic field; hence, the induced current will be in the same direction as the current which originally built the field. Since electrons continue to flow in the same direction as during discharge, the capacitor will charge in the opposite direction. Now with the capacitor charged again, although in the opposite direction, the process repeats. The capacitor discharges through the coil, changing electrostatic energy to magnetic energy. When the capacitor is completely discharged, all the energy is in the form of magnetic energy. As the magnetic field collapses, the capacitor is charged in its original direction, and all energy is again electrostatic. (See Figure 21.1.)

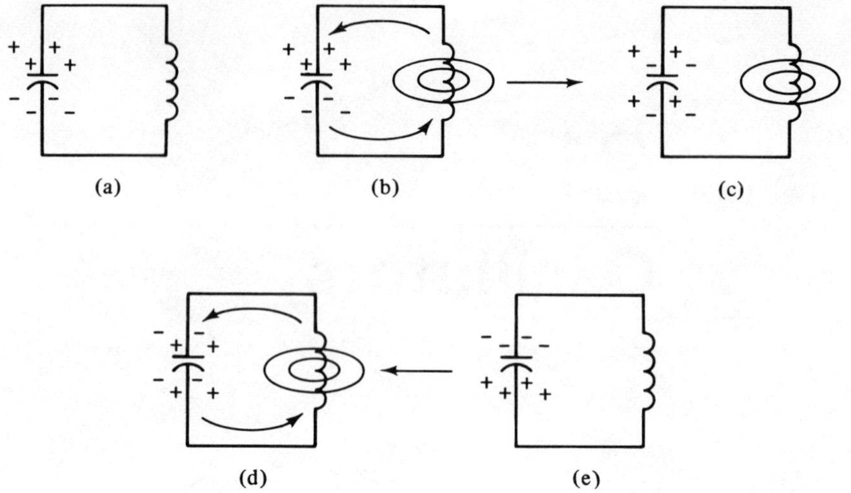

FIGURE 21.1 *(a) Charged capacitor (electrostatic energy); (b) discharge of capacitor building magnetic field; (c) discharged capacitor (magnetic energy); (d) collapse of magnetic field inducing current which charges capacitor; (e) capacitor charged in opposite direction (electrostatic energy).*

During one complete cycle, or oscillation, the capacitor, beginning in a charged state, discharges, charges in the opposite direction, discharges again, and charges in its original direction. Current through the coil flows first in one direction, then the other. Thus, we see that the circuit generates ac current and voltage. The frequency of the generated ac is the same as the resonant frequency of the parallel circuit. (See Figure 21.2.)

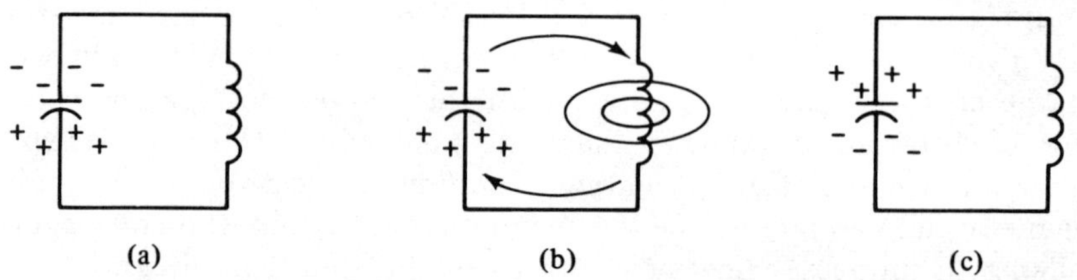

FIGURE 21.2 *(a) Capacitor charged in opposite direction (electrostatic energy); (b) current flow as capacitor discharges, then charges while magnetic field builds and collapses; (c) end of one complete cycle with capacitor in its original configuration.*

In the ideal LC circuit, this back and forth motion (oscillation) of the current, with energy changing from electrostatic to electromagnetic and back, would continue as long as the circuit existed. Practically, we know that any circuit will contain some resistance to current flow. This will cause energy to be dissipated in the form of heat, and the amplitude of the current will gradually decrease until the oscillations finally stop altogether. Obviously then, in order to use the parallel resonant LC circuit, frequently called a tank circuit or tank coil, as an oscillator, we will have to introduce additional energy into the circuit at just the right time to replace the energy dissipated as heat, and thus prevent the oscillations from dying out.

21.2 A SIMPLE ELECTRONIC OSCILLATOR

A simple electronic oscillator consists of a tank coil connected between grid and cathode of a triode with grid leak bias. (See Figure 21.3.) When

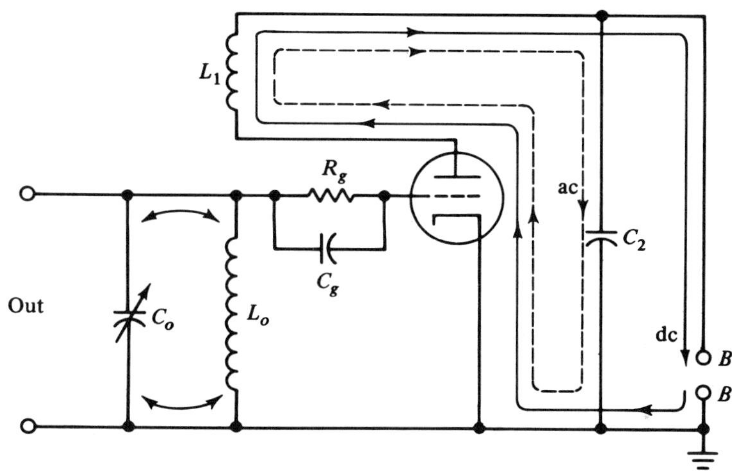

FIGURE 21.3 *Electronic oscillator.*

the tube conducts, a current flows through L_1. Mutual induction produces a current in L_o. The induced current in L_o charges capacitor C_o. When the top of capacitor C_o is positive, grid current will flow, charging grid leak capacitor C_g and thus establishing a bias voltage. The time constant for the grid leak circuit R_gC_g is long in comparison to the period of the oscillator L_oC_o, allowing C_g to maintain the bias.

Meanwhile, capacitor C_o is first discharging through L_o and then charging in the opposite direction. With the top of C_o negatively charged, no current flows in the triode, which is biased at cutoff. Capacitor C_o now begins discharging through L_o, making the grid less negative and building the magnetic field about L_o. When C_o is discharged, the magnetic field built about L_o starts collapsing, inducing a current which charges the top of C_o positively. This allows current to flow in the triode again. This plate current produces a magnetic field about coil L_1, which will induce a current in coil L_o. The current induced in coil L_o will be in such a direction as to oppose any building magnetic field; thus, the current due to mutual induction will be in the same direction as the self-induced current of the collapsing magnetic field (the two currents add). Coil L_1 provides positive feedback to the oscillator coil L_o. This positive feedback replaces energy losses due to resistance in the oscillator. Coil L_1 is said to "tickle" coil L_o in order to maintain oscillation in L_oC_o, so feedback coil L_1 is sometimes called a tickler coil. The amount of regenerative, or positive, feedback can be regulated by the size of coil L_1 and the distance between L_1 and L_o.

Capacitor C_2, incidentally, acts as a bypass capacitor for the ac signal, preventing it from passing through the dc power supply.

The frequency of oscillation can be determined if we recall that at resonance, the inductive reactance of the coil ($X_L = 2\pi fL$) equals the capacitive reactance of the capacitor ($X_C = 1/2\pi fC$). At resonance, $X_L = X_C$ or $2\pi fL = 1/2\pi fC$. Solving for frequency,

$$4\pi^2 f^2 LC = 1$$

$$f^2 = \frac{1}{4\pi^2 LC}$$

$$f = \frac{1}{2\pi\sqrt{LC}}$$

For example, if a 10-mH (millihenry) coil is in parallel with a 0.01-μF capacitor, the resonant frequency of the tank circuit is

$$f = 1/2 \times 3.14 \sqrt{10 \times 10^{-3} \text{ henry} \times 0.01 \times 10^{-6} \text{ farad}}$$
$$= 1/6.28 \sqrt{10^{-10}} = 1/6.28 \times 10^{-5} = 0.159 \times 10^{5} = 15{,}900 \text{ Hz.}$$

In the circuit just discussed, capacitor C_o can be adjusted to vary the resonant frequency. We will discuss several other types of variable frequency oscillators and one fixed frequency oscillator. All of them will require some form of regenerative feedback to maintain oscillation. Figure 21.4 illustrates the relationship of feedback, oscillator, and tube or transistor. The oscillation takes place in a parallel LC circuit. The triode or transistor amplifies the oscillation, a small portion of which is returned to the oscillator in the proper phase to maintain oscillation.

21.3 HARTLEY OSCILLATORS

The two Hartley oscillators in Figure 21.5 are both identifiable by the tapped coil. Notice, however, that the shunt-fed Hartley oscillator has

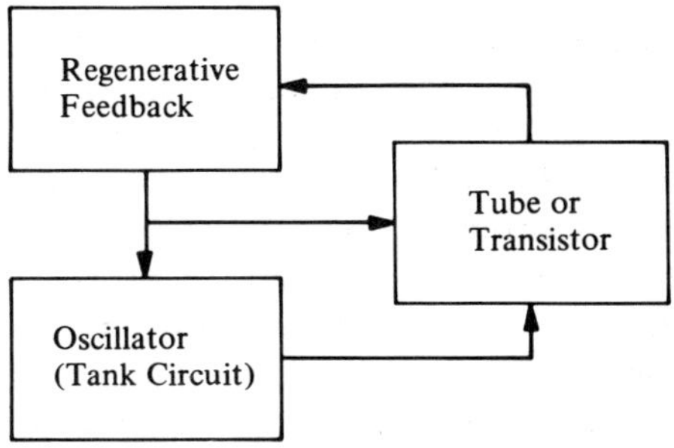

FIGURE 21.4

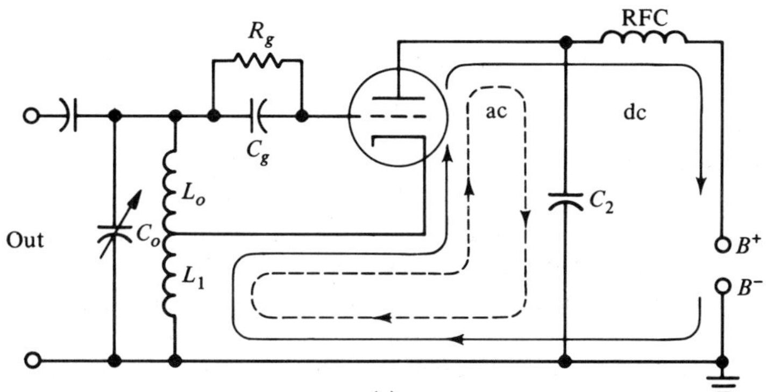

(a)

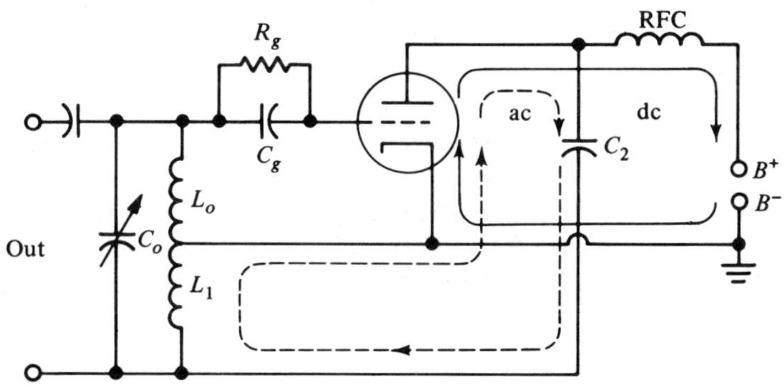

(b)

FIGURE 21.5 *(a) Series-fed Hartley oscillator; (b) shunt-fed Hartley oscillator*

a grounded cathode, whereas the series-fed Hartley oscillator has a grounded plate. In the series-fed oscillator, the dc path is through the coil L_1, which is in series with the triode. In the shunt-fed oscillator, the dc is shunted around L_1. In both circuits, the ac current in L_1 provides the regenerative feedback to maintain oscillation. As seen in the previous oscillator circuit, the grid leak bias provided by the grid capacitor and resistor is maintained, because the time constant of the R_gC_g circuit is longer than the resonant frequency of the tank circuit, and hence C_g does not have sufficient time to discharge while the top of C_o is positive. Capacitor C_2 is used as a bypass capacitor for ac, with the radio frequency choke coil (RFC) providing additional protection to help prevent any high-frequency ac from flowing in the dc power supply. If the variable capacitor C_o in the series-fed oscillator is not well insulated, its capacitance, and hence the resonant frequency, can be affected by the body capacitance of the person tuning the oscillator. On the other hand, in the cathode grounded shunt-fed oscillator with no dc flowing in L_1, body capacitance does not interfere with tuning.

The transistor equivalents of the two Hartley oscillators also have a center tapped coil in the parallel LC circuit. (See Figure 21.6.) In both transistor oscillators, resistor R_B provides a path for the flow of base current and controls the level of forward (emitter-base) bias. Base current produces voltage drops across R_E, the emitter-base junction, and R_B; the larger resistor R_B is, the larger is the voltage drop across it and hence the smaller the forward bias voltage across the emitter-base junction. Capacitor C_B also helps maintain the forward bias by preventing dc in the tank coil from reaching the base. Oscillations in the LC circuit provide the ac variations across the emitter base, which control the flow of collector current. This ac current flows through L_1 and provides the regeneration necessary for continued oscillation. Capacitors C_E and C_2 act as bypass capacitors. Resistor R_E acts as a stabilizing resistor in that, when collector current increases due to an increase in temperature, the resistance of R_E also increases, increasing the voltage drop across R_E and thus maintaining the operating level of the forward bias. Transistors, which have uniform temperature responses, can reduce the need for a stabilizing resistor.

21.4 COLPITTS OSCILLATORS

Regeneration can be accomplished in ways other than by mutual induction. The Colpitts oscillator, for example, uses a capacitor (C_1) to maintain oscillation. (See Figure 21.7.) With the exception of the means of introducing positive feedback, the circuit is similar to the shunt-fed Hartley

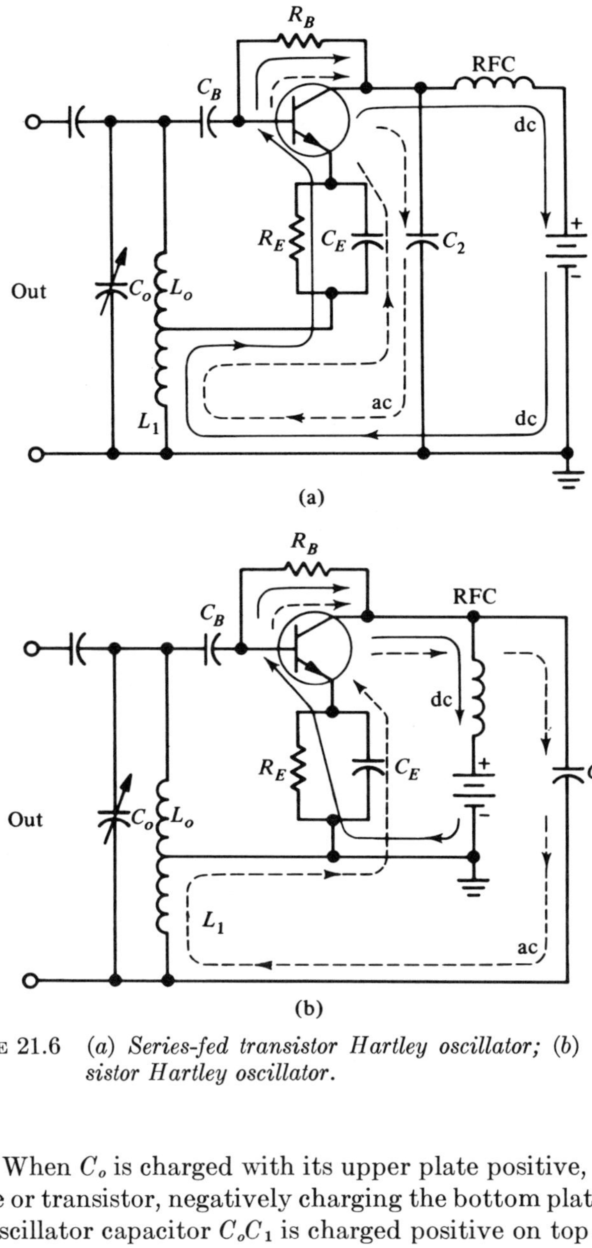

FIGURE 21.6 (*a*) *Series-fed transistor Hartley oscillator;* (*b*) *shunt-fed transistor Hartley oscillator.*

oscillator. When C_o is charged with its upper plate positive, current flows in the tube or transistor, negatively charging the bottom plate of C_1. Thus, the split oscillator capacitor C_oC_1 is charged positive on top and negative on the bottom. Additional electrons have been added to the bottom plate of C_1 to replace energy lost through resistance in the oscillator with additional electrostatic energy. The radio frequency choke coil (RFC) has enough resistance to serve also as the plate load or the collector resistor. (See Figure 21.8.)

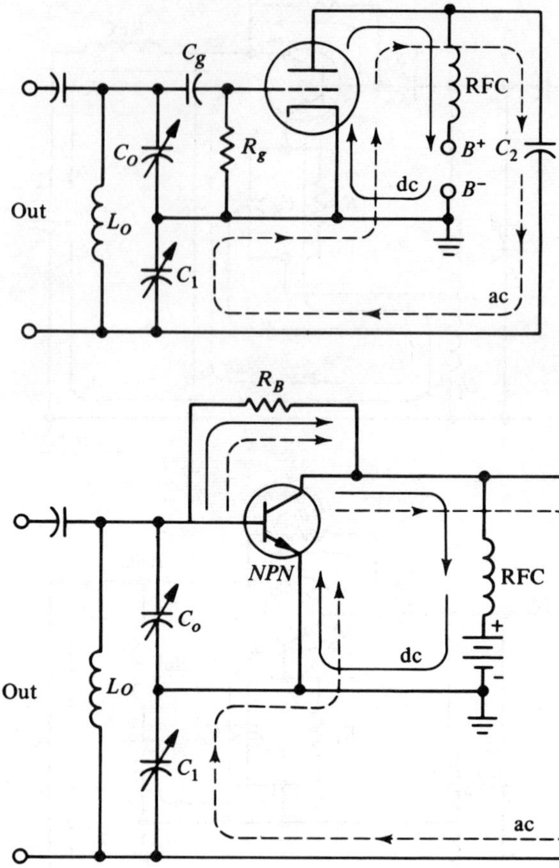

FIGURE 21.7 *Colpitts oscillators.*

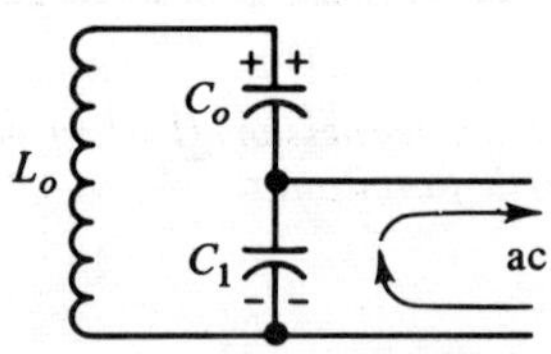

FIGURE 21.8

The Colpitts oscillator, like the Hartley, is self-starting. When electrons are first emitted from the cathode, some grid current will flow, charging C_g, which provides the bias, and introducing a small charge to the capacitor in the oscillator. If the gain of the tube or transistor is too small, not enough electrons will be sent to the oscillator to begin the oscillations necessary to produce more current in the tube or transistor and bring the amplitude of the ac signal up to the desired level. This problem of starting the oscillations is common to all types of oscillators.

Another difficulty occurs in maintaining the right amount of positive feedback. If the regenerative element in the oscillator is too small, there may not be enough positive feedback, in which case the oscillations would eventually die out or be reduced to a very low ac level. The VTVM (vacuum tube voltmeter) may be used to check the operation of the oscillator without having to directly observe its output. The potential across the grid resistor R_g (grid bias), or in a transistor circuit the potential across the base resistor R_B, is measured. If bias is steady, the oscillator is operating properly; a decreasing bias voltage would imply that the oscillations were dying out, thus allowing C_g to discharge through R_g. The reason the VTVM is used, rather than a dc voltmeter, is because the input impedance of the VTVM is much higher, and thus any appreciable current drain, which would affect the bias, is prevented.

21.5 TPTG OSCILLATOR

The tuned plate, tuned grid (TPTG) oscillator has its output taken across the tube, like an amplifier; in addition, it has a wide range of high-frequency response. (See Figure 21.9.) Regenerative feedback is obtained

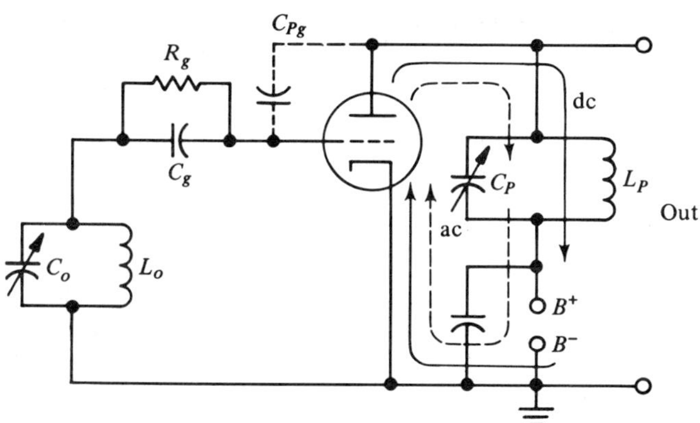

FIGURE 21.9 *TPTG (tuned-plate tuned-grid) oscillator.*

in the TPTG oscillator by returning part of the ac plate current to the grid via the interelectrode capacitance C_{pg} (plate to grid). When electrons are initially emitted from the cathode, grid bias is established and oscillations started, as described in previous circuits. When the top of the grid tank circuit is positive, the grid of the tube is above cutoff, and plate current flows making the top of the plate tank circuit negative. Also, as just mentioned, a small amount of plate current is returned to the grid via the interelectrode capacitance C_{pg}. The lower capacitive reactance ($X_C = 1/2\pi fC$) at higher frequencies explains why the TPTG is a good high-frequency oscillator. This current charges the top of C_{pg} negative, making the bottom which is attached to the top of the grid tank circuit positive, and thus regenerating the oscillator. Note that plate current can take either of two paths to return to ground: (1) through the plate tank circuit (the right branch) or (2) through the interelectrode capacitance C_{pg} and the grid tank circuit (the left branch). This second path provides the feedback.

Feedback can be controlled by controlling the amount of plate current in the plate tank circuit. To do this, the resonant frequency of the plate tank circuit is slightly larger than that of the grid tank, and both are slightly above the operating frequency. (See Figure 21.10.) Remember

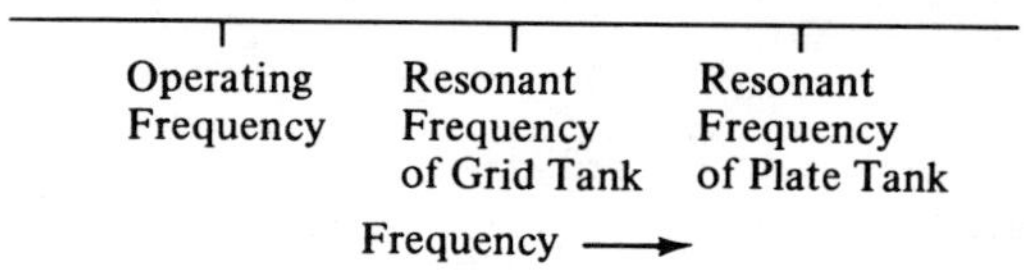

FIGURE 21.10

that a parallel LC circuit takes on the character of the smaller reactance, and its total impedance is at a maximum at resonance. The operating frequency is below resonance, and thus the grid tank is inductive; it offers inductive reactance to the feedback current. (See Figure 21.11.)

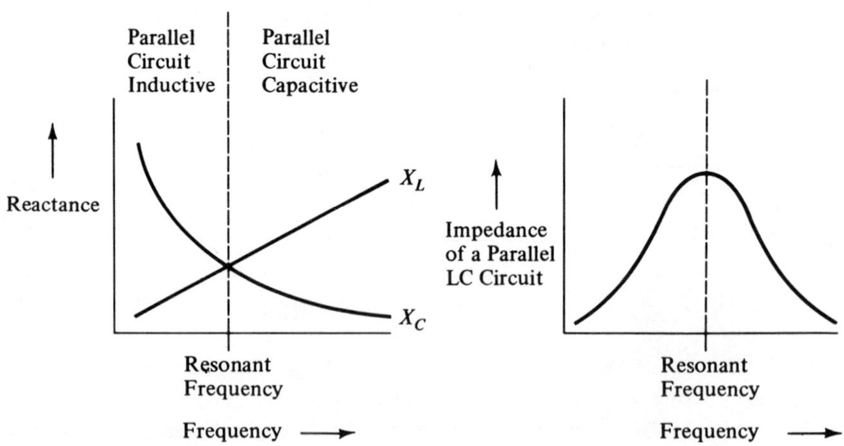

FIGURE 21.11

Since this inductive reactance is in series with C_{pg}, the total impedance offered to feedback current is the difference of the capacitive reactance due to C_{pg} and the inductive reactance due to the grid tank ($Z = X_C - X_L$). This means that the total impedance of the left branch is capacitive and is small enough to allow some feedback current to flow. When the impedance of the right branch is increased, less current will flow in the plate tank, and there will be more feedback current. Remembering that impedance is at a maximum at resonance, we can increase the impedance of the plate tank circuit by adjusting its variable capacitor to bring its resonant frequency closer to the operating frequency.

Note that the total impedance of the right branch is inductive (operating frequency is below resonance), whereas the total impedance of the left branch is capacitive. This means that the currents in the two branches are 180° out of phase, and the feedback is regenerative rather than degenerative.

2.16 CRYSTAL OSCILLATOR

The most common fixed frequency oscillator is the crystal oscillator. You may recall that we mentioned the piezoelectric effect when we talked about ways of obtaining a flow of electrons. When crystals of quartz or Rochelle salts have a voltage applied across them, they change shape. And when these same crystals are mechanically deformed by twisting or compressing, they produce a voltage drop across their surfaces. The amount of voltage dropped across the crystal depends on the crystal itself and on how it is cut.

Quartz is often used because of its high Q (20,000 or larger), that is, because of its narrow bandpass with respect to resonant frequency. Recall that $Q = f_r/B = X_L/R$. However, not all cuts of quartz crystal will produce a voltage drop across their surfaces. In addition, various cuts of quartz reflect both positive and negative temperature coefficients of frequency. The resonant frequency of X-cut crystals decreases as the temperature increases; the resonant frequency of Y-cut crystals increases as temperature increases. For work which requires that frequency not vary by even a few hertz (cycles each second), the crystals are kept in a small oven, where the temperature can be held constant. (See Figure 21.12.)

The crystal is either mounted between two metal plates or a metallic film is deposited on two of its surfaces. Now suppose we have a crystal which has been cut and mounted in such a way that it produces a voltage drop when distorted and is distorted when a voltage is applied to it. If an ac voltage, whose frequency is the same as the natural mechanical fre-

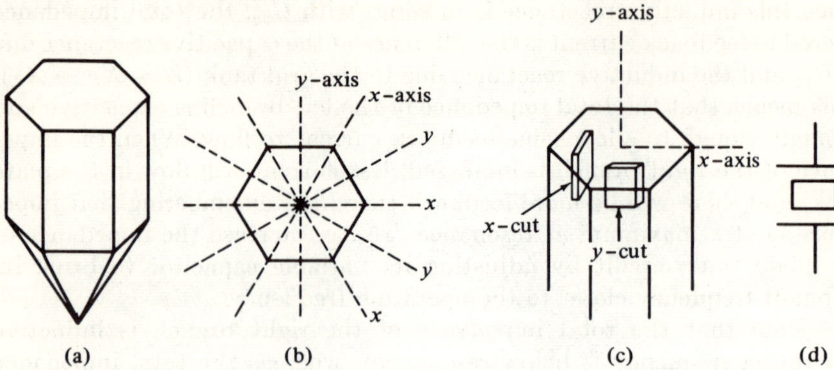

FIGURE 21.12 *(a) Quartz; (b) quartz cross section showing x- and y-axes; (c) x-cut (perpendicular to x-axis), y-cut (perpendicular to y-axis); (d) schematic of crystal.*

quency of the crystal, is applied to the crystal, it will begin vibrating, or oscillating, at this frequency. As it vibrates (or is distorted), it produces an ac voltage between its surfaces. The external voltage necessary to maintain oscillations is small in comparison to the voltage produced by the crystal. Thus, if the crystal were placed in the grid–cathode circuit of a triode and a small amount of the plate current were fed back to maintain oscillation, we would have a basic fixed frequency oscillator.

In the schematic seen in Figure 21.13, regenerative feedback is obtained through the plate-to-grid capacitance (C_{pg}), just as in the tuned plate tuned grid oscillator. The ammeter between the LC circuit and the

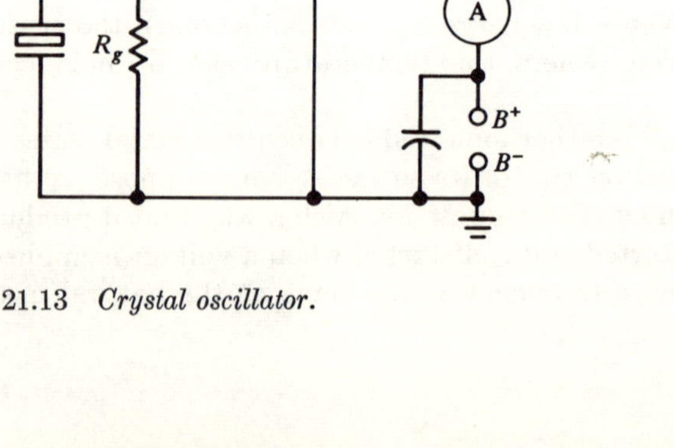

FIGURE 21.13 *Crystal oscillator.*

B^+ supply provides a means of adjusting the plate circuit oscillations to obtain positive feedback and thus maintain the crystal oscillations. The plate oscillator is a parallel LC circuit whose resonant frequency, which is varied by adjusting C_p, occurs when the ac plate current, as seen on the ammeter, is at a minimum. Since the impedance of the parallel LC circuit is at a maximum at resonance, output voltage would also be at a maximum at this point. (See Figure 21.14.) A portion of this ac signal is fed back to

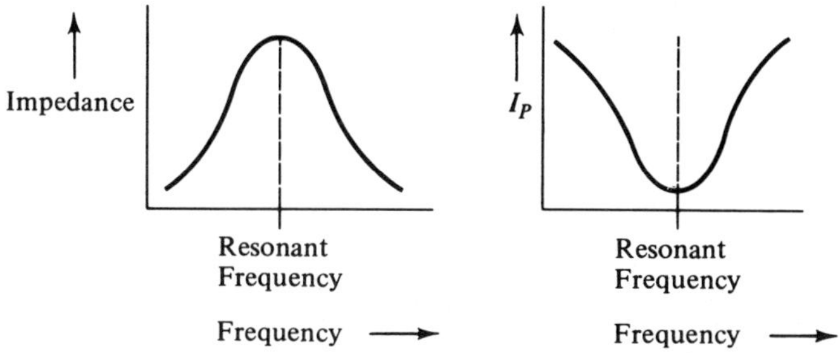

FIGURE 21.14 *Parallel LC resonant circuit.*

the crystal oscillator via the interelectrode capacitance C_{pg}. If the plate oscillator is kept slightly inductive (resonant frequency would have to be above the operating frequency, hence its capacitive reactance would be larger than its inductive reactance), then the feedback will be regenerative. This can be accomplished by adjusting C_P. If the plate oscillator is made slightly capacitive, the feedback will be degenerative. (See Figure 21.15.)

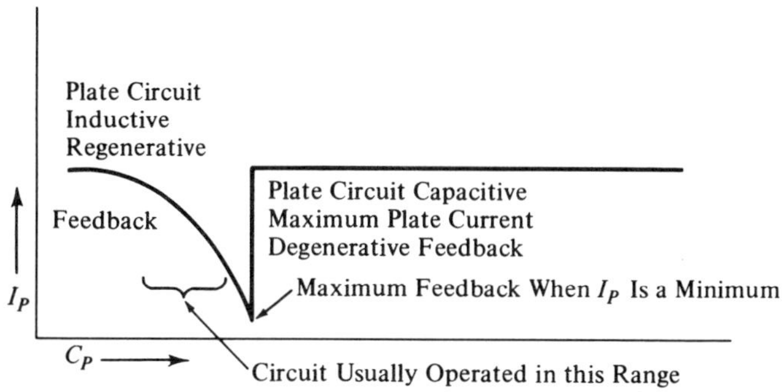

FIGURE 21.15 *Variations in plate current I_P due to adjustments in C_P.*

The phase relation between the inductive plate circuit and the crystal oscillator is such that the feedback voltage, applied to the crystal while it is enlarging, causes it to further increase its physical dimensions. Thus, the feedback is regenerative. Pressure, applied to the enlarged crystal by its holder, then causes it to contract, generating a large voltage. This generated voltage is large enough to overcome grid leak bias, driving the grid above the cutoff point for a short period of time and allowing plate current to flow. This ac plate current is fed to the plate oscillator to maintain its oscillations.

Excessive feedback would lead to the destruction of the crystal. Therefore, the plate oscillator is slightly detuned by not operating it at the point for minimum current (maximum feedback). This decreases the output voltage somewhat but also protects the crystal. The output is taken across the coil in the plate oscillator by transformer action.

Questions

(1) What point in the swing of a simple pendulum may be compared to the instant the capacitor is discharged and the magnetic field about the coil is at a maximum in a parallel LC oscillator?

(2) Is regenerative feedback present in a pendulum type clock?

(3) How may the piezoelectric effect be used in a phonograph?

Suggested Experiments

(1) Oscillator. The output of an oscillator may be observed and its frequency determined by comparing its output to that of a known signal using an oscilloscope. See the section on Lissajous figures in Chapter 28.

CHAPTER 22

Modulation

22.1 INTRODUCTION

The audible range of sound, produced by the human voice, varies from approximately 100 to 1200 Hz. The audio range of sounds, which can be perceived by the human ear, varies from 20 to 20,000 Hz. This extended range includes sound frequencies produced by instruments from the bass viol to the piccolo and the harmonics, or multiples, of these frequencies. Of course, sound waves can only be transmitted for very short distances by means of a loudspeaker. Even if the loudspeaker is replaced by an antenna, which transmits these AF in the form of electromagnetic waves rather than sound waves, the range of transmission would still be rather short because of low antenna efficiency at these frequencies. In addition, if all the AF transmitters had to operate over this same frequency range (20–20,000 Hz), it would be impossible to separate two signals coming from different sources.

A process called modulation is used to overcome these problems of AF transmission. Modulation is the variation in duration, amplitude, or frequency imposed on a RF signal, called the carrier, by a code or by audio signals. Because RF transmitting antennas are much more efficient in operation than AF antennas and because RF extend over such a broad range, making separation between different transmitters easy to maintain, RF waves are used to "carry" the audible signals. At the receiving end of the modulated RF, demodulation takes place, and the audible signal is retained.

22.2 PULSE MODULATION

Duration, or pulse, modulation cannot be used to transmit the AF. Instead, the duration of a RF transmitter's output signal is varied. Letters and symbols are represented by different durations of the RF signal and are thus transmitted in code form. Morse Code uses only two signals, one short (dit) and one long (dah). The characters of the alphabet are represented by different arrangements of these "dot" and "dash" signals. Following are code representations for a few letters in the English alphabet.

E dit
T dah
I didit (ditdit)
M dahdah

Notice that in combinations of several pulses, the short pulse is written di, not dit. The long signal (dah) is three times as long as the dit. The space between characters is the length of a dit, whereas the space between words is usually around seven times the length of a dit. Government regulations require an individual to be able to send and receive Morse Code before being granted permission to transmit over long distances. Detailed information about obtaining government permission to broadcast may be obtained from the American Radio Relay League.

We have already discussed several ways of obtaining RF signals using oscillators. The keying of an RF signal, that is, turning it on and off, can be accomplished with the circuit in Figure 22.1. An RF signal is

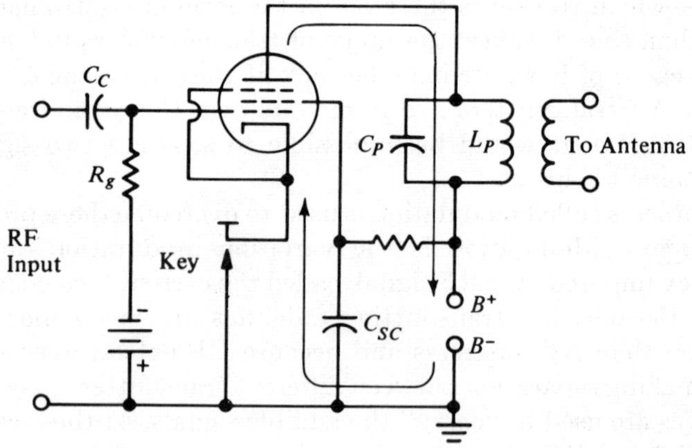

FIGURE 22.1 *RF keying circuit.*

placed on the grid of the pentode via coupling capacitor C_C. In addition to the grid leak bias established by R_gC_C, the grid has a fixed bias, supplied by the battery, which operates the pentode at class C (bias voltage below cutoff) for high efficiency. The switch, or key, in the cathode circuit controls the flow of plate current. When the key is closed, both ac and dc circuits are complete. RF current flows when the input is positive enough to drive the grid voltage above cutoff. The plate oscillator resonates with the input from the RF oscillator, and the output, when the key is closed, is at this same RF. There is no current, and hence no output, when the key is open. The output is taken across the coil by transformer coupling and placed on a transmitting antenna. The letter R, which is didahdit in Morse Code, would appear as in Figure 22.2 at the output of our circuit.

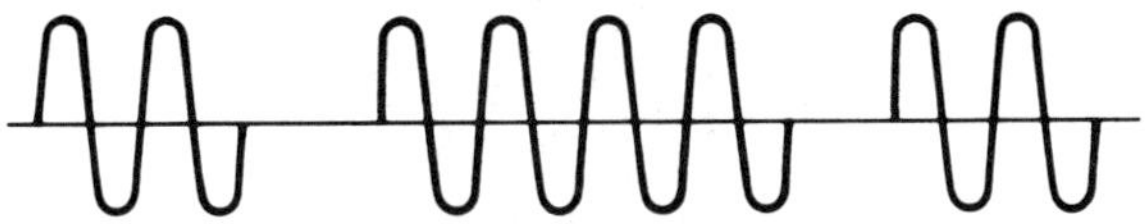

FIGURE 22.2

These pulse modulated signals can be transmitted great distances, because the distinction between signal and no signal is quite easily detected by a receiver, just as it is much easier for us to observe whether a light is on or off than to detect small changes in brightness.

22.3 AMPLITUDE MODULATION

One method of using RF signals as carriers to transmit the audio range is to vary the amplitude of the RF at the same rate at which the AF varies. Thus, amplitude modulation impresses an AF signal on an RF signal as shown in Figure 22.3. Notice that when the modulating signal is positive, the RF carrier amplitude increases, and when the modulating signal is negative, the RF carrier amplitude decreases.

22.4 PLATE AM

The RF plate modulated amplifier in Figure 22.4 illustrates how AM (amplitude modulation) can be obtained. In the circuit containing triode V_2, the AF input is felt on the grid and thus varies the flow of plate current through V_2. When the AF signal goes positive, the plate current through V_2 increases. This causes an increase in the voltage drop across the primary coil of transformer T_1. Transformer action couples this increased

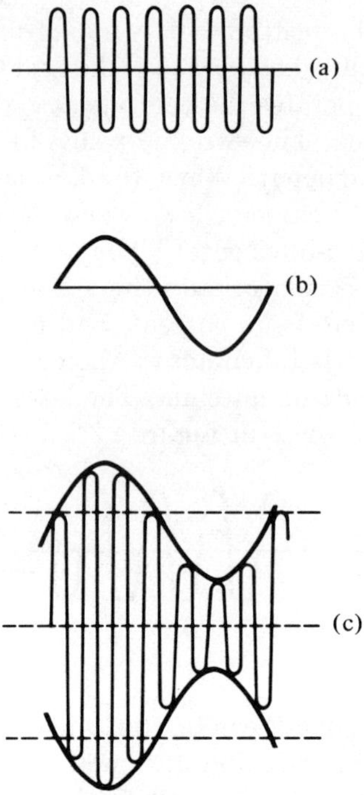

FIGURE 22.3 *(a) RF unmodulated carrier; (b) audio frequency modulating signal; (c) amplitude modulated carrier.*

voltage to the plate circuit of pentode V_1. Thus, the voltage applied to the plate of V_1 is the B^+ of the dc power supply plus the ac from the secondary of T_1. Increasing this supply voltage causes an increase in plate current and hence an increase in the amplitude of the RF voltage output. Thus, a positive going audio signal results in increased RF amplitude. Correspondingly, a negative going audio signal will cause a decrease in RF signal amplitude, and we have amplitude modulation.

22.5 GRID AM

Some other forms of amplitude modulation are suppressor grid AM, screen grid AM, control grid AM, and cathode AM. In each of these modulating techniques, an AF is coupled to a tube component by transformer action. Grid modulation, for example, is accomplished by the circuit of Figure 22.5. In this circuit, the grid bias is varied at an AF rate. The RFC (choke) coil is needed to keep the RF input out of the audio input. The triode is a class A amplifier. Thus, when the audio signal

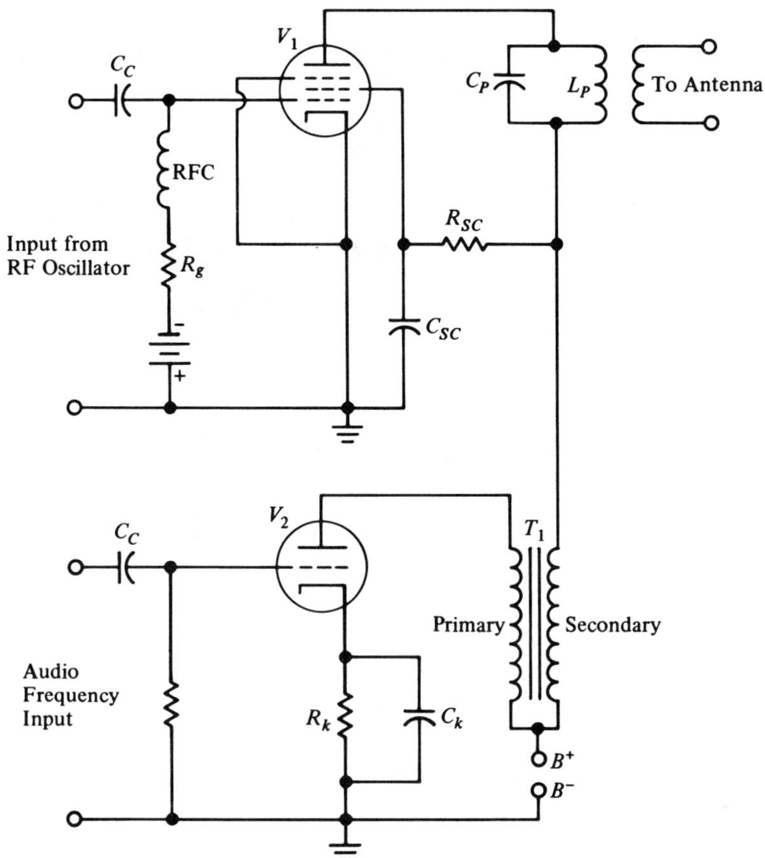

FIGURE 22.4 *Plate amplitude modulation.*

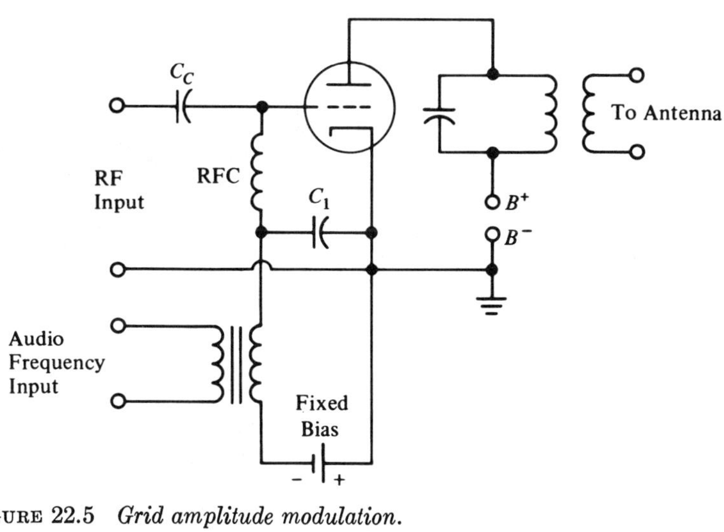

FIGURE 22.5 *Grid amplitude modulation.*

is placed on the grid, the tube conducts, and the changing grid bias modulates the amplitude of the RF carrier. The output is taken across the coil of an oscillator by transformer action; the oscillator provides a continuous RF output. Waveforms are shown in Figure 22.6.

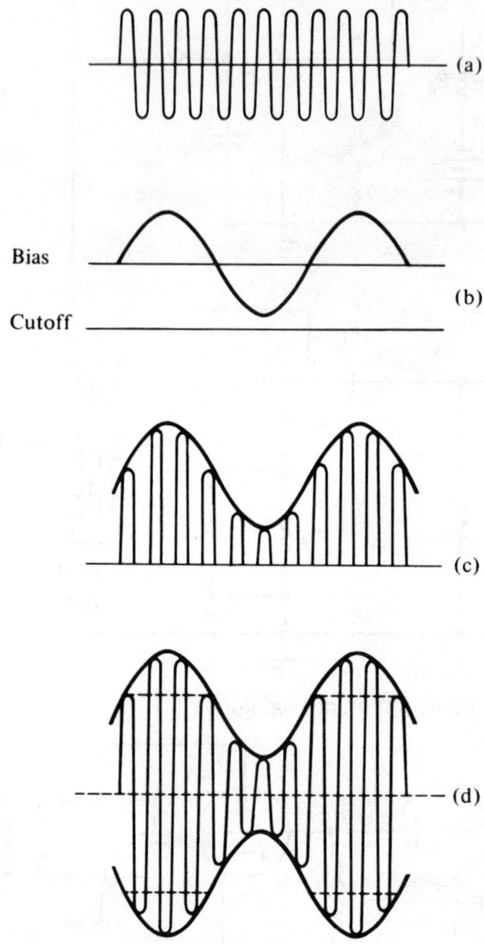

FIGURE 22.6 *(a) RF carrier; (b) audio frequency (triode biased above cutoff); (c) plate current; (d) output.*

22.6 AM BANDWIDTH

The RF carrier for amplitude modulation is at one specific frequency. When the AF signal is combined with this RF carrier, other frequencies are produced. These other frequencies can be found by adding or subtracting the frequencies of the RF carrier and the audio signal. When the two are added, we have the upper sideband frequencies. Upper sideband = RF + AF. The lower sideband frequencies, which also occur when the two signals are beat together, are found by subtracting. Lower sideband = RF − AF. (See Figure 22.7.)

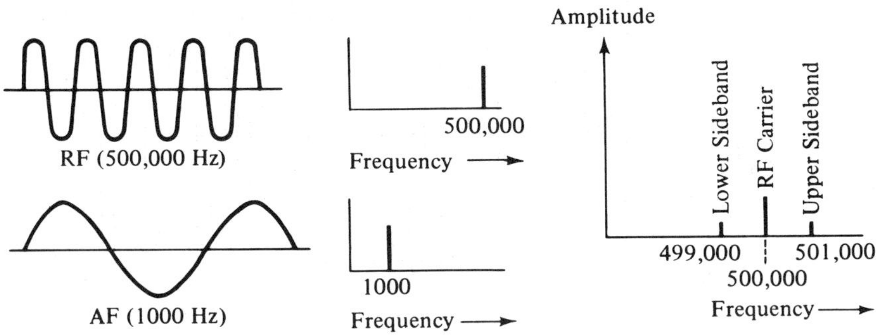

FIGURE 22.7

Suppose, for example, that the RF of an AM transmitter is 500,000 Hz, and a pure tone with AF of 1000 Hz is to be transmitted. The upper and lower sideband frequencies are:

$$\text{Upper Sideband} = 500{,}000 + 1000 = 501{,}000 \text{ Hz}$$
$$\text{Lower Sideband} = 500{,}000 - 1000 = 499{,}000 \text{ Hz}$$

Because of these sidebands' association with the carrier frequency, the transmission of an AM signal requires a band of frequencies. Therefore, more than one RF is allotted for transmission. The band of frequencies required in our example would be from 499,000 Hz to 501,000 Hz. Subtracting, this is a bandwidth of 2000 Hz. Notice that the bandwidth is twice the AF. Thus, any AF that falls within this bandwidth can be transmitted, that is, any AF of 1000 Hz or less. A typical AM broadcast station has a bandwidth of 10,000 Hz, 5000 Hz on each side of the carrier. Since audible sounds have frequencies ranging up to 20,000 Hz, the above broadcast bandwidth is too small for high-fidelity AF response, and many of the multiples, or harmonic frequencies, are missing from AM transmission.

22.7 AM PERCENT MODULATION

Recall that with pulse modulation, the signal could be easily distinguished due to its characteristic of being either on or off. Similarly, if an amplitude modulated signal varies from a maximum amplitude to zero amplitude, detection of the signal is much easier than if the change in the amplitude of the modulated signal is small. The percent of modulation, therefore, gives an indication not only of transmitter performance but also of relative ease of detection. (See Figure 22.8.)

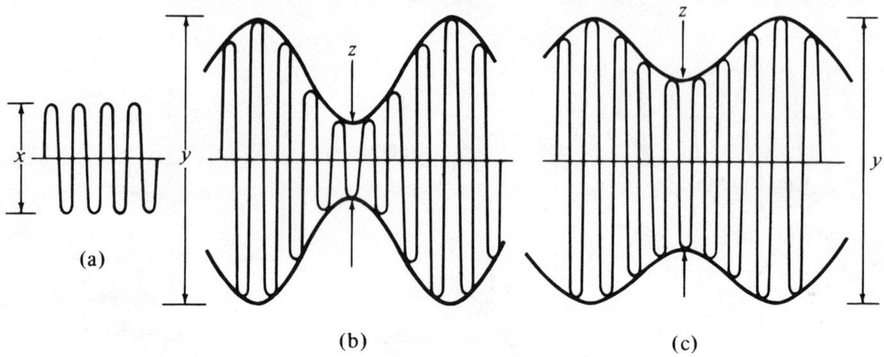

FIGURE 22.8 *(a) RF carrier; (b) good modulation (good detection); (c) poor modulation, under modulation (poor detection).*

Percent of modulation is found by comparing either the greatest increase in amplitude due to modulation or the greatest decrease in amplitude due to modulation to the unmodulated carrier. The increase in amplitude is $y - x$ and the decrease in amplitude is $x - z$ as seen in Figure 22.8. Thus, the percent of modulation can be found with either of the following equations: percent amplitude modulation (using increase in amplitude) $= [(y - x)/x]\ 100$; or percent amplitude modulation (using decrease in amplitude) $= [(x - z)/x]\ 100$.

Modulation of 100 percent will deliver maximum output for the transmitter; 100 percent modulation can be maintained if the gain of the AF signal is the same at all frequencies. In addition, notice that the peak AF voltage input must equal the peak RF voltage input. If the AF signal strength was less than the RF voltage, undermodulation would occur. (See Figure 22.9.)

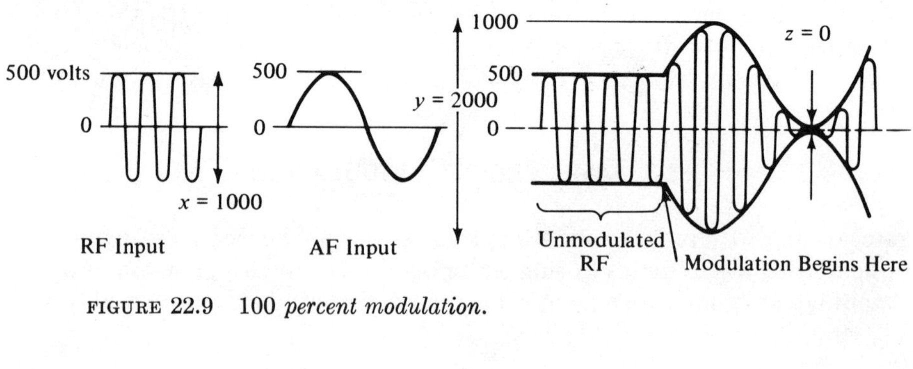

FIGURE 22.9 100 *percent modulation.*

22.8 FREQUENCY MODULATION

We now turn our attention to the third method of imposing AF on an RF carrier, called frequency modulation or FM. FM transmission is accomplished by varying the frequency of the RF carrier while its amplitude remains unchanged. In the waveforms of Figure 22.10, notice that the

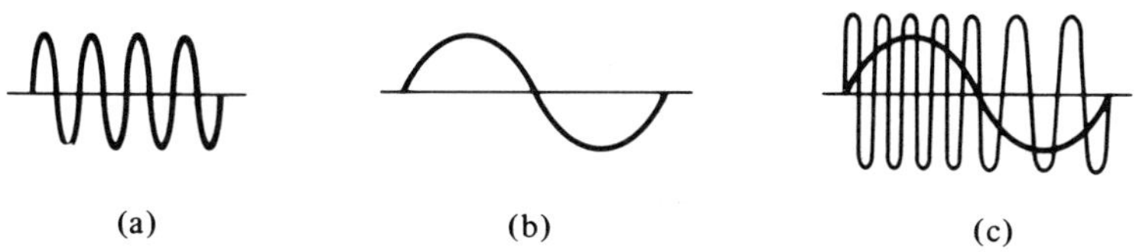

FIGURE 22.10 *(a) RF signal; (b) AF signal; (c) frequency modulated (FM) signal.*

modulated signal frequency increases when the audio signal goes positive, and the FM signal frequency decreases when the audio signal goes negative. The frequency of the audio signal determines the duration of the increases and decreases in the frequency of the carrier signal. The loudness (amplitude) of the audio signal determines by how much the carrier frequency increases and decreases. (See Figure 22.11.)

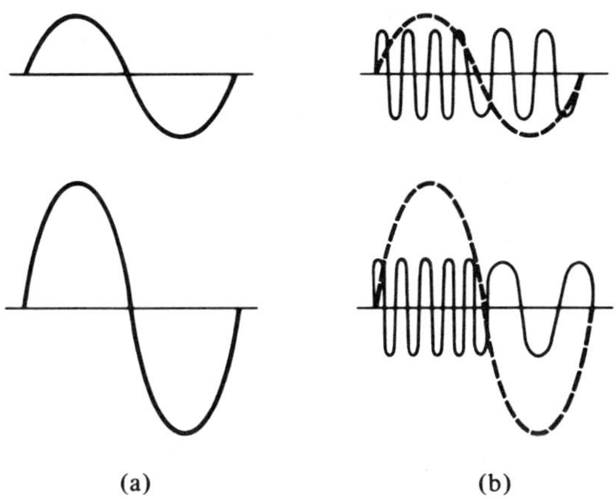

FIGURE 22.11 *(a) AF signal; (b) FM signal. (Note: The lower two graphs have the same frequency as above but amplitude of AF signal is increased.)*

In order to bring about these changes in the RF carrier, the resonant frequency of an oscillator must be made to change as the audio signal varies. Figure 22.12 illustrates one way of bringing about this type of

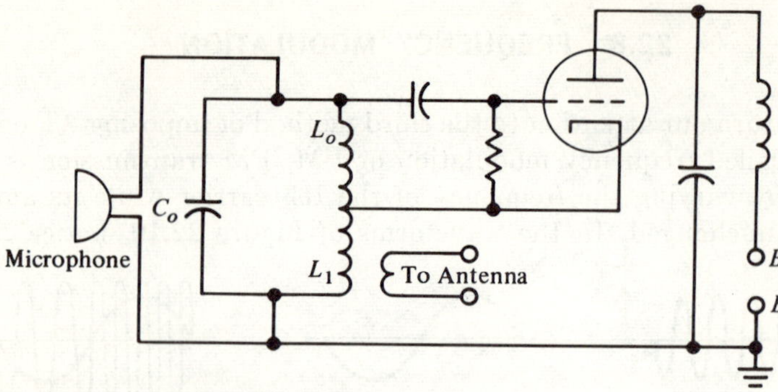

FIGURE 22.12 *Frequency modulation.*

modulation. It consists of a capacitance microphone coupled to a series-fed Hartley oscillator. The capacitance microphone is a capacitor with one flexible plate. Sound waves (AF waves), striking the microphone, cause the flexible plate to vibrate at this frequency. As the plate vibrates, the distance between the two plates changes, thus changing the capacitance ($C \alpha 1/d$). Since this capacitor (the microphone) is in parallel with the LC circuit in the Hartley oscillator, changes in the capacitance of the microphone cause changes in the total capacitance of the LC circuit. Recall for capacitances in parallel $C_T = C_{\text{Microphone}} + C_o$. The total capacitance of the LC circuit must be used to calculate the resonant frequency of the oscillator ($f = 1/2\pi \sqrt{LC}$). This means that changes in the capacitance of the microphone result in changes in the resonant frequency of oscillation, and this output RF signal (taken across L_1) thus varies as the audio signal varies. Note that the louder the signal, the greater will be the deflection of the flexible plate, causing a greater change in capacitance and a correspondingly greater variation in resonant frequency of the RF signal. Hence the amplitude of the audio signal determines by how much the FM signal changes.

A more common method of modulating the frequency of an RF carrier is with a reactance tube connected in parallel to an oscillator tank. (See Figure 22.13.) The reactance tube circuit is not an amplifier. It replaces the microphone in the previous circuit, and its purpose is to present a variable reactance in parallel with the LC oscillator circuit. When the AF input varies, the reactance of V_1 varies. This, in turn, varies the resonant frequency of the oscillator. The FM output is taken across the coil of the oscillator by transformer action.

In the transistor FM circuit, the audio signal varies the forward bias, thus causing corresponding changes in collector-base voltage (reverse bias). These variations in collector-base voltage change the output capacitance of the transistor.

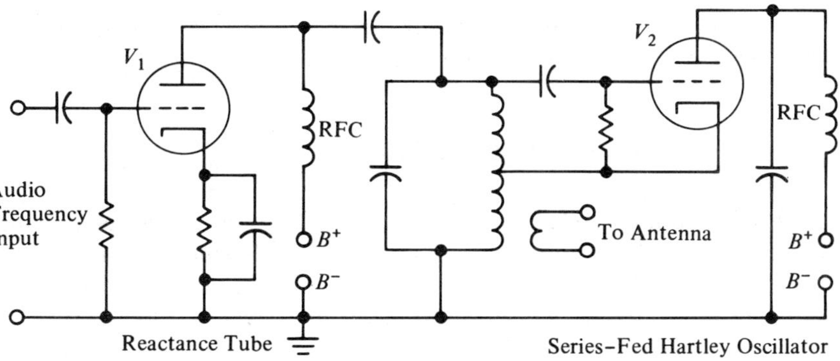

FIGURE 22.13

Remember that the depletion zone between semiconductor materials becomes larger, when the reverse bias voltage increases. (See Figure 22.14.) An increased distance (larger depletion zone) between emitter

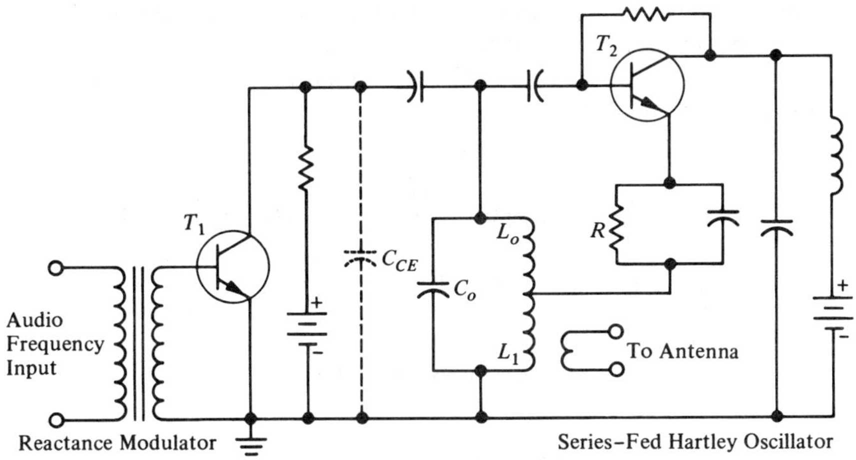

FIGURE 22.14

and collector would decrease the collector-emitter capacitance (C_{CE}). When a positive going AF signal increases the forward bias of T_1, the collector-emitter voltage also increases, and the collector-emitter capacitance decreases. Since C_{CE} is in parallel with the LC oscillator circuit, the total capacitance is $C_T = C_{CE} + C_o$. If C_{CE} decreases, then C_T decreases, and the resonant frequency of the oscillator increases ($f\uparrow = 1/2\pi\sqrt{LC\downarrow}$). Thus, a positive going audio signal produces an increase in RF oscillations; similarly, a negative going audio input will produce a decrease in RF oscillations. And the circuit output is FM (frequency modulated).

Notice again the greater the amplitude of the positive going AF signal, the greater will be the increase in RF carrier frequency.

22.9 FM BANDWIDTH

The FM sidebands include more than just the sum and difference of the carrier RF and the pure AF. Since multiples, or harmonics, of the AF are carried by an FM signal, they must also be added to, and subtracted from, the RF carrier. This produces a much broader bandwidth than found in AM. Commercial FM bandwidth is 150,000 Hz as compared to the AM bandwidth of 10,000 Hz. The inclusion of the harmonics in the bandwidth for FM transmission produces a truer sound reproduction than is possible with AM, but it also means that there must be a larger broadcast band assigned to FM than to AM. Although eight multiples of any AF will produce nearly all tonal characteristics, the number of harmonics produced on either side of the carrier frequency may sometimes be greater than eight. To accomodate the varying number of sidebands and eliminate interference from stations whose bandwidths might extend near one another, guard bands of 25,000 Hz are set up on each side of the bandwidth. (See Figure 22.15.)

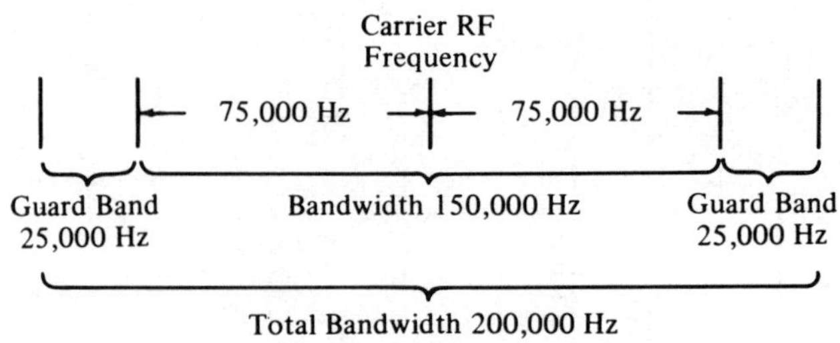

FIGURE 22.15

22.10 FM PERCENT MODULATION

Recall that the percent of modulation of an AM signal varied from zero to 100 percent as the amplitude of the modulating AF signal varied from zero to the amplitude of the unmodulated RF carrier. The percent of modulation of an FM signal is obtained from a knowledge of bandwidth. If the positive going AF signal input causes the carrier frequency to increase by 75,000 Hz, and the negative going AF causes it to decrease by 75,000 Hz, then 100 percent modulation exists. This can be expressed by the following equations: percent frequency modulation = [(Maximum frequency of output − carrier frequency)/75,000] × 100; percent frequency modulation = [(carrier frequency − minimum frequency of output)/75,000] × 100.

For example, if the broadcast frequency of an FM station is 100 MHz (megahertz) and the maximum output frequency is 100,025,000 Hz, then the percent of frequency modulation is:

$$\text{percent frequency modulation} = \left[\frac{(100{,}025{,}000 - 100{,}000{,}000)}{75{,}000}\right] \times 100$$

$$= \left(\frac{25{,}000}{75{,}000}\right) \times 100 = 33.3 \text{ percent.}$$

The broad bandwidths of 200,000 Hz required for commercial FM broadcasting, as compared to 10,000 Hz for AM, necessitates a much broader FM broadcast band. The commercial FM broadcast band is from 88,000,000 Hz to 108,000,000 Hz, a range of 20,000,000 Hz. By way of comparison, the commercial AM broadcast band is from 550,000 Hz to 1,600,000 Hz, a range of 1,050,000 Hz. Observe that the FM broadcast band is nearly twenty times as broad as the AM band. We might also calculate the maximum possible number of stations in one location broadcasting either AM or FM. Since the commercial FM broadcast bandwidth is 20,000,000 Hz and each FM station is allotted 200,000 Hz, we can determine the maximum number of FM stations by dividing: 20,000,000 Hz/200,000 Hz for each station = 100 FM stations. Now dividing the commercial AM broadcast bandwidth by the bandwidth allotted each AM station, we have 1,050,000 Hz/10,000 Hz for each station = 105 AM stations. Notice that the maximum possible number of FM and AM stations is nearly the same.

A spectrum of frequencies indicates how narrow the broadcast bands are. (See Figure 22.16.)

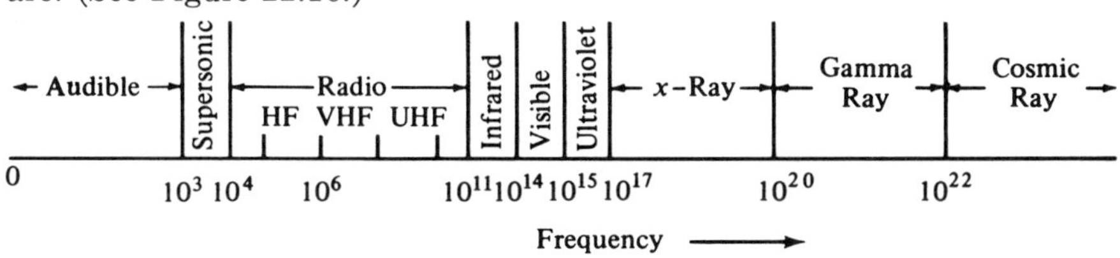

FIGURE 22.16 *Spectrum of frequencies.*

Questions

(1) Can all languages be coded for pulse modulation?
(2) Describe the types of modulation.

Suggested Experiments

(1) Pulse Modulation. You may wish to learn Morse Code and investigate becoming a "Ham" radio operator.

(2) Amplitude Modulation. A wave pattern of an amplitude modulated signal just before transmission may be observed with an oscilloscope. The pickup coil is placed near the output tank coil of the transmitter. (See Figure 22.17.)

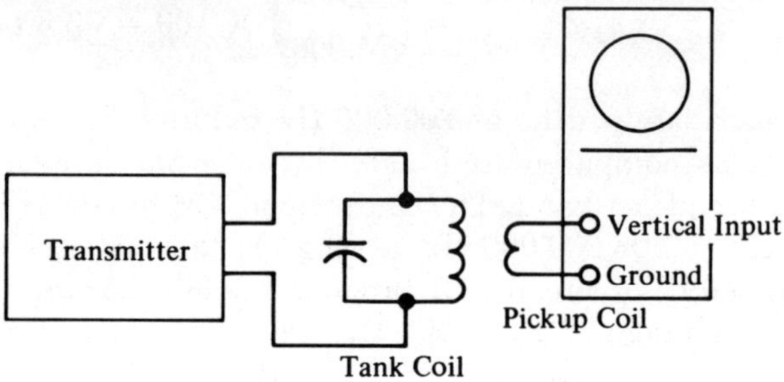

FIGURE 22.17

CHAPTER 23

Transmission Lines and Antennas

23.1 OPEN TRANSMISSION LINES

We have briefly discussed the electronics associated with AM and FM transmitters. Now we will consider how this RF energy is first transmitted to the antenna and then radiated into space in the form of electromagnetic energy to eventually be received by an AM or FM radio.

Remember when we discussed transformers, we mentioned that they may be used to step up voltage and step down current and thus reduce the power loss ($P = I^2R$) in transmitting ac over long lines. RF signals from AM or FM transmitters must be carried over transmission lines to the antenna with a minimum of power loss in order to provide maximum output power at the antenna. To obtain maximum power transfer from transmission lines to antenna, the output impedance of the transmission lines must equal the input impedance of the antenna.

Consider the transmission lines from an RF generator. (See Figure 23.1.) The instant the transmitter begins to send out a signal, the current and voltage, leaving the transmitter and going down the line, are in phase and determine the input impedance of the line ($Z_{in} = V/I$). Neglecting small resistive losses in current and voltage, the output impedance at any point along the line will equal the input impedance ($Z_{out} = Z_{in} = V/I$).

When the current and voltage reach the end of open transmission lines, the potential difference (voltage) between the lines is at a maximum, but the current flowing between the wires across the open space is, of course, zero. Recall that up to this point (end of line) the current and voltage were in phase. This out-of-phase relation between current and voltage at the end of the transmission lines will be reflected back along

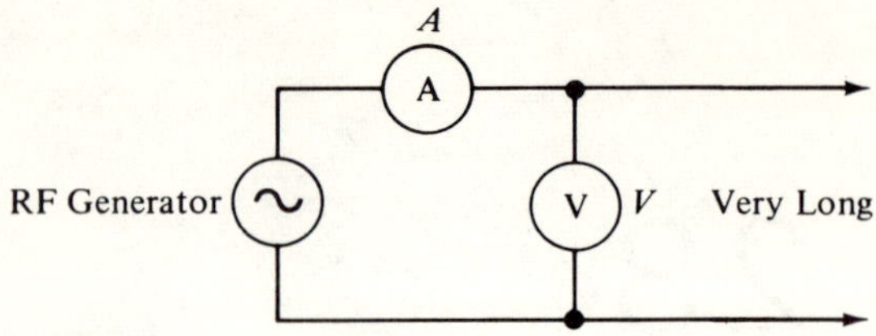

FIGURE 23.1 *Transmission lines*

the lines, setting up what are called standing waves. At the open end of the line, the opposition to current flow is at a maximum; one-fourth of a wavelength back along the line, the opposition to current flow is at a minimum. (See Figure 23.2.)

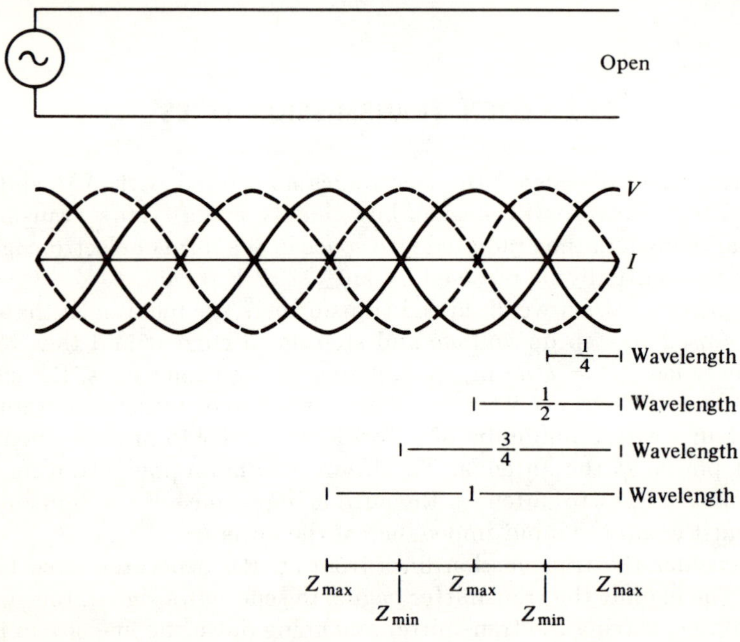

FIGURE 23.2

We want to place an impedance, which will allow maximum power transfer, across the open end. Since closing the ends of the lines with the impedance will prevent standing waves from being set up, it would initially appear that in order to have maximum power transfer we need only place a resistor across the open end of the lines equal to V/I (the input impedance). However, the fallacy here is that the opposition to current flow along RF transmission lines is not purely resistive. The changing magnetic field associated with ac flowing in the lines produces inductive reactance. And the two transmission lines, being separated by an insulator, or dielectric, also display capacitive reactance. These two reactances are in addition to the relatively small internal resistance to current flow in the line. Any impedance matching device to be placed across the wires must take these reactances, as well as the resistance, into account.

The maximum opposition to current flow at the ends of the open transmission lines may be thought of as a capacitor and a coil in parallel. Recall that a parallel LC circuit has its maximum impedance at resonance. Thus, the impedance matching device should be a parallel LC circuit, with its resonant frequency equal to the RF carrier frequency, and a small resistance in series with it. (See Figure 23.3.)

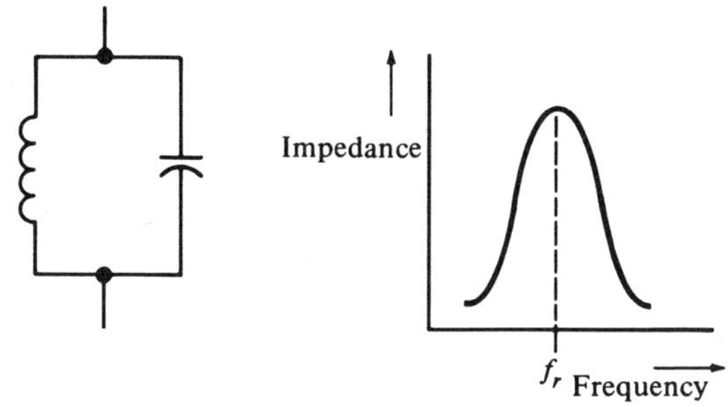

FIGURE 23.3 *Maximum impedance at resonance.*

23.2 CLOSED TRANSMISSION LINES

If, instead of being left open, the ends of the transmission lines are shorted, the voltage across the ends will be zero, and the current flowing across the ends will be at a maximum. The opposition to current flow will be at a minimum at the end and at a maximum one-fourth of a wavelength back from the end. (See Figure 23.4.)

Thus, a one-quarter wavelength section of shorted transmission line has a minimum impedance at the bar and a maximum impedance at the open end. (See Figure 23.5.) These characteristics of the one-quarter wavelength section permit it to be used as an insulator for the lines or as an impedance matching device. When the open ends are placed across the transmission lines, their high impedance essentially insulates the lines. Remember that the length of the ⊔ must be one-quarter wavelength of the transmitted signal. (See Figure 23.6.)

As mentioned, these one-quarter wavelength ⊔'s can also be used as impedance matching devices. The impedance near the open end of the ⊔

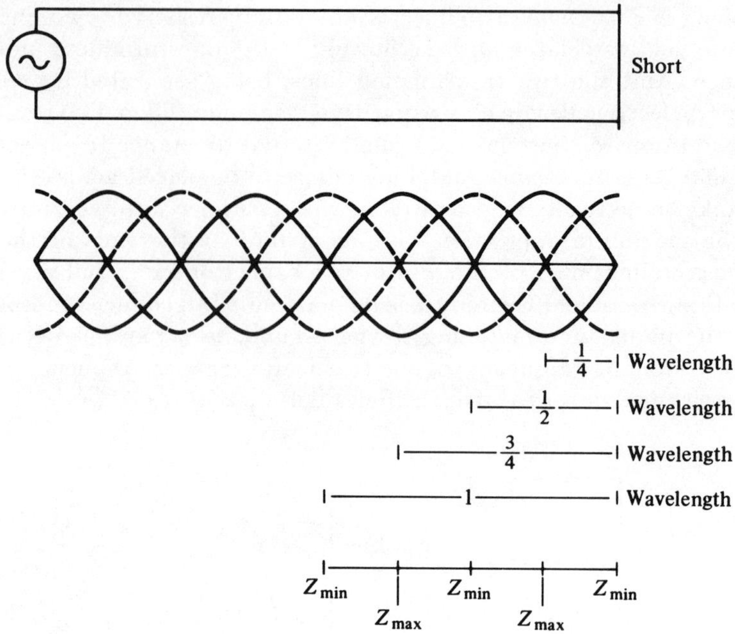

FIGURE 23.4

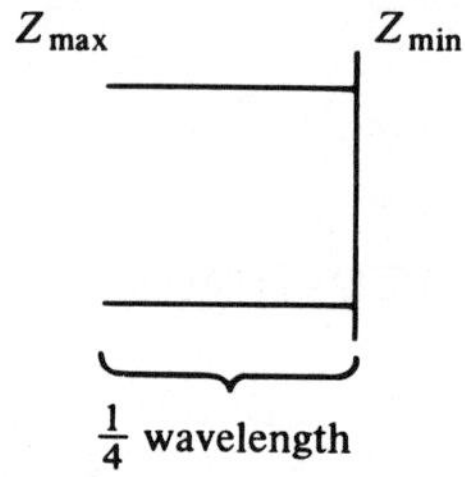

FIGURE 23.5

is quite large. It decreases as we move toward the closed end. Thus, the impedance of a transmission line may be matched by connecting it to points on the one-quarter wavelength section which offer the same impedance as the line. Another pair of taps closer to the closed end will offer a lower impedance output. (See Figure 23.7.)

23.3 DIPOLE ANTENNA

The last one-quarter wavelength section of open-ended transmission lines can be used as an antenna by bending these sections back to an angle of

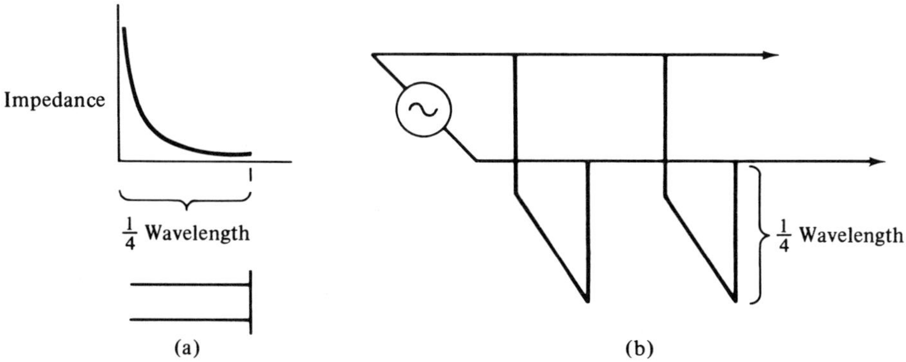

FIGURE 23.6 *(a) Impedance versus length of shorted transmission line; (b) one-quarter wavelength ⊔'s used as insulators.*

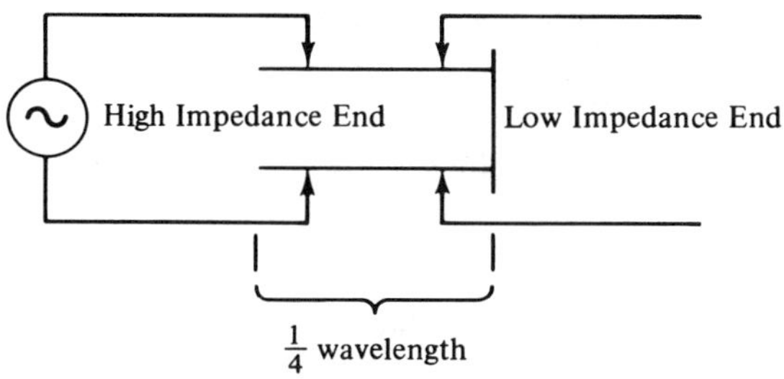

FIGURE 23.7 *One-quarter wavelength ⊔'s used as impedance matching device.*

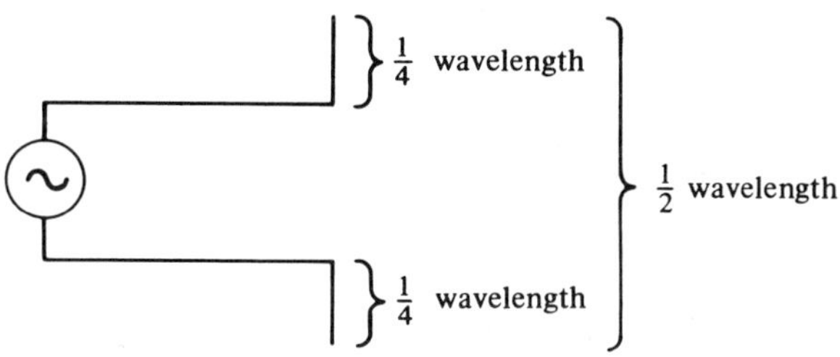

FIGURE 23.8 *One-quarter wavelength section used as a dipole antenna.*

90° with the lines. (See Figure 23.8.) Referring to the diagram of standing waves reflected from the open end of a transmission line (Figure 23.2), we see that impedance is at a minimum one-quarter wavelength back from the end. Thus, in the dipole antenna, impedance is at a minimum across the two bends in the wires. A minimum impedance containing both capacitive and inductive reactances is obtained with a series resonant circuit containing a coil and a capacitor. (See Figure 23.9.) Since capacitive and inductive reactances in a series circuit are 180° out of phase and

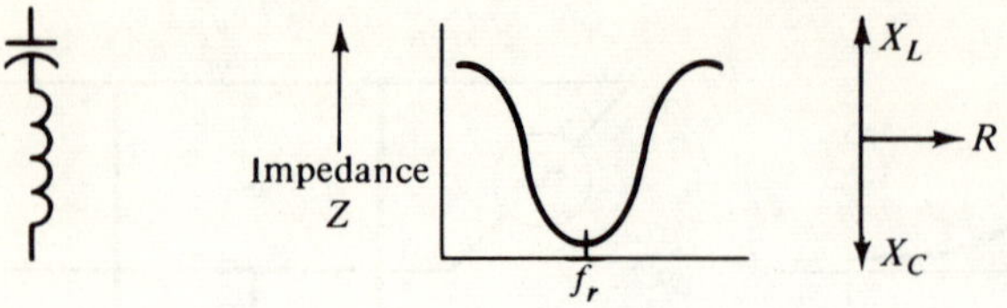

FIGURE 23.9 *Minimum impedance at resonance.*

since $X_C = X_L$ at resonance, the only opposition to current flow at the input of the antenna (the bends in the wire) is resistive.

The resistance of the antenna can be calculated using the power equation $P = I^2R$. Thus, $R = P/I^2$, where P is the power radiated from the antenna, and I is the effective ac current. For most half wavelength dipole (two poles) antennas, this resistance is about 73 ohms; that is, a resistor of 73 ohms connected across the "bends" of the open-ended transmission lines would dissipate the same amount of power as the dipole antenna. Since the opposition to current flow of the dipole antenna is 73 ohms, then the input impedance of the transmission lines should also be 73 ohms for maximum power transfer.

23.4 WAVELENGTH

The wavelength, represented by the symbol λ (Lambda), of any RF transmitter can be calculated from the frequency or it can be measured experimentally. The study of optics tells us that the speed of light (c), or the speed of any other type of electromagnetic radiation including RF signals, is equal to the product of the frequency of the wave (f) and the length of the wave (λ). In equation form, $c = f\lambda$. The speed of light or of RF radiation is 186,000 mi/sec = 300,000,000 m/sec. This is the speed of electromagnetic radiation in air; it will actually be somewhat slower along the conductor, but for our purpose, we can assume that it remains constant. If the above equation is solved for wavelength by dividing both sides by frequency f, we obtain $\lambda = c/f$ or $\lambda =$ (300,000,000 m/sec)$/f$. Suppose, for example, that the output of an RF transmitter is 3000 MHz. Then the length of one wave would be

$$\lambda = \frac{c}{f} = \frac{300{,}000{,}000}{3{,}000{,}000{,}000} = 0.1 \text{ m} = 10 \text{ cm}.$$

A one-quarter wavelength section of transmission line for this transmitter would be $\frac{1}{4}\lambda$, that is, one-fourth of 10 cm or 2.5 cm long (approximately one inch).

To determine wavelength experimentally, we connect the RF transmitter to two parallel wires (transmission lines with an open end). Connect two uninsulated wires to the terminals of a small flashlight bulb, suspend the bulb by these wires between the transmission lines, and move it up and down the lines. The light from the bulb will be brightest at points where the current is at a maximum. The distance between successive bright spots will be one half of a wavelength. (See Figure 23.10.)

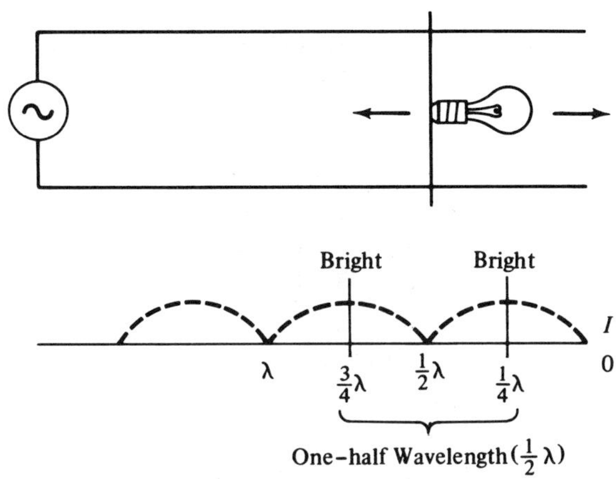

FIGURE 23.10

23.5 HERTZ DIPOLE ANTENNA

The half-wave dipole antenna, sometimes referred to as a hertz antenna, has a voltage and current distribution such that the voltage is at a maximum at the ends of the antenna and at a minimum at the two bends (the center); current is just the opposite (minimum at the ends, maximum at the center). Note that, at one instant, end A of the dipole is positive and end B is negative. On the second half cycle, A is negative and B is positive. (See Figure 23.11.)

At any instant, an electrical field exists between the positive and negative ends of the dipole. An electrical field cannot be seen, but it can be mapped by placing a positive test charge (positive ion) in the field and observing its motion. In this electrical field, produced by the dipole antenna, the test charge will have two forces acting on it. It will be repelled by the positive end and attracted by the negative end of the dipole. The path that these forces cause the charge to take shows the direction of the electrical field. A mapping of the electrical field along a hertz antenna at one instant is shown in Figure 23.12.

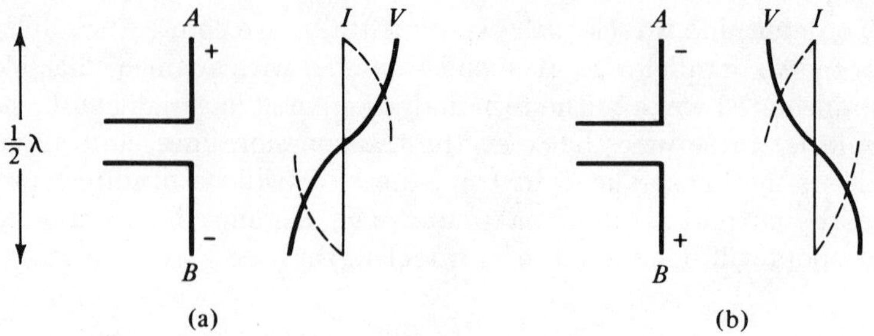

FIGURE 23.11 *Voltage and current waveforms in a hertz antenna. (a) First half cycle; (b) second half cycle.*

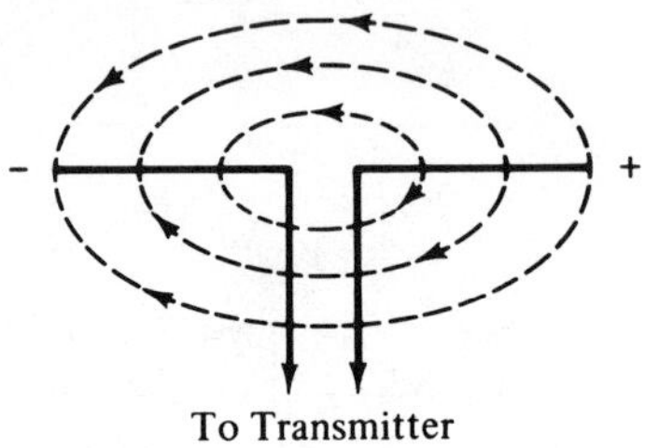

FIGURE 23.12 *Electrical field of hertz dipole antenna at one instant.*

Since there are electrons moving in the dipole antenna, we know there is also a magnetic field present. This magnetic field circles the antenna, whereas the electrical field extends from end to end between the dipoles. Thus, the electrical and magnetic fields are perpendicular to each other. (See Figure 23.13.)

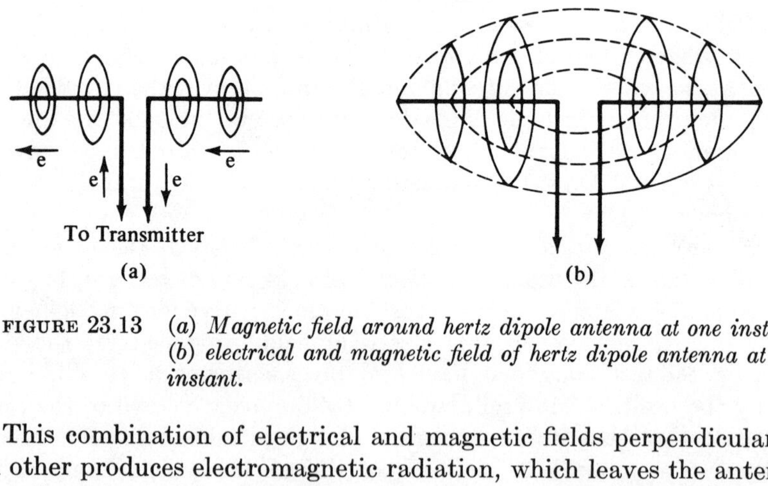

FIGURE 23.13 *(a) Magnetic field around hertz dipole antenna at one instant; (b) electrical and magnetic field of hertz dipole antenna at one instant.*

This combination of electrical and magnetic fields perpendicular to each other produces electromagnetic radiation, which leaves the antenna and radiates into space, traveling at the speed of light (186,000 mi/sec).

The radiation pattern for the antenna varies with variations in the RF placed on the antenna by the transmitter. For example, if the RF signal is a sine wave, the building and collapsing of the electrical and magnetic fields during one half cycle appear in Figure 23.14. These two fields com-

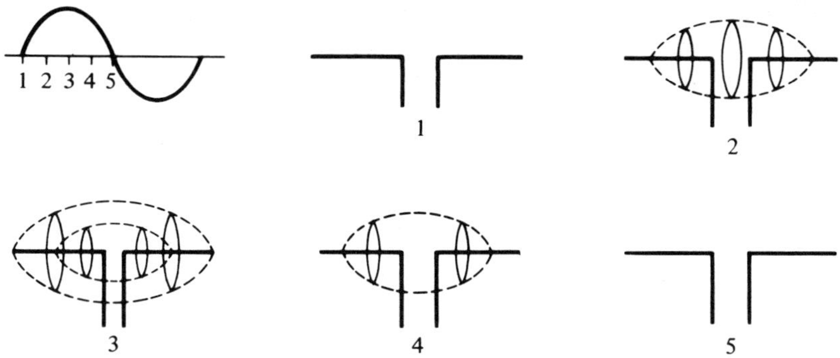

FIGURE 23.14 *Building and collapsing of electrical and magnetic fields during one half cycle.*

bine at each instant to produce electromagnetic energy. The electromagnetic radiation given off at each instant moves away from the antenna in all directions. Thus, the electromagnetic field about the antenna after one half cycle looks much like a donut with the arms of the antenna popping through the hole. (See Figure 23.15.)

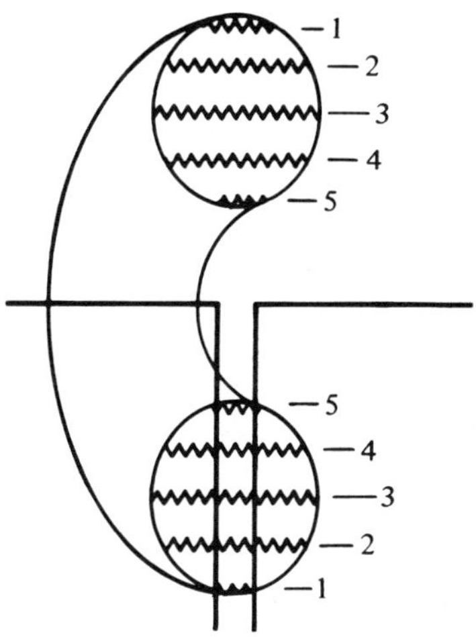

FIGURE 23.15 *Electromagnetic radiation emitted from antenna during one half cycle.*

In reality, the radiation pattern will not be this symmetrical due to interference from objects around the antenna. A specialized receiver, called a field strength meter, is used to map these unsymmetrical electromagnetic fields.

A vertical one-quarter wavelength Marconi antenna is a variation on the half-wave hertz dipole antenna just discussed. To obtain the Marconi antenna, the lower half of the hertz antenna is put in the ground, leaving only one-quarter wavelength to radiate energy into space. The other one-quarter wavelength (the grounded half of the antenna) radiates energy into the earth. (See Figure 23.16.)

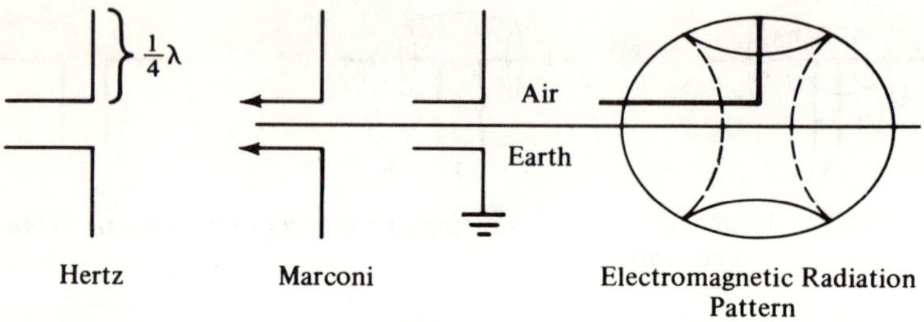

FIGURE 23.16

23.6 RADIATION PATTERNS

Standard broadcast stations have a ground wave pattern, which extends approximately 100 miles from the antenna in all directions. The electromagnetic energy which is radiated into space may take either of two types of paths. Those waves that travel along or toward the surface of the earth are called space waves. Those that travel upward toward the sky are called sky waves. The distance that space waves travel depends on the height of the antenna with respect to the surrounding countryside. In addition, the effective distance traveled is limited by the curvature of the earth's surface.

The sky waves of some RF can travel around the earth by a process of refraction and reflection. If you are familiar with the phenomenon of mirages and know how a bouncing ball behaves, the concepts of refraction and reflection will be easy to follow.

Mirages occur when the air near the surface of the earth is thinner (less dense) than the air above the surface. This situation can be caused by the earth's heating of the air at its surface. When light travels through layers of air with different densities, its path is changed slightly, producing the mirage as shown in Figure 23.17. This same process of refraction bends electromagnetic sky waves back toward the earth. The RF waves are refracted by the ionosphere, a layer of ionized gas molecules beginning approximately 40 miles above the surface of the earth and extending to about 350 miles above the surface. (See Figure 23.18.) The positive and negative ions, which make up the ionosphere, are produced from gas molecules bombarded by cosmic radiation from outer space.

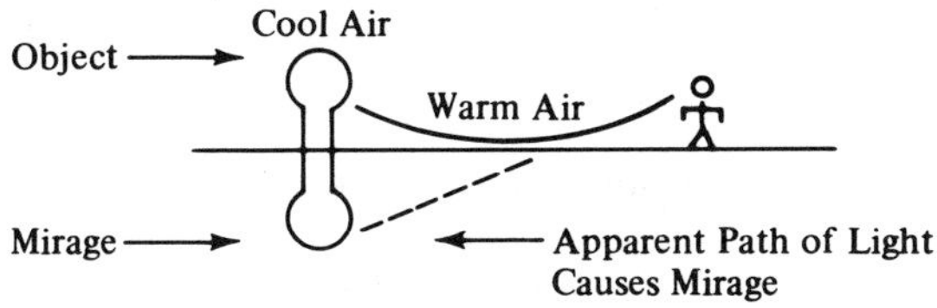

FIGURE 23.17

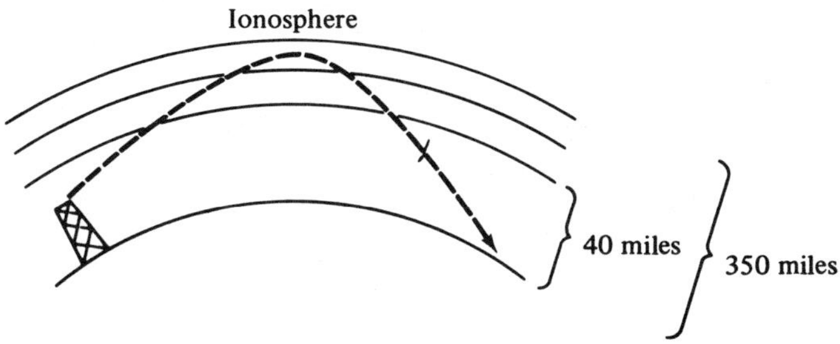

FIGURE 23.18 *Refraction of RF signal by ionosphere.*

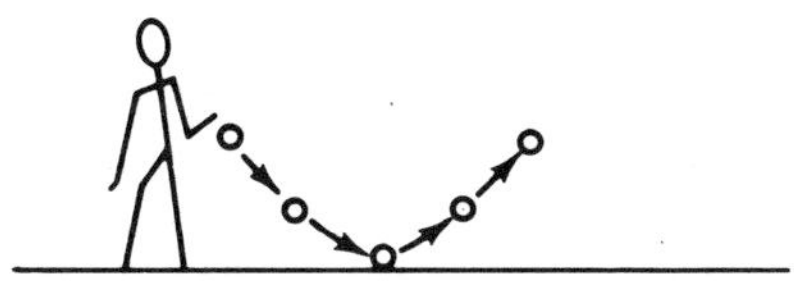

FIGURE 23.19 *"Reflection" of ball from a smooth surface.*

A ball hitting a hard surface at an angle bounces off in the opposite direction. In much the same manner, when a refracted sky wave strikes the hard surface of the earth, it is reflected back toward the sky. By repeated refractions and reflections, the signal is sent around the earth. (See Figures 23.19 and 23.20.)

Combining the three radiation patterns (ground, space, and sky), we observe that there are still some locations on the surface of the earth where the RF electromagnetic signal is not received. These locations are called the skip distance. (See Figure 23.21.)

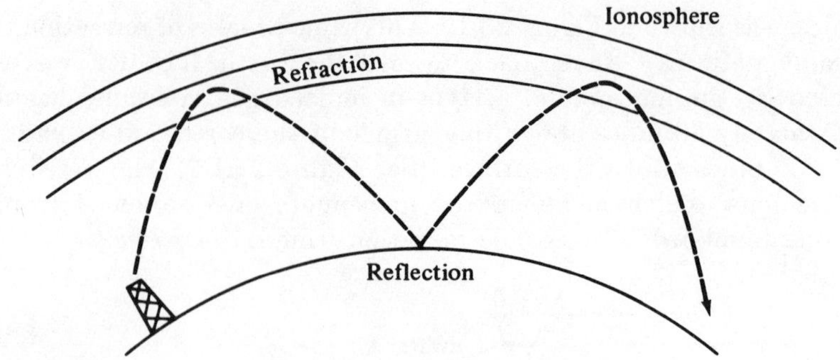

FIGURE 23.20 *Refraction and reflection of electromagnetic energy.*

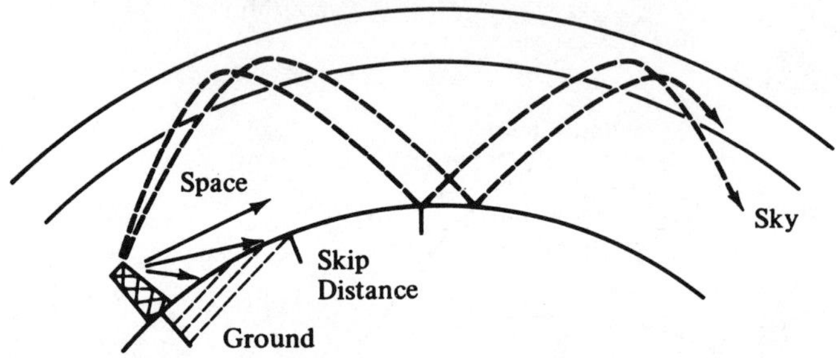

FIGURE 23.21 *Skip distance for low-frequency electromagnetic radiation.*

As the frequency of the RF signal increases into the FM range and beyond, the ionosphere will not effectively refract the electromagnetic waves. Thus, the only useful radiation from the antenna is the space wave. The signal from high-frequency transmitters, such as FM or radar, can be received only if the antenna is within the "line of sight" of the receiver; that is, it must be possible to draw a straight line through space connecting antenna and receiver.

Questions

(1) How is a shorted transmission line similar to a vibrating guitar or violin string?

(2) RF radiation is refracted, or bent, by the ionosphere. What everyday phenomena give evidence that visible light is refracted by the atmosphere?

CHAPTER 24

Detectors

24.1 DIODE DETECTORS

The detection of the electromagnetic radiation, which has been amplitude modulated and then transmitted into space, can be accomplished by crystal diode detectors, grid leak detectors, regenerative detectors, and heterodyne detectors. After we consider some of the circuits used to detect these signals, we will add other circuitry to amplify the signal strength to an audible level.

In the diode detector, when the electromagnetic radiation cuts across the receiving antenna, a small voltage is induced across the antenna, causing current to flow down the antenna lead to ground. (See Figure 24.1.) As the signal passes to ground, it travels through the primary of an RF transformer, which couples the signal to a parallel LC circuit, so tuned that its resonant frequency is the same as the frequency of the incoming RF signal. By changing the value of the variable capacitance in the LC circuit, different RF frequencies from different broadcast transmitters may be received. The diode (or crystal) acts as a half-wave rectifier to the input RF signal by conducting only when the signal goes positive. The RC filter, consisting of capacitor C and resistor R, acts as a demodulator. The time constant ($1\ \text{TC} = R \times C$) is long in comparison to the time for one cycle of the RF signal. Because of the slow discharge of C through R, voltage is maintained across the load when the RF signal drops to zero. Caution must be exercised in choosing the time constant so as to avoid distorting the AF signal imposed on the RF input. The time constant should be short in comparison to the period of the AF; otherwise, the capacitor would charge to a peak value and then would be so slow to

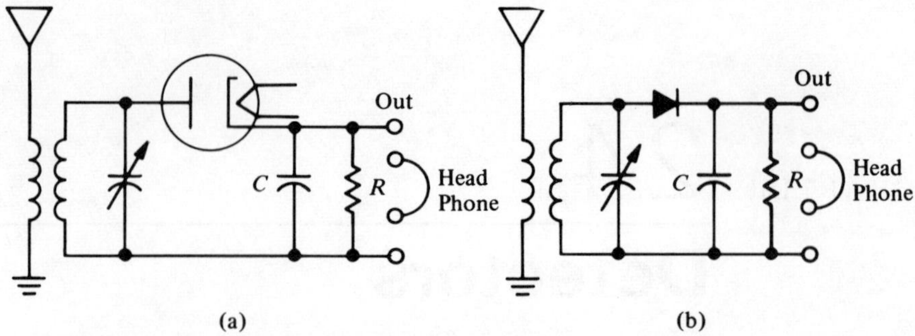

FIGURE 24.1 *(a) Diode detector; (b) crystal detector.*

discharge that it would miss intermediate variations in AF amplitude. (See Figure 24.2.)

We want the AF output to be distorted as little as possible. The degree of exactness of reproduction of the AF by a detector or receiver is referred to as fidelity. The output from the diode or crystal detector

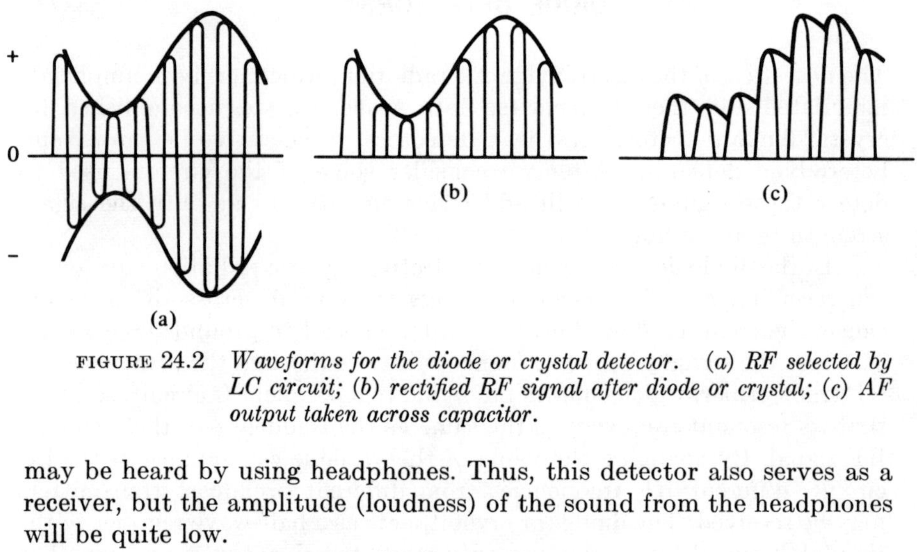

FIGURE 24.2 *Waveforms for the diode or crystal detector. (a) RF selected by LC circuit; (b) rectified RF signal after diode or crystal; (c) AF output taken across capacitor.*

may be heard by using headphones. Thus, this detector also serves as a receiver, but the amplitude (loudness) of the sound from the headphones will be quite low.

24.2 GRID LEAK DETECTOR

The grid leak detector provides both detection and amplification, thus giving a much stronger AF output. (See Figure 24.3.) The single triode in this circuit will both detect and amplify. The cathode and grid act as a diode detector. The grid, although small in surface area, acts as the plate of this detection section by attracting electrons from the cathode when the incoming RF signal is positive. Thus, the cathode–grid circuit

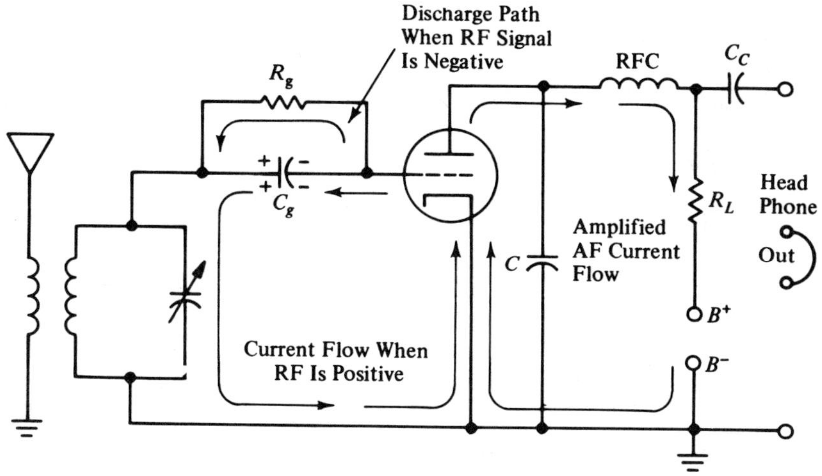

FIGURE 24.3 *Grid leak detector.*

does the job of a half-wave rectifier as the diode did in the diode detector. Next, the RF signal must be demodulated, or filtered, to obtain an AF output. This is accomplished by the grid resistor and capacitor. Again, the time constant of the R_gC_g circuit must be long in comparison to the period of the RF signal, and short in comparison to that of the AF signal. The filtering action of the R_gC_g circuit places an AF signal on the grid which is then amplified by the triode. (See Figure 24.4.)

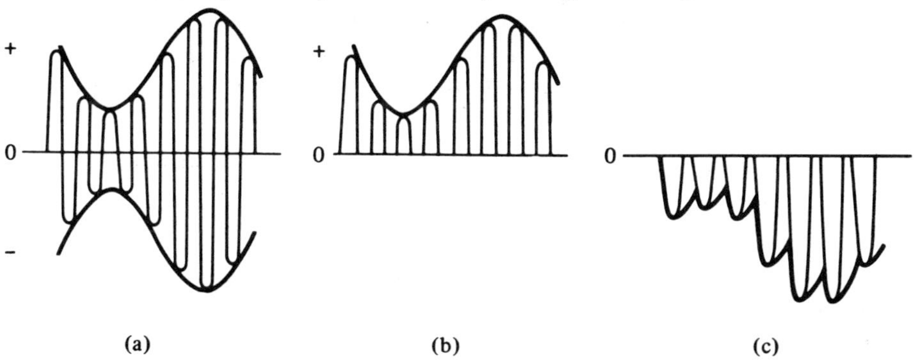

FIGURE 24.4 *Waveforms for the grid leak detector. (a) RF selected by LC circuit; (b) half-wave rectification of grid current by cathode grid; (c) AF signal placed on grid after demodulation by R_gC_g circuit.*

In addition to the amplified AF signal, there will also be some RF, which goes on the grid and is amplified. The purpose of capacitor C is to bypass, or shunt, this RF to ground. Further removal of RF is accomplished by the radio frequency choke coil (RFC). The output AF signal is taken across the load resistor R_L.

The common-base transistor detector is similar to the grid leak detector in that detection takes place in the base-emitter circuit and amplification in the base-collector circuit. (See Figure 24.5.) When the

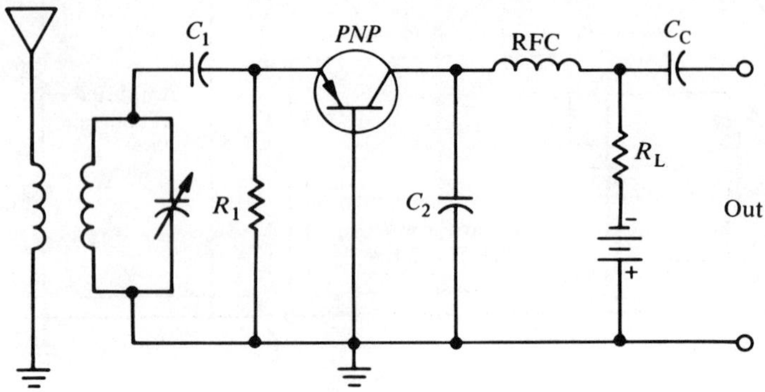

FIGURE 24.5 *Common-base detector.*

incoming RF signal is positive, current flows through the base-emitter circuit charging capacitor C_1. When the RF signal goes negative, C_1 discharges through R_1. The time constant for R_1C_1 is long for RF signals, so C_1 does not completely discharge. Thus, capacitor C_1 and resistor R_1 act as a demodulator for the RF signal, producing an AF signal. Since R_1C_1 also produces the emitter-base bias, this AF signal is applied to the emitter-base forward bias. The collector current then varies with the AF signal, that is, it varies at an audio rate, giving an AF output. Any remaining RF signal is removed from the output by bypass capacitor (C_2) and RF choke coil (RFC).

24.3 PLATE DETECTOR

In the plate detector, we again have detection and amplification by a single triode. (See Figure 24.6.) The difference between the plate detector and the grid leak detector is that in the plate detector, the triode first amplifies the RF signal and then detects the AF. Cathode resistor R_k establishes the bias for the triode at cutoff. When the incoming RF signal is placed on the grid, plate current flows during the positive half cycle. This rectifies the plate current. The top plate of capacitor C is connected to the plate of the triode and is thus positively charged; the bottom plate, connected to the cathode, is negatively charged. When plate current flows, its RF component, meeting the high impedance of RFC, is bypassed through C. The RF current makes the top of C less positive, and C dis-

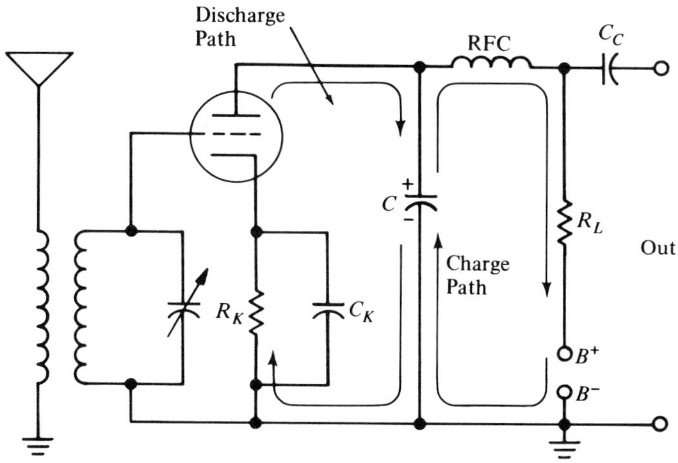

FIGURE 24.6 *Plate detector.*

charges through the triode. When the RF signal is zero, C charges through R_L. The time constant for the charging path is long in comparison to the period of the RF, but short in comparison to the period of the AF. Thus, the voltage across C varies at an AF rate. The final output, taken across R_L, is inverted due to triode amplification. (See Figure 24.7.)

24.4 REGENERATIVE DETECTOR

The regenerative detector utilizes the principle of positive, or regenerative, feedback to improve signal strength and is consequently very sensitive to weak incoming signals. It is also quite selective as to the signal chosen for demodulation. The regenerative detector is like a grid leak detector except for the addition of feedback coil L_2. (See Figure 24.8.) That RF signal, which is not filtered by R_gC_g but which goes on the grid and is amplified by the triode, passes through plate coil L_2 and provides positive feedback to the original signal voltage in oscillator L_oC_o. This increases the total voltage to be detected and amplified by the triode. The amount of feedback can be controlled by moving coil L_2 or by placing a variable resistor across L_2 which allows part of the signal to bypass L_2. Control of feedback is important in the detector, because excessive positive feedback causes oscillations in L_oC_o, which produce howls and whistles in the earphones. In addition, the coupling between L_o and L_1 is such that the energy of these oscillations in L_oC_o can be transmitted by the receiving antenna, producing these same howls and whistles in other receivers. To avoid this, the detector is tuned to the desired station and

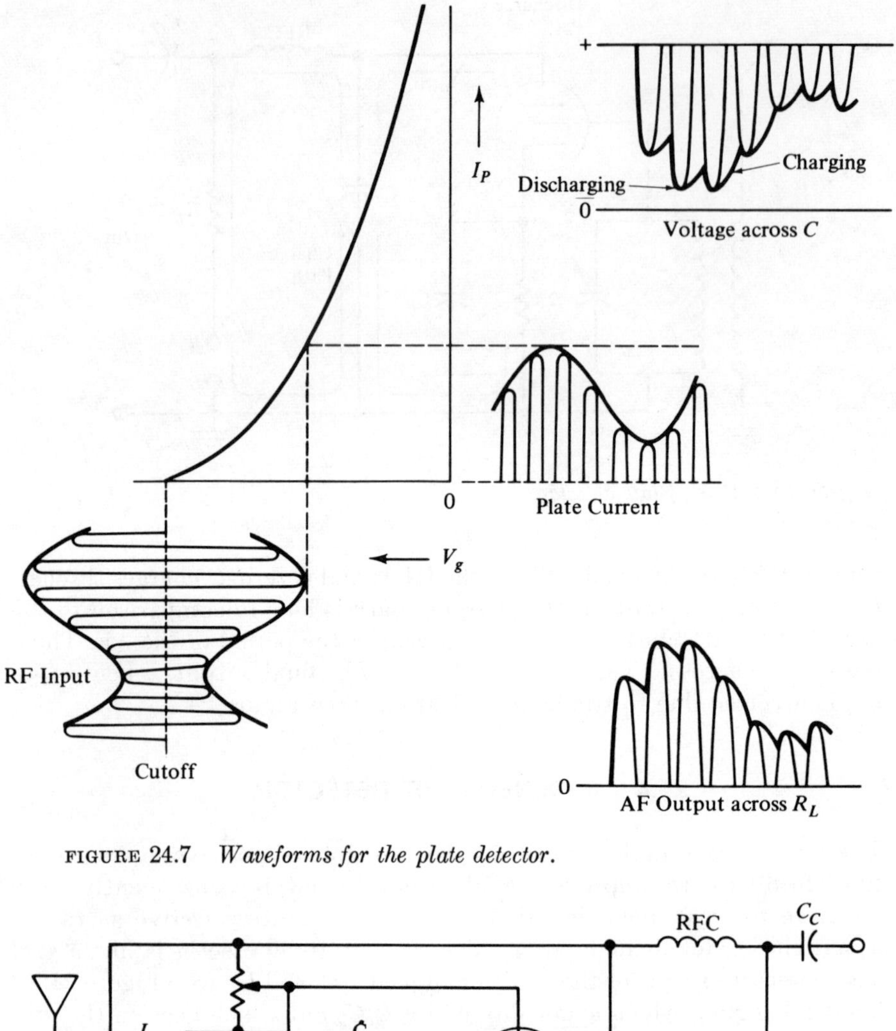

FIGURE 24.7 *Waveforms for the plate detector.*

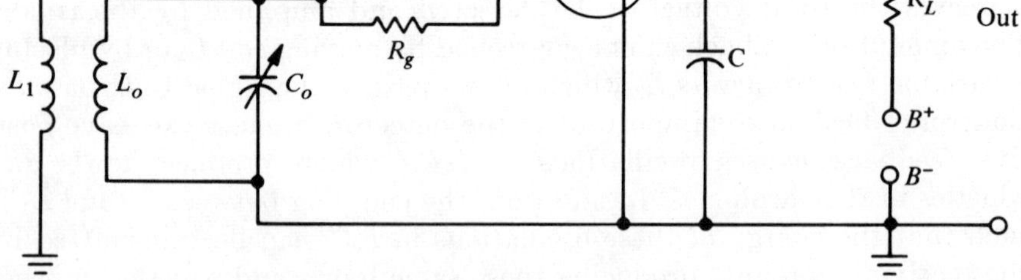

FIGURE 24.8 *Regenerative detector.*

then the regeneration level is adjusted to the point just before the howls and whistles begin.

24.5 FM SLOPE DETECTOR

All of the detectors thus far discussed have demodulated AM input signals. One method of demodulating FM waves is with a slope detector. This circuit selects the incoming RF, converts the frequency modulated signal to a frequency and amplitude modulated signal, and then rectifies and filters the AM to obtain an AF output. (See Figure 24.9.)

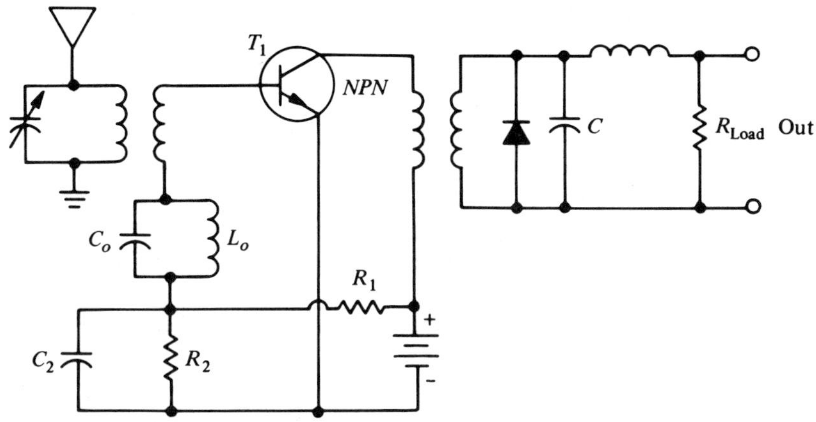

FIGURE 24.9 *FM slope detector.*

The resonant circuit L_oC_o, which is tuned above the carrier frequency, produces a large voltage output when the frequency of the incoming signal from the antenna is near resonance for L_oC_o. When the incoming signal frequency decreases, the voltage across resonant circuit L_oC_o also decreases. This varying voltage across L_oC_o is applied to the emitter-base forward bias, which is provided by voltage divider R_1 and R_2. (R_2 is the base resistor, and C_2 is its bypass capacitor.) The base-emitter bias has the voltage from L_oC_o added to it, and hence the forward bias voltage changes at the RF rate. This causes changes in the gain of the *NPN* transistor. Therefore, the RF output of the collector varies not only in frequency, but also in amplitude. This output signal is transformer coupled to a crystal detector where it is rectified and demodulated. The time constant of R_LC is long in comparison to the period of the RF, but short in comparison to the period of the AF. Thus, the slow discharge of C across R_L produces an audio output, which is quite similar to the original AF input to the FM transmitter. (See Figure 24.10.)

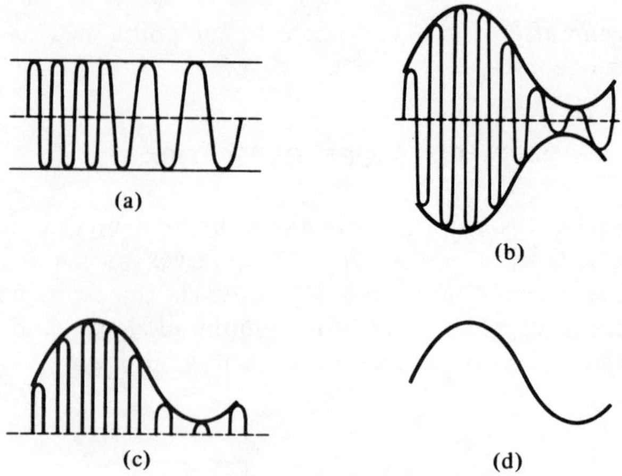

FIGURE 24.10 *Waveforms for the FM slope detector. (a) RF input (frequency nodulated); (b) output of T_1 (frequency and amplitude modulated); (c) rectification by crystal detector; (d) AF output across R_{Load}.*

24.6 FM LIMITER AND DISCRIMINATOR

The FM discriminator circuit, which is another method of demodulating FM signals, requires an FM input which has frequency variations only, with no variations in amplitude. If some amplitude variations have developed in the FM signal between transmission and reception, they must be eliminated by means of a limiter circuit.

The triode limiter uses a variable grid leak bias, which automatically adjusts to any changes in amplitude of the FM signal, and thus provides a constant amplitude for the FM output. (See Figure 24.11.) The grid leak bias is provided by the signal's driving the grid positive for a short time, allowing grid current to flow, which charges C_g and establishes the bias. When the amplitude of the signal increases, it drives the grid more positive, causing more grid current to flow and making the bias more negative. Each time the grid goes positive, the plate current reaches saturation (maximum value), and thus the amplitude of the FM output is held at this constant level. The plate LC circuit provides a continuous output and couples the FM signal to the discriminator. Now the job of the discriminator is to change the RF FM signal to AF. (See Figure 24.12.)

The FM discriminator contains identical diodes (D_1 and D_2), equal resistors ($R_1 = R_2$), equal capacitors ($C_1 = C_2$). The discriminator is sensitive to frequency changes in the FM signal with respect to the RF carrier. When the frequency of the input signal is greater than the carrier frequency, the AF output of the discriminator is positive; and when the

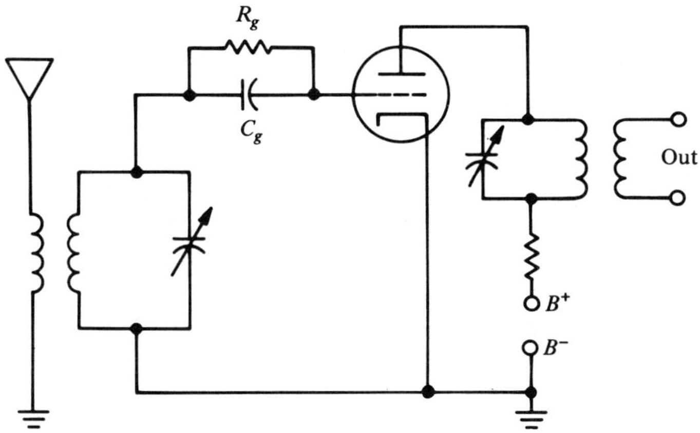

FIGURE 24.11 *Grid leak bias limiter.*

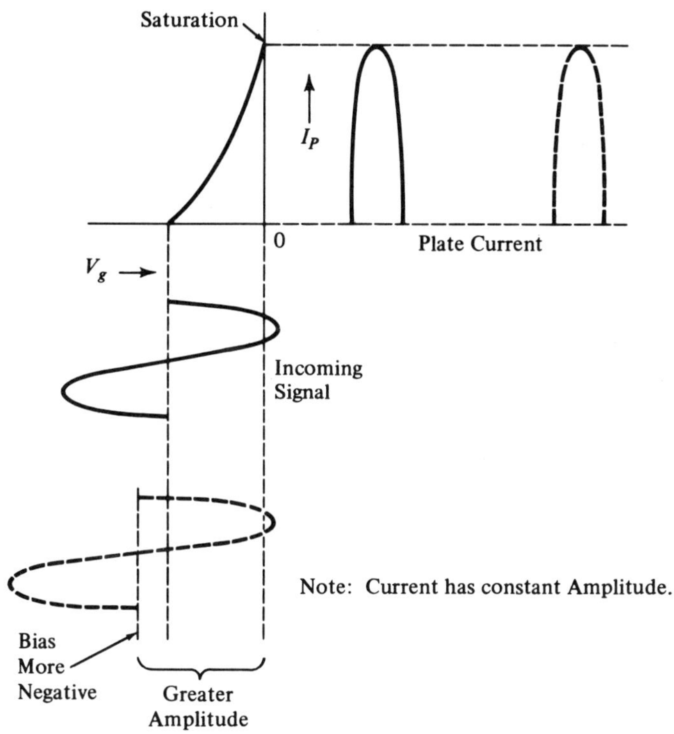

FIGURE 24.12 *Waveforms for the grid leak bias limiter.*

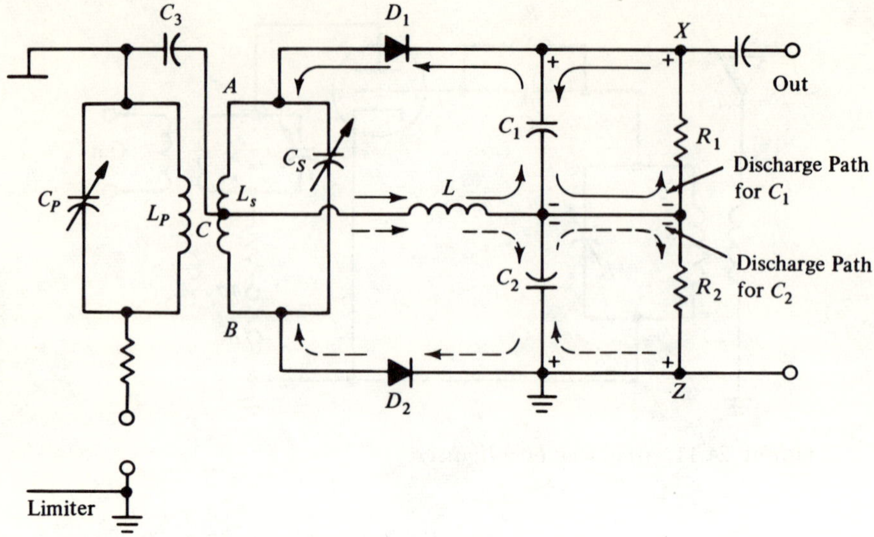

FIGURE 24.13 *Discriminator—FM demodulator.*

input signal frequency is less than the carrier frequency, the AF output is negative. (See Figure 24.13.)

The LC circuits L_PC_P (the primary tank) and L_SC_S (the secondary tank) are both tuned to the carrier frequency, which is sometimes called the rest frequency. First, we assume that the incoming signal is at the carrier frequency. This input signal is coupled to the discriminator in two ways: through transformer action in the tuned circuits and through coupling capacitor C_3, which is connected to a center tap on L_S. There is a phase relation between these two voltages applied to the discriminator. To discover this relation, we must recall two ideas. First, the current and voltage of the parallel LC circuit at resonance are out of phase; second, the current in a capacitor leads the voltage by 90° (*ICE*). Since the current in L_PC_P is in phase with the current in C_3 and the voltage across L_PC_P is the same as that across L_SC_S, we have the phase relations shown in Figure 24.14. The important observation to make here is that the voltage coupled through C_3 and the voltage coupled through the tank circuits are 90° out of phase at resonance (at the carrier frequency).

The voltage coupled by C_3 is applied to both diodes D_1 and D_2. But the transformer coupled voltage is divided by the center tap on L_S, half is applied to D_1, half to D_2. Since A is positive when B is negative, the voltage from A to C is just the opposite of the voltage from B to C, that is, they are out of phase. Each, in turn, is 90° out of phase with the voltage coupled by C_3. The total voltage applied to each diode is shown in Figure 24.15. Notice that these two total voltages are of the same magnitude.

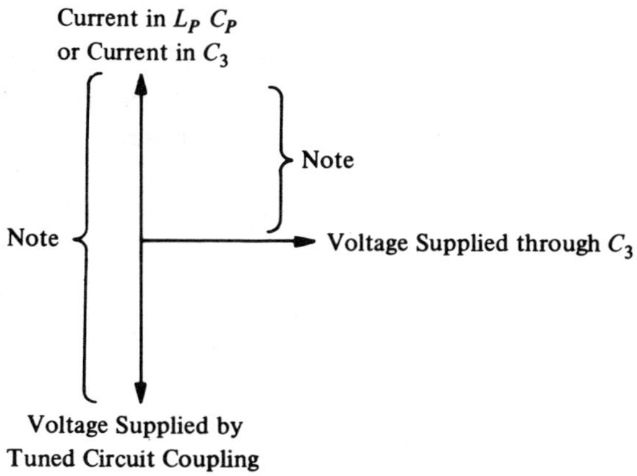

FIGURE 24.14 *Input at carrier frequency. (Note: Current in C_3 leads voltage by 90°; current and voltage in LC tank circuit are out of phase.)*

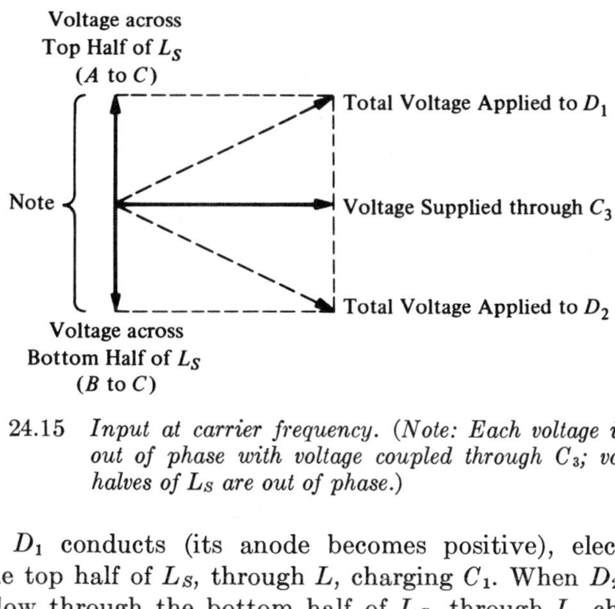

FIGURE 24.15 *Input at carrier frequency. (Note: Each voltage in L_S is 90° out of phase with voltage coupled through C_3; voltage across halves of L_S are out of phase.)*

When D_1 conducts (its anode becomes positive), electrons flow through the top half of L_S, through L, charging C_1. When D_2 conducts, electrons flow through the bottom half of L_S, through L, charging C_2. Because the total applied voltages are equal and the charging circuits are identical, the capacitors charge to the same level. Capacitors C_1 and C_2 discharge through R_1 and R_2, respectively, when the diode in its charging path is not conducting. They charge at the RF rate, but the time constant of the discharge paths is long with respect to the RF period and short with respect to the AF period. Thus, the voltages felt across R_1 and R_2 are equal in magnitude (due to the slow discharge), but opposite in direction (X is positive and Z is positive). They cancel, making the output voltage zero when the input is at the carrier frequency.

Now suppose the input signal frequency goes above the carrier frequency. At the higher frequency, the characteristic of the parallel circuit $L_P C_P$ is capacitive, so the phase angle between the transformer coupled voltage and the voltage coupled through C_3 becomes less than 90°. This affects the total voltage applied to each diode as shown in Figure 24.16. We see that the total voltage applied to D_1 is now greater

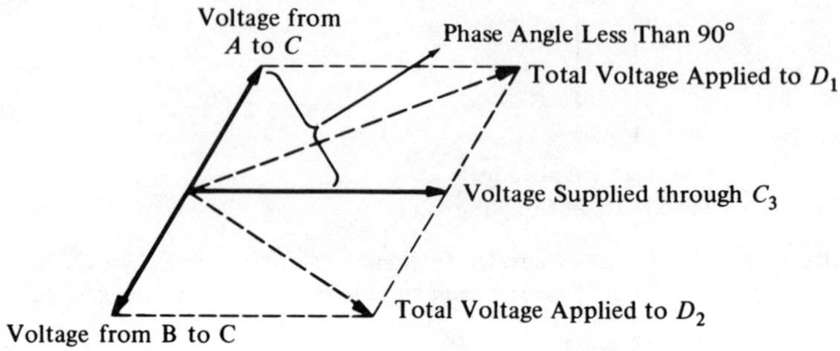

FIGURE 24.16 *Input frequency above carrier frequency.*

than the total voltage applied to D_2. C_1 is thus charged to a higher level than C_2. Point X is more positive than point Z, because D_1 conducted more electrons than D_2. Subtracting the voltage across YZ (across R_2) from the larger voltage across XY (across R_1), the final AF output is positive with respect to ground.

Similarly, when the input frequency drops below carrier frequency, C_2 will be charged to a higher level than C_1. (See Figure 24.17.) Point Z

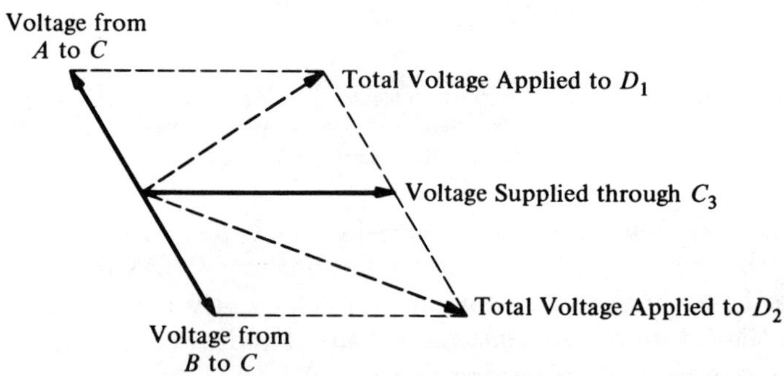

FIGURE 24.17 *Input frequency below carrier frequency.*

will be more positive than point X; in other words, X is more negative than Z, which is at ground potential. Subtracting the voltage across XY from the larger voltage across YZ, we have the AF output voltage, which is negative since X is negative with respect to ground. (See Figure 24.18.)

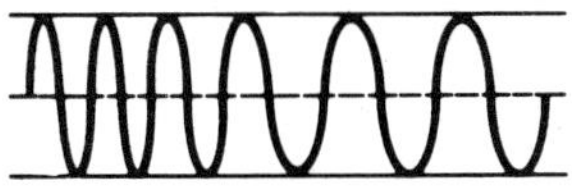

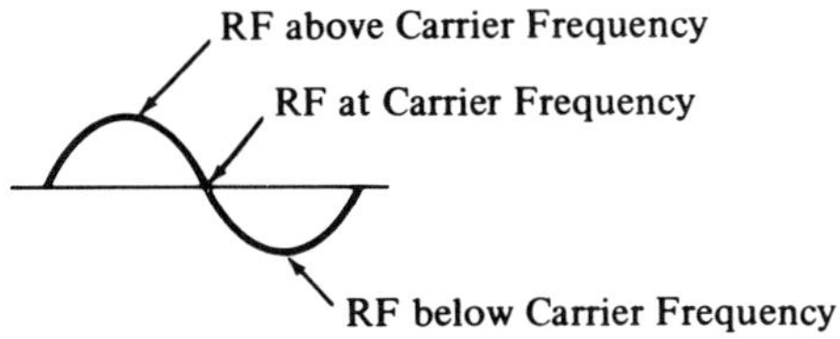

FIGURE 24.18 *(a) FM input; (b) AF output (across R_1 and R_2).*

24.7 RATIO DETECTOR

A variation on the FM discriminator, which does not require a limiter stage to precede demodulation, is the ratio detector. (See Figure 24.19.)

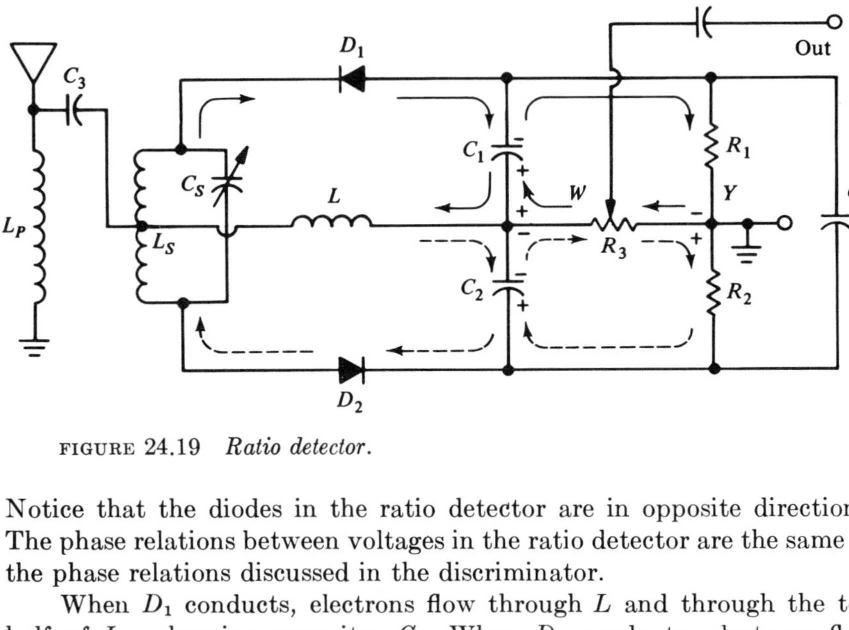

FIGURE 24.19 *Ratio detector.*

Notice that the diodes in the ratio detector are in opposite directions. The phase relations between voltages in the ratio detector are the same as the phase relations discussed in the discriminator.

When D_1 conducts, electrons flow through L and through the top half of L_S, charging capacitor C_1. When D_2 conducts, electrons flow through the bottom half of L_S and through L, charging capacitor C_2. At the carrier frequency, the voltages supplied to D_1 and D_2 are equal in magnitude and the charging paths are identical, so C_1 and C_2 charge to the same level. C_1 and C_2 both discharge through R_3 at an AF rate (C_1 discharges through R_1 and R_3; C_2 through R_2 and R_3). The voltages dropped across R_3 are equal in magnitude but opposite in direction. Thus they cancel, and the total voltage dropped across R_3, which is the output voltage, is zero at the carrier frequency.

When the frequency of the input goes above the carrier frequency, the voltage applied to D_1 increases and that applied to D_2 decreases, just as in the discriminator. Thus, C_1 charges to a higher level than C_2, and point Y is more negative than point W, or W is more positive than Y, which is at ground potential. The output AF voltage, taken across R_3 between W and Y, is positive when the input frequency is above rest frequency.

Similarly, when the frequency of the input goes below the carrier frequency, the voltage applied to D_2 increases and that applied to D_1 decreases. Capacitor C_2 charges to a higher level than C_1, and point Y on R_3 will thus be more positive than point W, or W is negative with respect to Y (ground potential). Thus, the AF output is negative when the input frequency is below carrier frequency.

The purpose of capacitor C_4 is to prevent any variations in amplitude of the FM signal from reaching output resistor R_3. (See Figure 24.20.) C_4 has a large capacitance and thus low capacitive reactance at high fre-

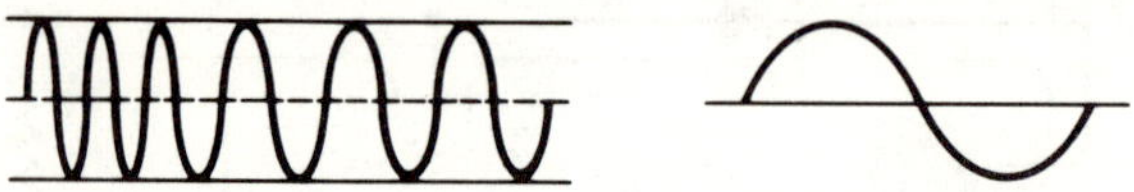

(a) (b)

FIGURE 24.20 *(a) FM input to ratio detector; (b) AF output across* R_3.

quencies. Suppose the amplitude of the FM signal suddenly increases, charging capacitors C_1 and C_2 to a higher level. During the discharge of C_1 and C_2, a part of this additional charge will go on C_4, and the remainder goes through R_3, thus preventing the amplitude of the voltage dropped across R_3 from varying due to variations in the amplitude of the FM input. C_4 then takes the place of a limiter.

Questions

(1) In what ways does the diode detector differ from the half-wave rectifier?

(2) Describe the process of FM detection.

Suggested Experiments

(1) Detector–Receiver. You may want to build a simple AM diode detector–receiver or even a more elaborate receiver. The pleasure of receiving the first signal from a receiver you have built is not soon forgotten.

CHAPTER 25

Receivers

25.1 PULSE RECEIVERS

The problem of amplifying the signal enough so that it may be heard by means other than with a headphone is solved by the receiver. We will consider pulse duration receivers, AM receivers, and FM receivers. Since we have already discussed circuits which demodulate AM and FM, these receivers will essentially just apply amplifiers to strengthen the signal. The pulse modulated transmitter, you recall, required the use of Morse Code to send intelligible information, because the output of this transmitter was simply an on-and-off RF signal. Receiving and amplifying this signal can, of course, be accomplished by an antenna coupled to an RF amplifier. The problem is that RF is beyond the limits of the human ear's audible range, so an AF must be obtained from the RF.

The FM demodulator we discussed earlier compared signal frequencies to the carrier frequency in order to obtain an AF output. This same principle can be applied to the pulse modulated signal. An oscillator, which is in the receiver for pulse modulated signals, generates a signal whose frequency is slightly larger than the frequency of the RF pulse input to the receiver. When the two signals are combined, we can obtain the sum or the difference of the two frequencies. If the difference falls in the audible range, then we have a new output pulse signal which can be heard.

This principle of combining two different frequencies and detecting their differences is often used to tune musical instruments. When two notes of nearly the same frequency are played simultaneously, a beat frequency, which is actually the difference of the two frequencies, is heard.

As the two notes are tuned so their frequencies are nearly the same (that is, the musicians are playing the same note or tone), the number of beats heard each second, or the beat frequency, decreases to zero. The farther apart the two tones, the greater is the beat frequency. For instance, if you play G (392 Hz) and A (440 Hz), the difference, or beat frequency, is 48 Hz, which is still in the audible range.

Similarly, if the pulse transmitter radiates an RF signal of 650 kHz and the beat frequency oscillator in the receiver generates an RF signal of 651 kHz, the difference is 1 kHz, which is within the audible range.

The block diagram in Figure 25.1 illustrates the relationship of circuits making up a pulse receiver. The input to the RF amplifier is tuned

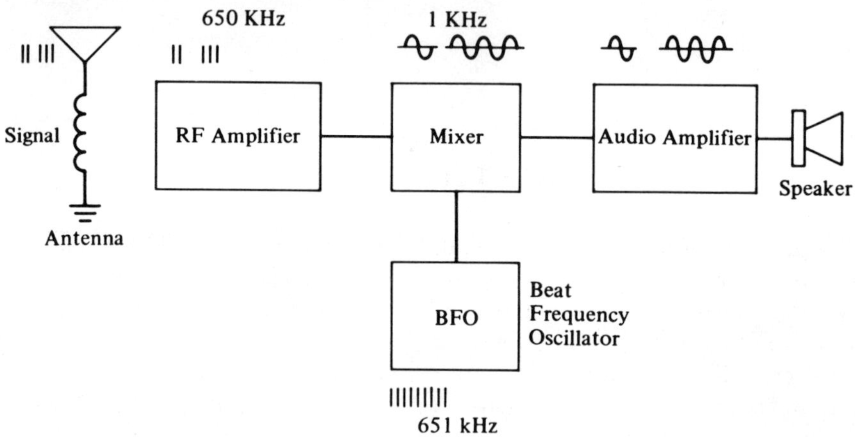

FIGURE 25.1 *Pulse detector.*

to the transmitting frequency, in this case 650 kHz. The beat frequency oscillator is tuned to 651 kHz and the mixer to 1 kHz. The procedure of combining, or mixing, two signals of similar frequencies and using the difference is called heterodyning. A crystal mixer with its input and output waveforms is shown in Figure 25.2.

25.2 TRF RECEIVERS

A receiver for AM signals might consist of an RF amplifier, a detector to change RF to AF, and an audio amplifier as shown in the block diagram of a tuned radio frequency (TRF) receiver. (See Figure 25.3.) This receiver may use any of the detectors already mentioned: plate detector, regenerative detector, and so on. It is important that the RF amplifier

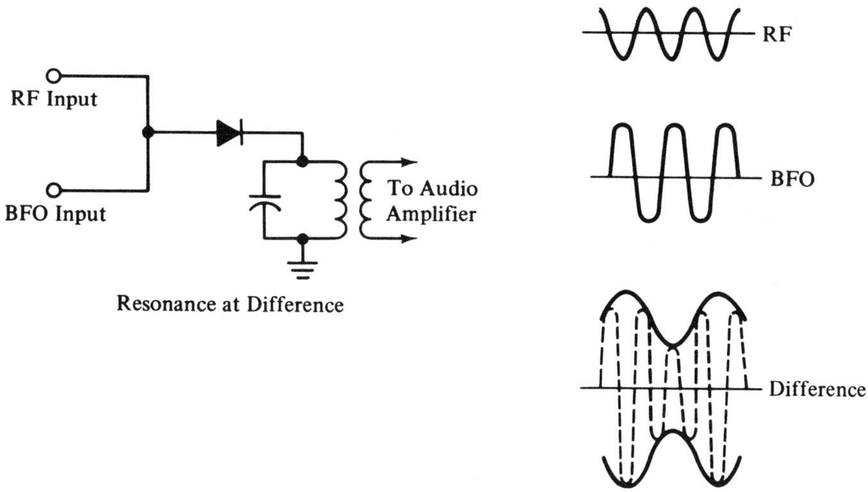

FIGURE 25.2 *Crystal mixer.*

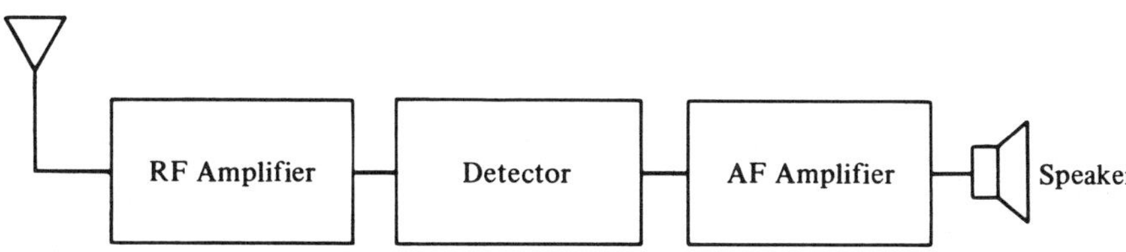

FIGURE 25.3 *TRF (tuned radio frequency) AM receiver.*

and the detector both be carefully tuned to the frequency of the transmitted RF signal. This may be done by connecting the two tuning capacitors together, using the same rotor for both. If changing from one incoming station to another is accomplished by a very small rotation of the tuning capacitor, fine tuning for a single station becomes impossible. This can be remedied by placing a small capacitor in parallel with the tuning capacitor on the RF input which, in effect, spreads the bandwidth and thus facilitates fine tuning. (See Figure 25.4.)

25.3 AM SUPERHETERODYNE RECEIVER

We can apply the heterodyne principle to AM receivers to facilitate amplification of RF signals before detection and subsequent AF amplification. This type of AM receiver, which is widely used today, is called a superheterodyne receiver. The block diagram in Figure 25.5 for the super-

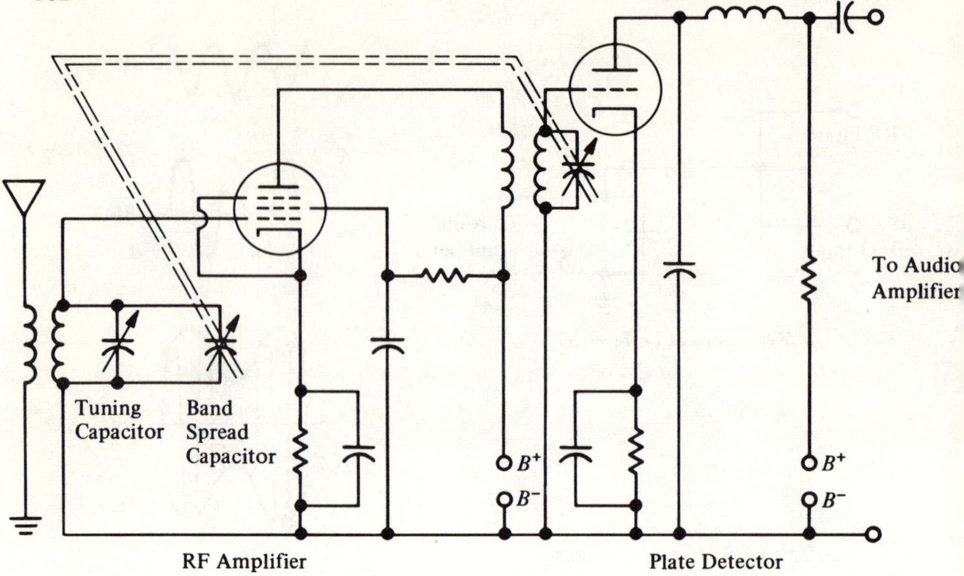

FIGURE 25.4

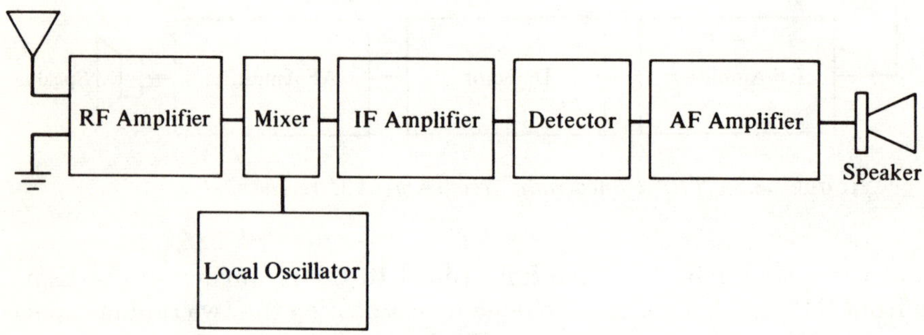

FIGURE 25.5 *Superheterodyne AM receiver.*

heterodyne receiver introduces a new circuit, the IF (intermediate frequency) amplifier.

The IF (intermediate frequency) amplifier amplifies the AM beat frequency signal produced by the mixer. This signal is, of course, the difference between the RF signal and the local oscillator frequency. Since this intermediate beat frequency will be much lower than the RF signal frequency, it is easier to amplify. Thus, the mixer and IF amplifier replace several stages of RF amplification, which would be required to increase signal strength in the TRF receiver. Remember that since the RF amplifier has a very narrow frequency response, each of the RF amplifier stages in a TRF receiver must be tuned to exactly the same frequency. This is difficult to do even when a common shaft, or rotor, is used for all the tuning capacitors. With the superheterodyne receiver, the IF frequency is always the same, 455 kHz. Therefore, the coupled stages of IF amplifiers do not require readjustment for each new broadcast frequency. Tuned circuit coupling, consisting of two parallel LC circuits as primary and secondary, is normally used for the IF stages of the superheterodyne receiver.

The implication of a constant IF frequency of 455 kHz is that the local oscillator frequency must change with each change in broadcast frequency in such a way that their difference always remains 455 kHz. In this receiver, the local oscillator frequency is kept 455 kHz above the broadcast frequency. For example, if you want to listen to a station which broadcasts at a frequency of 840 kHz, then the local oscillator in your superheterodyne receiver will produce a frequency of 1.295 MHz. The difference, which is produced in the mixer, in the broadcast RF and the local oscillator frequency is 1,295,000 − 840,000 = 455,000 Hz.

Tuning the local oscillator to keep it exactly 455,000 Hz above the broadcast RF signal over the entire AM bandwidth is called tracking. Tracking is accomplished by connecting the variable capacitor in the RF tank circuit to the variable capacitor in the local oscillator's tank circuit. When capacitors are connected by a common rotor, they are said to be ganged together. The connection must keep the capacitance of the local oscillator circuit smaller than that of the RF tank, in order for it to have a higher frequency output. $f\uparrow = 1/2\pi\sqrt{LC\downarrow}$. Since capacitance is directly proportional to area, the capacitance of the capacitor in the local oscillator can be kept smaller by making its plates smaller. $C\downarrow = A\downarrow/d$. (See Figure 25.6.) Another technique, used to provide the frequency difference

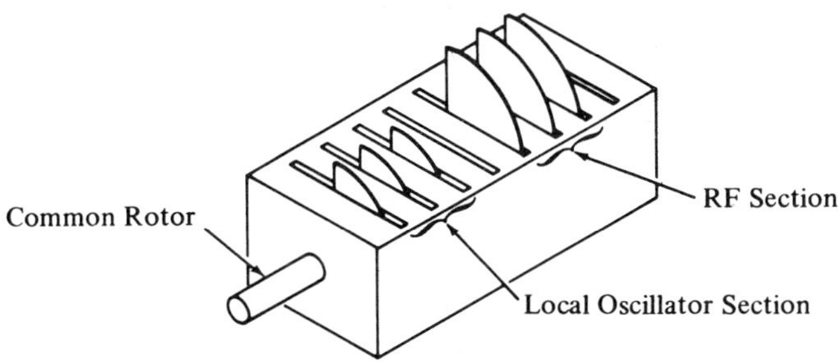

FIGURE 25.6 *Ganged capacitors of different size.*

between the RF input and the local oscillator, is to make the sections of the ganged capacitors the same size, but to place a small capacitor in series with the local oscillator tank circuit. Remember that the total capacitance in a series circuit is smaller than the largest capacitance. Therefore, this small capacitor, sometimes called a padder, decreases the capacitance of the local oscillator section of the ganged capacitor. (See Figure 25.7.)

FIGURE 25.7 *Local oscillator tank circuit with series capacitance.*

25.4 PENTODE MIXER

A pentode mixer is often used in AM receivers. The function of the pentode mixer, just like the crystal mixer already discussed, is to combine the RF signal (840 kHz in the above example) with the signal from the local oscillator (1.295 MHz) to obtain an IF signal (455 kHz), which is then fed to an IF amplifier. (See Figure 25.8.) The RF tuning capacitor is ganged with the capacitor in the Hartley local oscillator. The RF signal is placed on the control grid, and the local oscillator signal is placed on

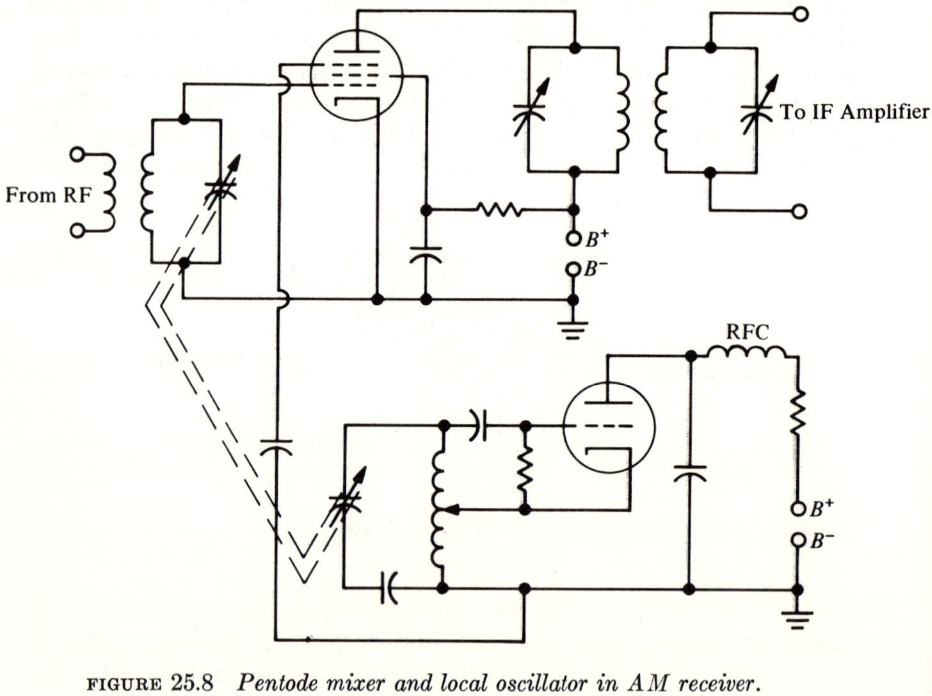

FIGURE 25.8 *Pentode mixer and local oscillator in AM receiver.*

the suppressor grid. Both signals vary the cathode-to-plate current. Thus, the output of the pentode mixer combines the two signals, and its frequency is the difference of their frequencies. The plate tank coil is tuned to the difference of the two signal frequencies, that is, to 455 kHz, and this IF signal is tuned circuit coupled to an IF amplifier.

25.5 PENTAGRID CONVERTER

Many superheterodyne receivers combine the mixer and the local oscillator in one tube. The tube has five grids, and the mixer circuit is called a pentagrid converter. Elimination of a separate tube for the local oscillator saves space and money in the manufacture of the receiver. The action of the pentagrid converter is much the same as that of the pentode mixer and separate local oscillator just studied. The circuit shows the same ganging of RF and local oscillator capacitors with the IF tank circuit tuned to the difference of the two frequencies. (See Figure 25.9.)

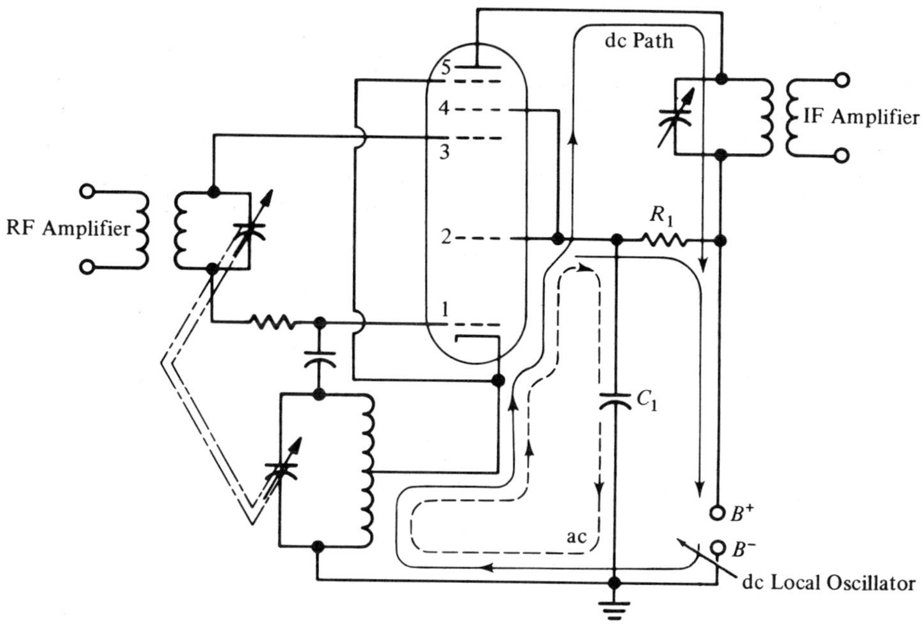

FIGURE 25.9 *Pentagrid converter.*

The signal from the local oscillator is placed on grid 1; grid 2 serves as its plate. Bypass capacitor C_1 provides a path for feedback to the bottom of the local oscillator. The RF signal is placed on grid 3. Grid 4 is kept less positive than the plate by R_1 and serves to maintain the cathode-to-plate current. Capacitor C_1 also serves as a bypass capacitor for grid 4. Grid 5, which is connected to the cathode, reduces the speed of electrons hitting the plate and returns secondarily emitted electrons to the plate to prevent a space charge. The IF oscillator also serves as the plate load.

25.6 AUTOMATIC VOLUME CONTROL

Changes in ionospheric conditions, the occasional presence of sun spots, moving objects between transmitter and receiver, and many other factors can change the strength of the incoming signal on the antenna of the receiver. These variations would then be amplified, and the output of the receiver would get louder or fade, which would require constant adjustment of the volume control. To eliminate this continual adjustment of the volume, necessitated by external conditions, we introduce automatic volume control (AVC). The manual control to adjust to desired listening level will not be eliminated. The AVC is simply a feedback method of maintaining the desired listening level.

The key to controlling gain in any amplifier is the bias level of the control grid. If the grid is quite negative, the incoming signal will not increase plate current as much as if the grid were less negative. AVC controls gain by varying grid bias so that when an unusually strong signal is received by the antenna, the bias is made more negative. (See Figure 25.10.) The output of the diode detector is a rectified AF voltage taken

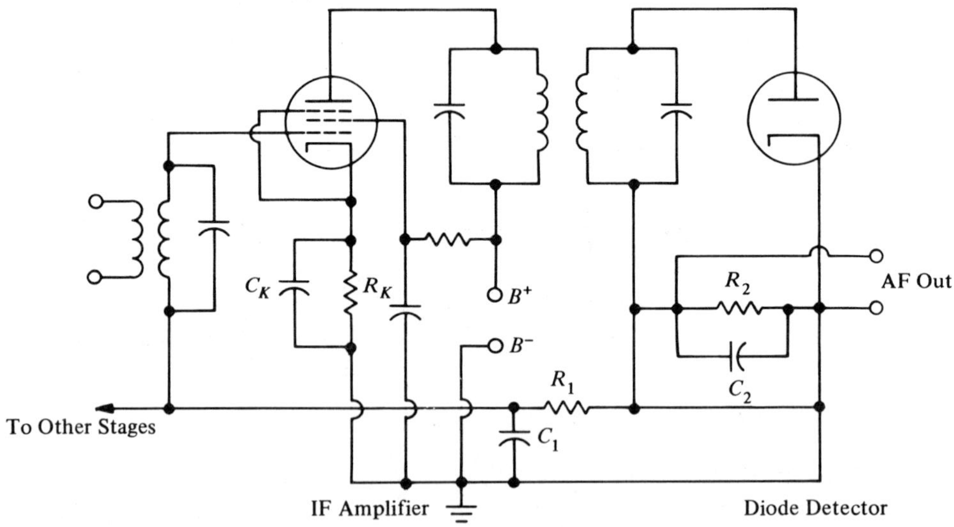

FIGURE 25.10 *AVC (automatic volume control).*

across R_2C_2. Since the diode only conducts when its cathode is negative, the voltage drop across R_2 is negative. This negative voltage is placed across R_1C_1 and filtered further to produce a negative dc voltage which is applied to the control grids of the tubes in the various stages of the IF amplifier. Thus, when the signal strength increases, the voltage across R_2 becomes more negative. This causes a more negative bias to be placed on the grids of the preceding stages of amplification and thus reduces gain.

25.7 FM RECEIVER

The block diagram of Figure 25.11 for the FM receiver is similar to the AM superheterodyne receiver. We have discussed the action of all these

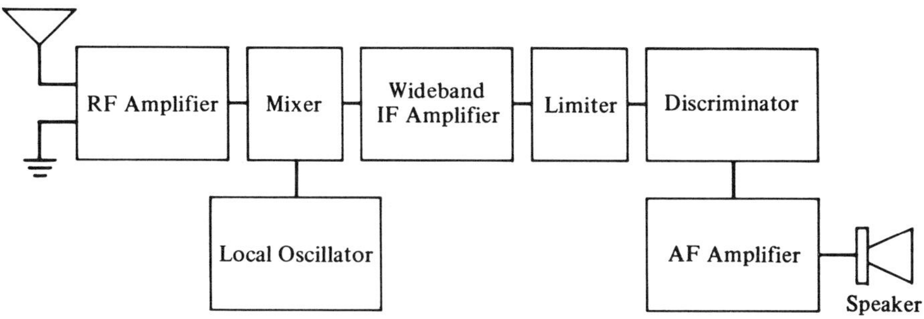

FIGURE 25.11 *FM superheterodyne receiver.*

circuits including the limiter and discriminator, which, of course, differ from the AM detector stage. The local oscillator is ganged to the tuning capacitor in the RF amplifier as before, and it generates a signal whose frequency is above that of the RF from the antenna. The mixer essentially subtracts the smaller antenna frequency from the oscillator frequency, and the result is applied to the IF amplifier. Since FM has a varying input frequency, this difference, supplied by the mixer, will also vary over the FM bandwidth. Thus, the IF amplifier must respond to frequencies over a range of 200 kHz. The frequency range for the IF amplifier in a FM receiver is from 10.6 MHz to 10.8 MHz. In order to obtain this range from the mixer, the local oscillator frequency must be 10.7 MHz (the middle frequency of this range) above the carrier frequency of the FM. For example, if a FM station is broadcasting at 98.0 MHz, the local oscillator must generate a frequency of 98.0 MHz + 10.7 MHz = 108.7 MHz.

Since the bandwidth of FM signals is so broad (200 kHz in comparison to 10-kHz bandwidth of an AM station), the IF amplifier, in order to cover such a wide band of frequencies, will have a rather small gain at each stage. (See Figure 25.12.) An alternative is to tune each stage to a small segment of the 200,000-Hz bandwidth, thus allowing greater gain at each stage. (See Figure 25.13.)

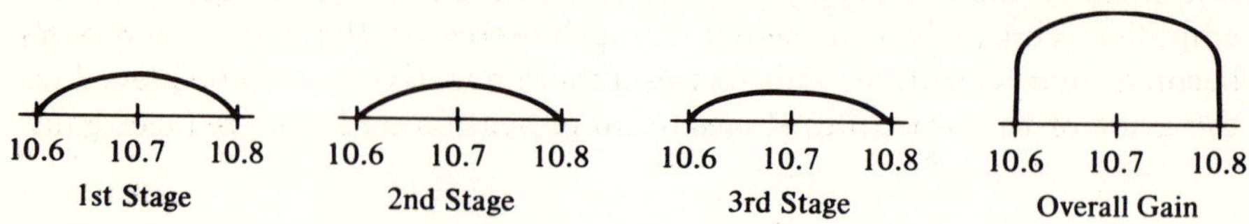

FIGURE 25.12 *Gain in various stages and overall gain of an IF wide band amplifier.*

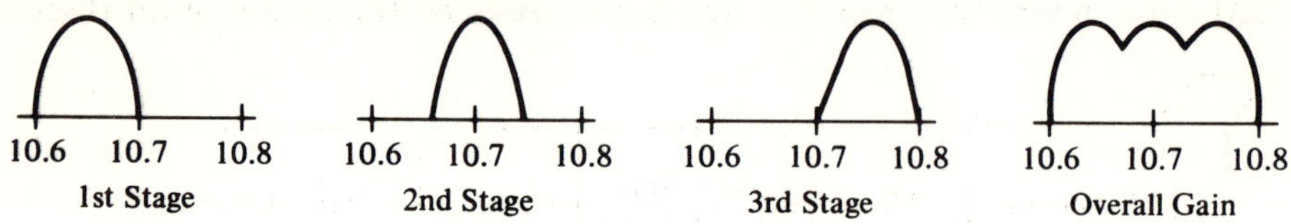

FIGURE 25.13 *Gain in various stages and overall gain in IF wide band amplifier.*

Questions

(1) The AM broadcast band is from 550 kHz to 1440 kHz. What is the band of frequencies of the local oscillator in an AM radio?

(2) Since the quality of FM reception is superior to AM reception in that FM is more interference free than AM, and FM has a wider range of audible response than AM, why are there more AM stations than FM in the United States?

CHAPTER 26

Multivibrators

26.1 ONE-SHOT MULTIVIBRATOR

The ac generator and the various oscillator circuits all produced a sinusoidal output. The amplitude of the positive going signal was equal to the amplitude of the negative going signal, and the wavelength for each half cycle remained constant. (See Figure 26.1.) We now want to consider a

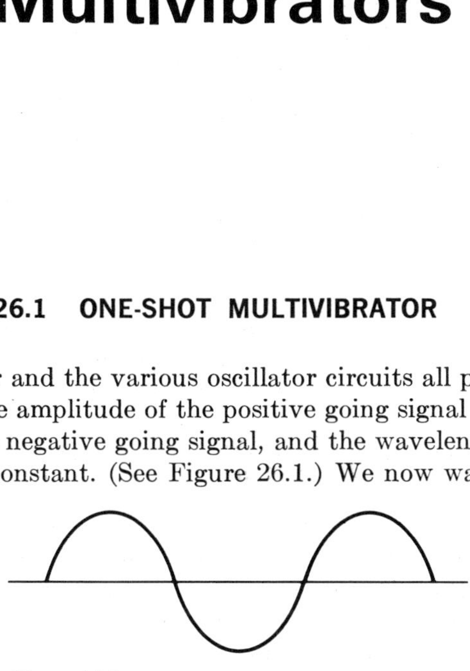

FIGURE 26.1 *Sinusoidal wave.*

few oscillators which do not produce such a uniform sinusoidal output. These oscillators are generally referred to as relaxation oscillators. Their outputs may be rectangular, square, or triangular. (See Figure 26.2.)

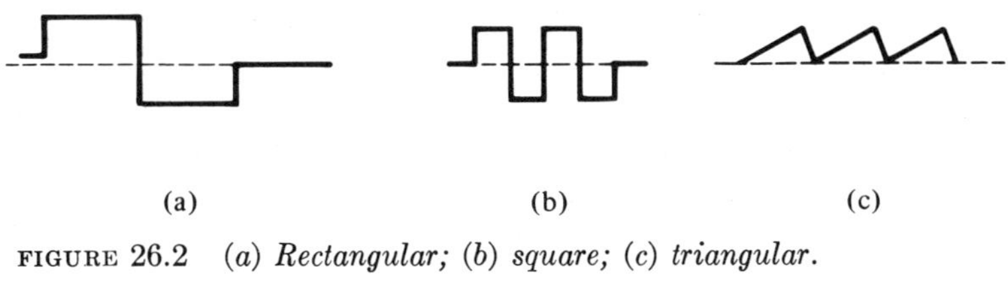

FIGURE 26.2 (*a*) *Rectangular;* (*b*) *square;* (*c*) *triangular.*

The first type of relaxation oscillator to be discussed is the multivibrator. The one-shot multivibrator requires a trigger voltage to operate for one cycle. Then, for the circuit to produce another cycle, the trigger voltage must be applied again. In the circuit in Figure 26.3, which

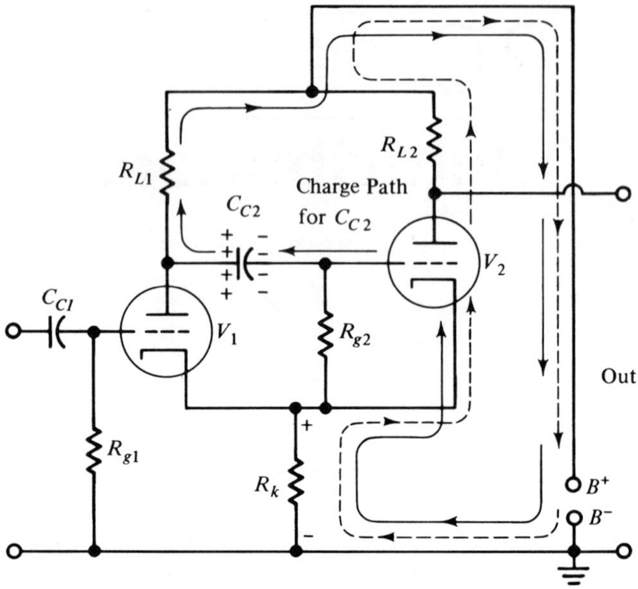

FIGURE 26.3 *One-shot multivibrator (V_2 conducting).*

resembles a two-stage RC coupled amplifier, either tube V_1 or tube V_2 is conducting. Notice that the grid of V_2 is connected to the cathode of V_2 via grid resistor R_{g2}. Since no current flows through R_{g2}, the grid of V_2 is at the same potential as the cathode, that is, it has zero bias. With the grid of V_2 at zero bias, the tube will conduct. Current passes through R_{L2}, the power supply, and cathode resistor R_K. When plate current flows in V_2, the plate voltage of V_2 decreases. The voltage drop across R_K, when V_2 is conducting, is sufficient to bias V_1 to cutoff, because the grid of V_1 is at ground potential (the bottom of R_K). Since no current flows in V_1, the plate voltage of V_1 is at near B^+ potential. The grid of V_2 is at the same potential as the cathode of V_2 (slightly positive). Thus, grid current flows, and capacitor C_{C2} becomes charged with the side of C_{C2} connected to the plate of V_1 more positive than the side connected to the grid of V_2. The potential across C_{C2} will be the difference between the voltage on the plate of V_1 and the voltage on the grid of V_2. Both voltages are positive, but the plate of V_1 is more positive than the grid of V_2. Load resistors R_{L1} and R_{L2} are equal.

Now we will trigger the one-shot multivibrator with a positive signal placed on the grid of V_1. The positive trigger must be sufficiently large to cause V_1 to conduct. When V_1 begins conducting, plate current flows, and the plate voltage of V_1 decreases. (See Figure 26.4.) With the

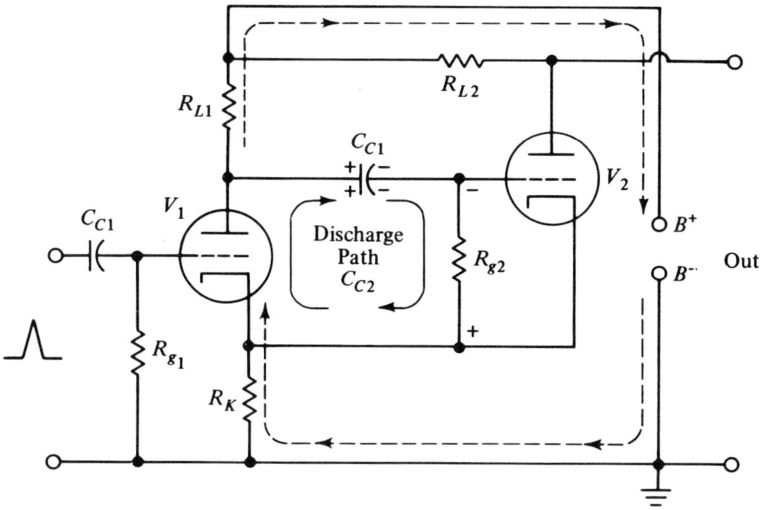

FIGURE 26.4 *One-shot multivibrator* (V_1 *conducting*).

plate of V_1 now below B^+ potential, C_{C2} begins to discharge through R_{g2} and V_1, which is conducting. This discharge current produces a voltage drop across R_{g2} and hence supplies a negative grid bias to V_2. With the grid of V_2 now negative, the plate current in V_2 decreases rapidly, causing a decrease in the voltage drop across R_K. This decreased voltage drop across R_K, which occurs when V_1 begins conducting, makes the cathode of V_1 less positive with respect to the grid and allows the tube to continue to conduct even after the trigger voltage has passed. With C_{C2} discharging, V_2 reaches cutoff and stops conducting; when the plate current in V_2 stops, the plate voltage will be near B^+ potential. (See Figures 26.5 and 26.6.)

Notice that the time constant for the discharge of C_{C2} through R_{g2} determines how long V_1 will conduct. For, as C_{C2} discharges, the voltage drop across R_{g2}, which supplies the bias to V_2, becomes less negative. When C_{C2} is discharged, the bias is above cutoff, and tube V_2 begins to conduct.

When plate current begins to flow in V_2, the plate voltage of V_2 decreases to the level it maintained before the trigger voltage, and the voltage drop across cathode resistor R_K increases. With the cathode of V_1 thus made more positive, the grid of V_1 appears more negative with respect to the cathode. In fact, cutoff is reached, and V_1 ceases to conduct. As the plate voltage of V_1 increases toward B^+ potential, C_{C2} begins to

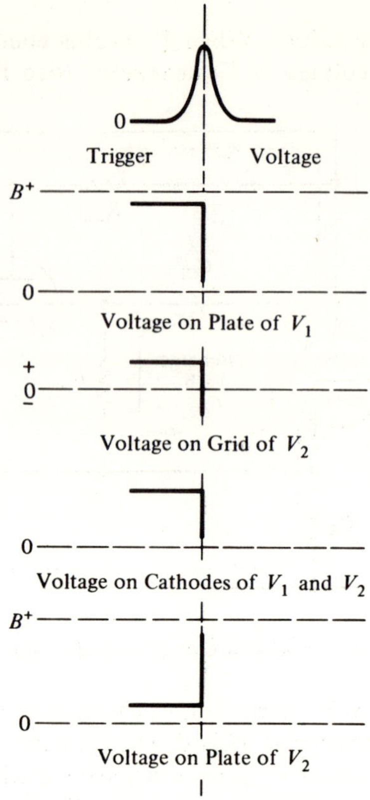

FIGURE 26.5 *Changes that occur when trigger voltage is applied to grid of* V_1.

charge, its left plate positive and its right plate negative. For the process to repeat itself, another trigger voltage must be applied to the grid of V_1. The charging of C_{C2} through R_{L1}, R_K, and V_2 is much faster than the discharging due to the smaller resistance of the charging path. (See Figure 26.7.)

A transistor one-shot multivibrator operates much like the vacuum tube one-shot multivibrator just described. It too requires a trigger voltage to cycle the circuit. (See Figure 26.8.) Before any signal is introduced to transistor T_1, the base emitter of T_2 is forward biased and T_2 is conducting, charging C_{C2}. When a positive trigger voltage is applied to T_1, the voltage drop across R_{B1} produces a forward base-emitter bias, and T_1 begins to conduct. When T_1 conducts, its collector voltage decreases, allowing C_{C2} to discharge through R. The flow of current through R produces a reverse bias voltage on the base emitter of T_2, and transistor T_2 is cut off. As C_{C2} continues to discharge, this reverse base-emitter bias

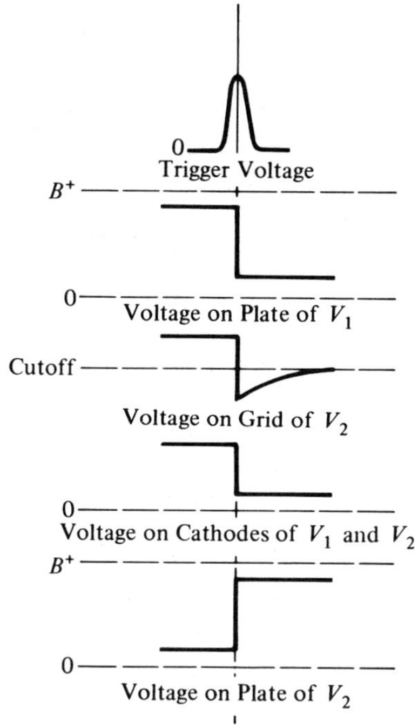

FIGURE 26.6 *Changes that occur while V_1 conducts and C_{C2} discharges.*

on T_2 decreases until finally the base emitter of T_2 becomes forward biased, and T_2 begins to conduct again. When T_2 conducts, the voltage drop across R_E makes the emitter of T_1 positive with respect to the base, and T_1 is cut off. The cycle will repeat only if another trigger voltage is introduced.

26.2 FREE RUNNING MULTIVIBRATOR

The free running multivibrator does not require an external trigger voltage to begin operation as did the one-shot multivibrator. Thus, the square-wave output of the free running multivibrator does not stop after one cycle, but continues as long as voltage is supplied to the circuit, and the components continue to function. (See Figure 26.9.) Because triodes V_1 and V_2 are not identical, one of the tubes will begin conducting before the other when the power supply, not shown, is turned on. Referring to the instant when full voltage is first applied to the circuit as t_1, we will suppose that V_1 conducts more than V_2 at instant t_1. As the plate current of V_1 increases, its plate voltage drops from near B^+ potential. This drop in plate voltage of V_1 is coupled to the grid of V_2 by coupling capacitor

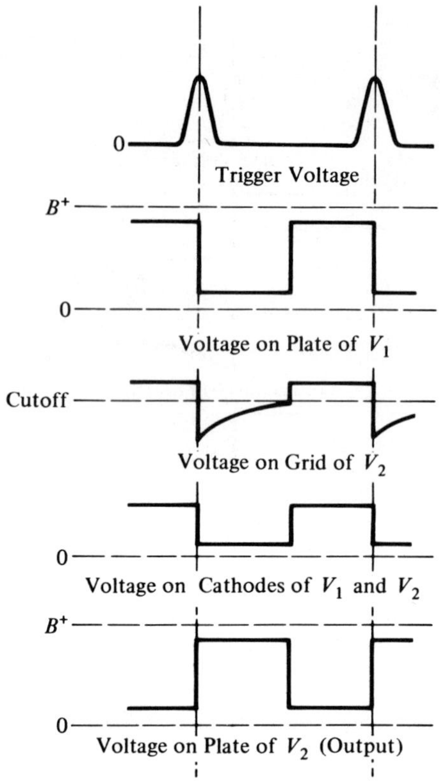

FIGURE 26.7 *Voltage during one cycle of one-shot multivibrator.*

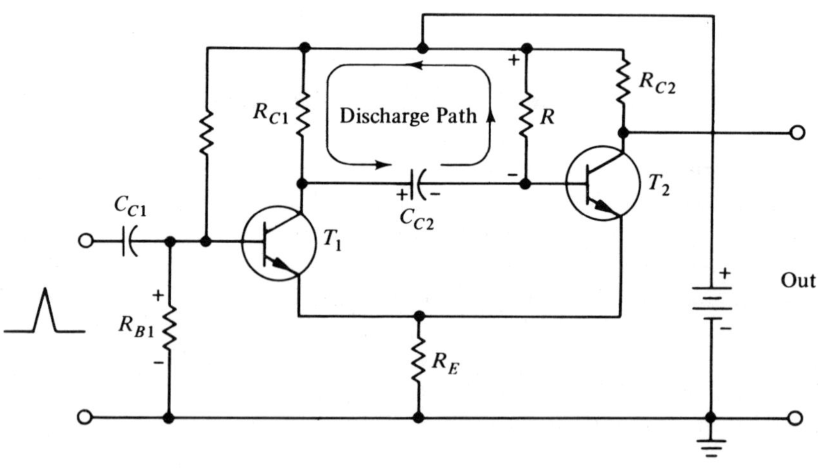

FIGURE 26.8 *Transistor one-shot multivibrator.*

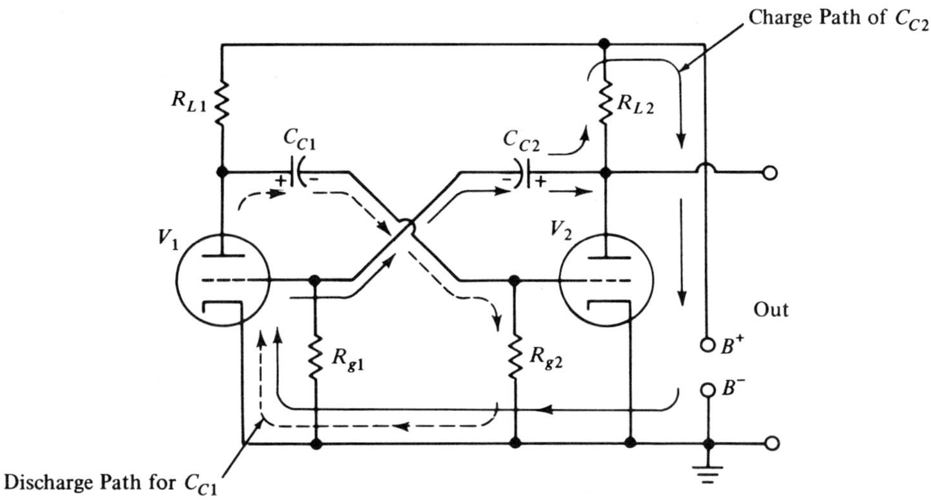

FIGURE 26.9 *Free running multivibrator (charging of C_{C2} and discharging of C_{C1} while V_1 conducts).*

C_{C1}, driving the grid of V_2 below cutoff and preventing tube V_2 from conducting. With no current in V_2, the plate voltage is at near B^+ potential. This rise in V_2 plate voltage is coupled to the grid of V_1 by C_{C2}, making the grid of V_1 more positive and causing V_1 to conduct even more. (See Figure 26.10.)

With the grid of V_1 positive, grid current flows charging C_{C2}. At the same time C_{C1}, whose left plate was initially at near B^+ potential, is discharging because of the drop in the plate voltage of V_1. As C_{C2} charges and C_{C1} discharges, we leave instant t_1 and move toward instant t_2. The time duration between instants t_1 and t_2 is determined by the time constant of $R_{g2}C_{C1}$, since C_{C1} discharges through R_{g2}. When capacitor C_{C1} has discharged sufficiently, the grid of V_2 goes above cutoff. With the plate of V_2 very positive and the grid just above cutoff, plate current begins to flow in V_2. At this instant we mark time t_2. (See Figure 26.11.)

Now the cycle reverses itself. The increase in plate current in V_2 causes a decrease in plate voltage, which is coupled to the grid of V_1 by capacitor C_{C2}. This results in a decrease of grid voltage in V_1 to below cutoff, and V_1 stops conducting. The plate voltage of V_1 rises to near B^+ potential, and this increase causes a further increase in the grid voltage of V_2 due to the presence of coupling capacitor C_{C1}. As the grid of V_2 becomes slightly positive, grid current flows charging C_{C1}. At the same time, capacitor C_{C2} is discharging because of the drop in plate voltage of V_2. (See Figure 26.12.) The time constant for the discharge of C_{C2} through R_{g1} determines the duration of the second half cycle. If the two time

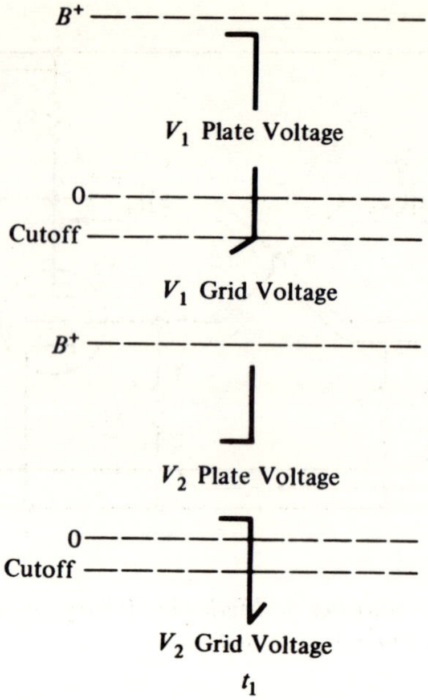

FIGURE 26.10

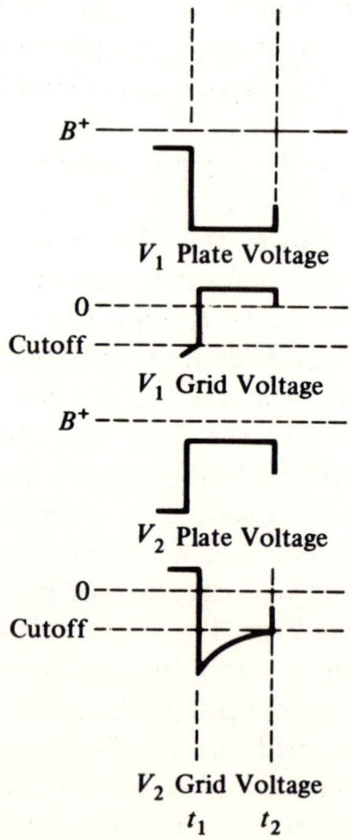

FIGURE 26.11

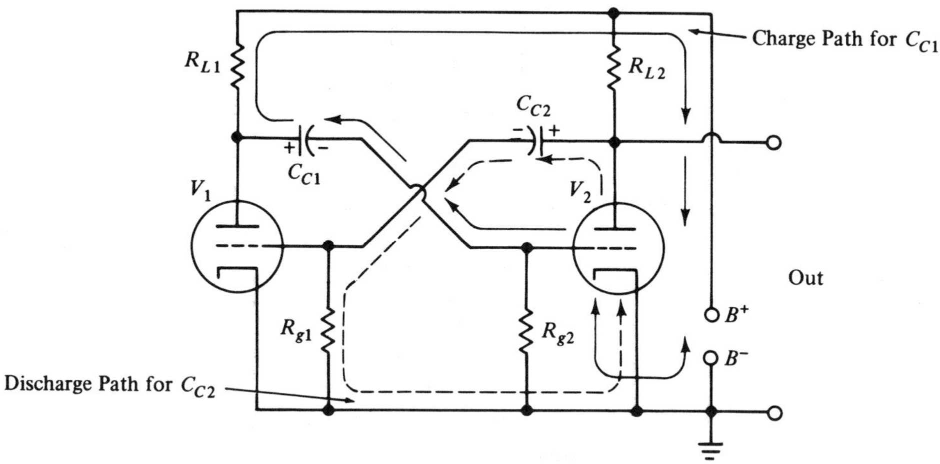

FIGURE 26.12 *Free running multivibrator (charging of C_{C1} and discharging of C_{C2} while V_2 conducts).*

constants (one for the discharge of C_{C1} through R_{g2} and the other for the discharge of C_{C2} through R_{g1}) are equal, then the length of each half cycle will be the same, and the output will be a square wave. If one of the time constants is longer than the other, the output will be unsymmetrical. (See Figure 26.13.)

To improve the stability of the free running multivibrator, the grid resistors may be returned to B^+ instead of ground. (See Figure 26.14.) The grid return to B^+ causes capacitors C_{C1} and C_{C2} to discharge much more rapidly during their respective half cycles. Thus, the instant at which the grids go above cutoff is much better defined, since it occurs on a more nearly vertical portion of the discharge curve. (See Figure 26.15.)

A transistor version of the free running multivibrator has many of the same characteristics as the vacuum tube type. (See Figure 26.16.) When transistor T_1 conducts, its collector voltage decreases, causing capacitor C_{C1} to discharge through R_{B2}, the battery, and the emitter collector of T_1. This discharge places a reverse bias on the emitter base of T_2, and T_2 does not conduct. As C_{C1} discharges, C_{C2} is charging through R_{C2} and the emitter base of T_1. When C_{C1} is discharged, the emitter base of T_2 becomes forward biased, and T_2 conducts. The second half cycle is similar to the first with the roles of T_1 and T_2 reversed.

26.3 FLIP-FLOP MULTIVIBRATOR

The bistable, or flip-flop, multivibrator has two stable states of operation. The circuit remains in one state of operation until a trigger voltage forces

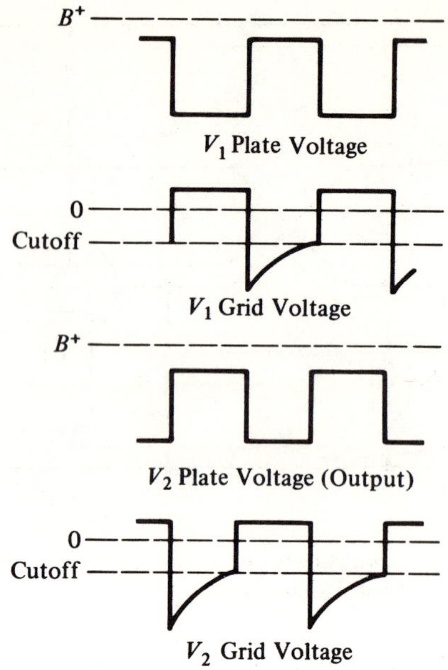

FIGURE 26.13

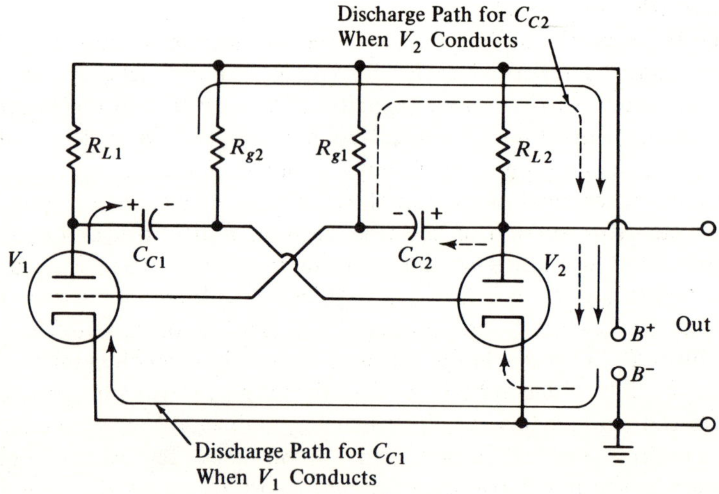

FIGURE 26.14 *Free running multivibrator with positive grid return.*

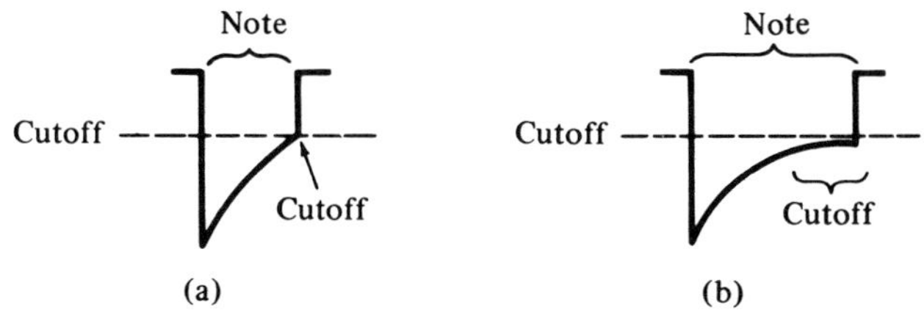

FIGURE 26.15 *(a) Positive grid return (good definition at cutoff); (b) ground grid return (poor definition of cutoff). (Note: Faster discharge in (a) implies steeper rise in curve; slower discharge in (b) implies more gradual rise in curve.)*

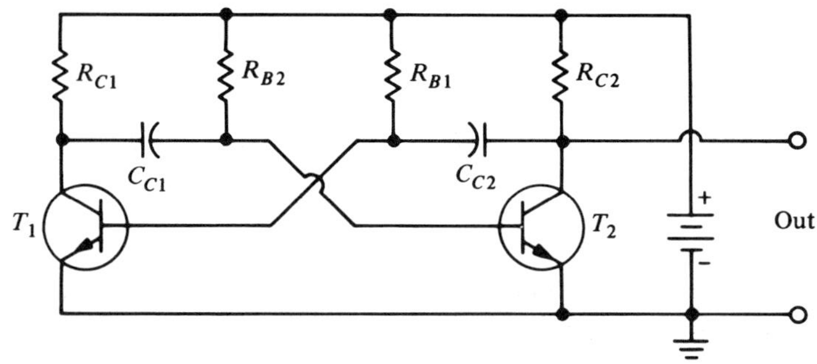

FIGURE 26.16 *Transistor free running multivibrator.*

it into the second. Then it does not change back to the first state until triggered again. The implication is that there are no time constants, associated with the charging and discharging of capacitors, which change the mode of operation. (See Figure 26.17.)

Before a trigger voltage is applied across input resistor R_{in}, one of the tubes, for example, V_1, starts conducting due to inequalities in the two tube circuits. The plate current of V_1 increases, and its plate voltage decreases. The decrease in V_1 plate voltage causes a decrease in current from C^- flowing through the voltage divider network of R_2 and R_{g2}. This means the top of R_2 is less positive (more negative), making the grid of V_2 more negative. When the grid of V_2 becomes more negative, current in V_2 is decreased, and V_2 plate voltage increases. This increased plate voltage for V_2 causes an increase in current from C^- flowing through the voltage divider network of R_1 and R_{g1}. This makes the top of R_1, and hence the grid of V_1, even more positive, and V_1 becomes an even better conductor. When V_2 reaches cutoff, there is no further increase in voltage on the plate of V_2 or the grid of V_1.

The introduction of a positive trigger voltage on the grids of both V_1 and V_2 at time t_1 has little effect on V_1, which is already conducting. But this positive voltage on the grid of V_2 causes its grid bias to go above

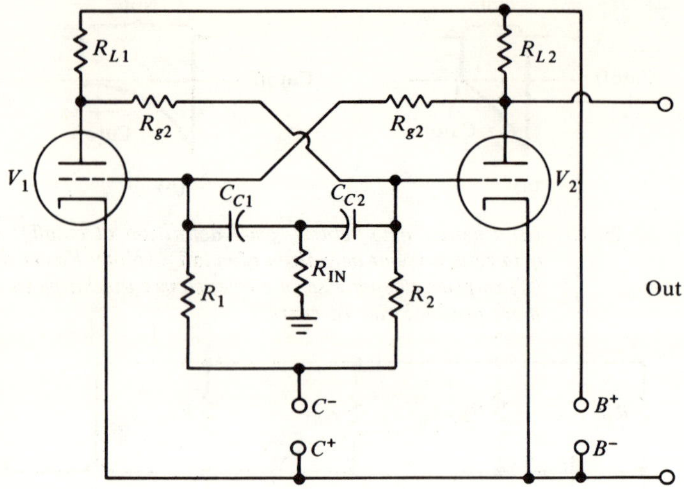

FIGURE 26.17 *Bistable, or flip-flop, multivibrator.*

cutoff, and V_2 conducts. The rapid rise in current in V_2 causes the V_2 plate voltage to decrease. This decrease in the plate voltage of V_2 is applied to the grid of V_1 via voltage divider network R_1 and R_{g1}. A more negative grid bias for V_1 causes a decrease in plate current and an increase in V_1 plate voltage. This increased plate voltage of V_1 is applied to the grid of V_2 by voltage divider network R_2 and R_{g2}, causing V_2 to conduct even more and driving the grid of V_1 toward cutoff and the plate voltage of V_1 toward B^+ potential.

The process reverses itself again when a second trigger voltage is applied. Since the grid of V_2 is already positive, this second trigger will not affect V_2 but, when applied to the grid of V_1, it drives V_1 above cutoff. V_1 begins to conduct, and the cycle already described repeats. (See Figure 26.18.)

A transistor bistable, or flip-flop, multivibrator "flips" when one transistor circuit is conducting and "flops" when the other transistor conducts. (See Figure 26.19.) If transistor T_1 is conducting, its collector current increases. The resulting decrease in collector voltage is applied to the base of T_2 by voltage divider R_2 and R_{B2}. This decreases the forward bias of T_2 until T_2 does not conduct. Since T_2 is not conducting, its collector voltage increases, and this increase is applied to the emitter-base forward bias of T_1 by voltage divider R_1 and R_{B1}. When a trigger voltage is applied across R_{in}, it does not aid T_1, which is already conducting, but it increases the emitter-base forward bias of T_2, causing T_2 to begin conducting. When T_2 conducts, the resulting decrease in collector voltage, which is applied to the emitter base of T_1, causes T_1 to stop conducting.

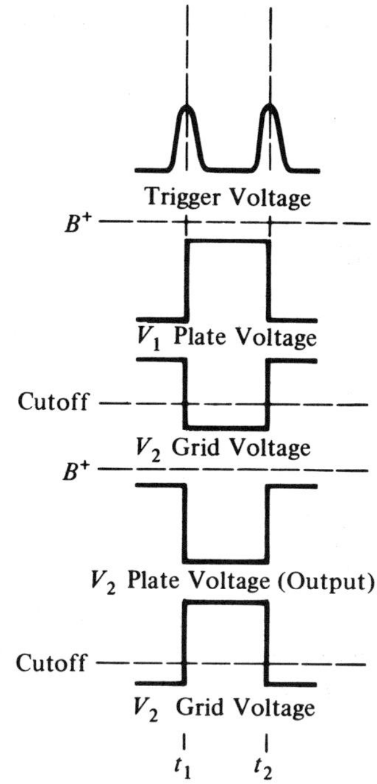

FIGURE 26.18

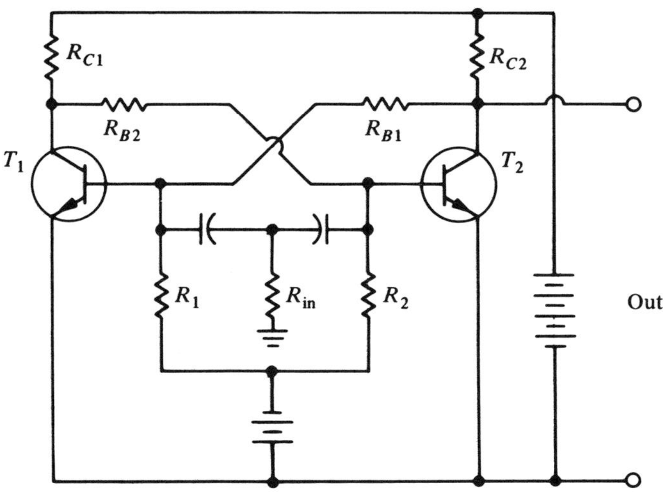

FIGURE 26.19 *Transistor bistable, or flip-flop, multivibrator.*

Another trigger voltage will be required to make T_1 conduct again, cutting off T_2.

Questions

(1) How may the length of one half cycle be varied in the one-shot multivibrator?

(2) In what sense is the free running multivibrator just the opposite of a full-wave rectifier?

(3) In what way does the bistable, or flip-flop, multivibrator display a memory?

Suggested Experiments

(1) Many audio oscillators also have a square-wave output. You may connect the square-wave generator to an oscilloscope to observe the symmetry of the square wave at different frequencies.

CHAPTER 27

Sawtooth Generators

27.1 MANUALLY OPERATED SAWTOOTH GENERATOR

A sawtooth voltage output is the product of a sweep generator. (See Figure 27.1.) The sweep to which the name refers is the sweep of an

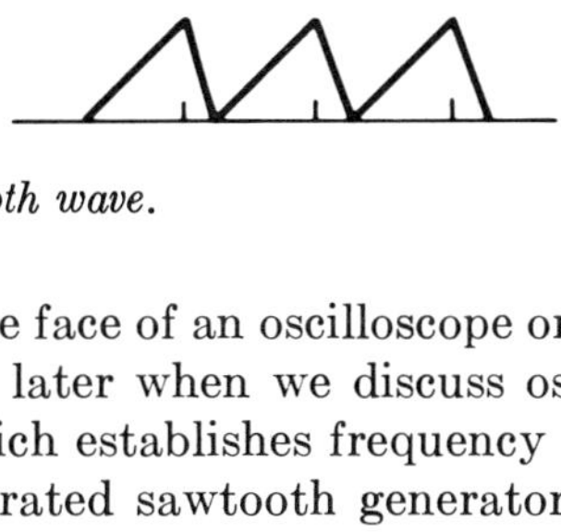

FIGURE 27.1 *Sawtooth wave.*

electron beam across the face of an oscilloscope or TV tube. The sawtooth voltage, as we will see later when we discuss oscilloscopes, has a linear increase in voltage, which establishes frequency control.

The manually operated sawtooth generator in Figure 27.2 consists of a switch, a capacitor, a resistor, and a battery. When the switch is in

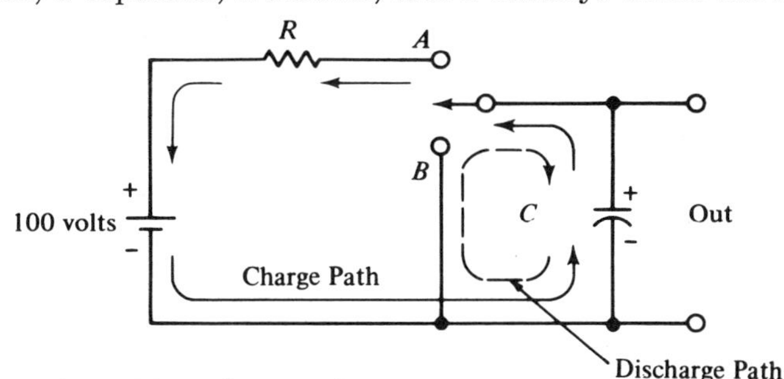

FIGURE 27.2 *Manual sawtooth generator.*

position A, the capacitor charges. The time for it to completely charge is five time constants [5 TC = $5(R \times C)$]. When the switch is moved to position B, the capacitor discharges. The complete discharge of the capacitor occurs almost instantaneously, because the time constant for the discharge path is approximately zero (no resistance). (See Figure 27.3.)

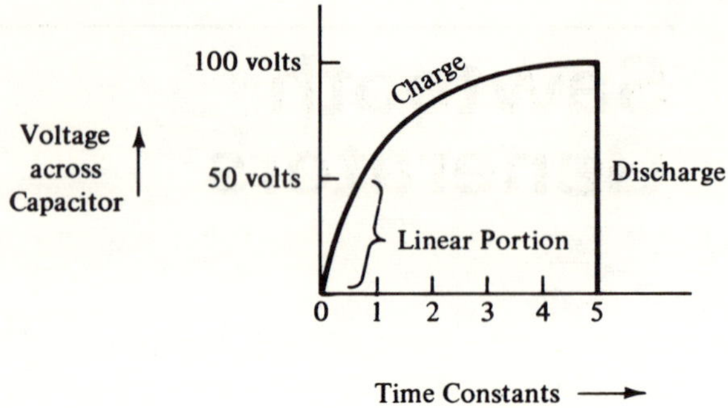

FIGURE 27.3 *Charge and discharge of capacitor.*

If we limit the charging of the capacitor to the linear portion of the curve, that is, we only allow it to charge for a fraction of one time constant and then cause it to discharge, the output across the capacitor looks more like a sawtooth. (See Figure 27.4.)

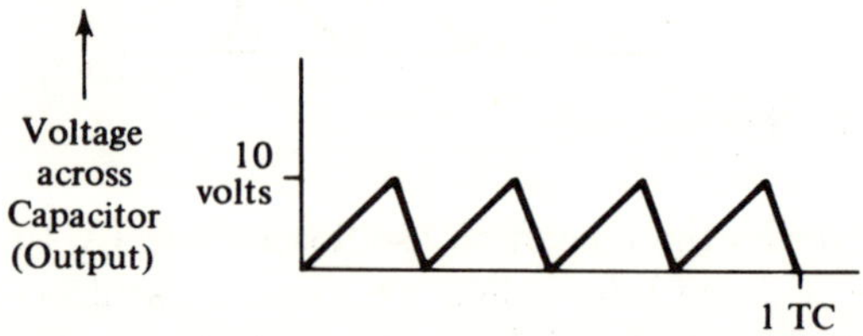

FIGURE 27.4

27.2 THYRATRON SAWTOOTH GENERATOR

To eliminate manual switching, a triode or a thyratron tube may be used. The thyratron tube is a gas filled, hot cathode triode. (See Figure 27.5.) The presence of gas in the tube is indicated by the dot on its schematic. When the potential between the cathode and plate of the thyratron reaches a certain value, the gas in the tube ionizes (breaks down into electrons and positive ions), and the tube conducts very rapidly. Below the ionization voltage, the tube does not conduct. The grid of the thyratron furnishes a fine control of the ionization voltage.

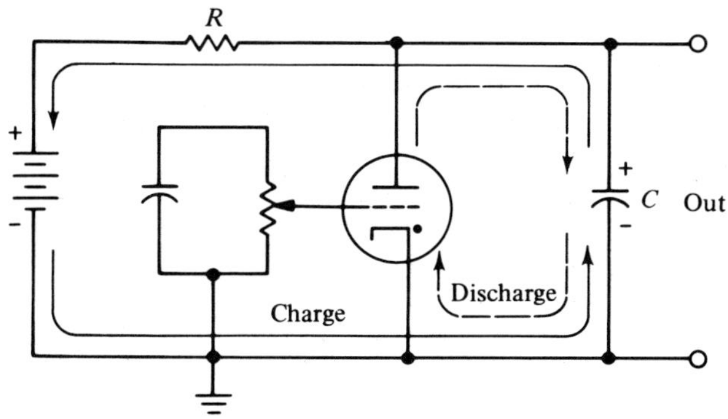

FIGURE 27.5 *Thyratron sawtooth generator.*

When the circuit is complete, the capacitor charges through the resistor. When the potential across the capacitor reaches the ionization potential of the thyratron, the tube conducts, discharging the capacitor rapidly and thus reducing the potential across both the capacitor and the thyratron. At the lower voltage, the gas within the tube deionizes (electrons and positive ions recombine), and the thyratron ceases to conduct. The ionization potential of the thyratron is usually set at the value of the capacitor voltage after it has been charging for approximately 1/10 of one time constant. (See Figure 27.6.)

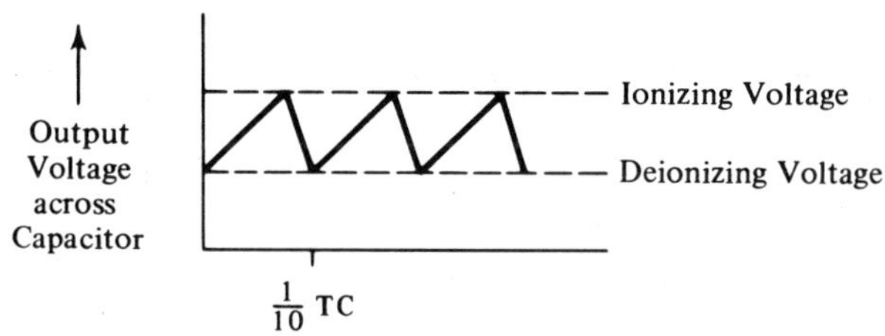

FIGURE 27.6

Temperature factors along with instabilities in bias and supply voltages make for erratic response from the thyratron sawtooth generator. A synchronizing bias voltage is used to improve the regularity, or response, of this generator. This synchronizing bias voltage may be supplied by any number of voltage generators, for example, by an oscillator or by a free running multivibrator. The sync signal makes the grid less negative at specific time intervals. This change in grid voltage causes the thyratron to "fire," or conduct, at that instant. (See Figure 27.7.)

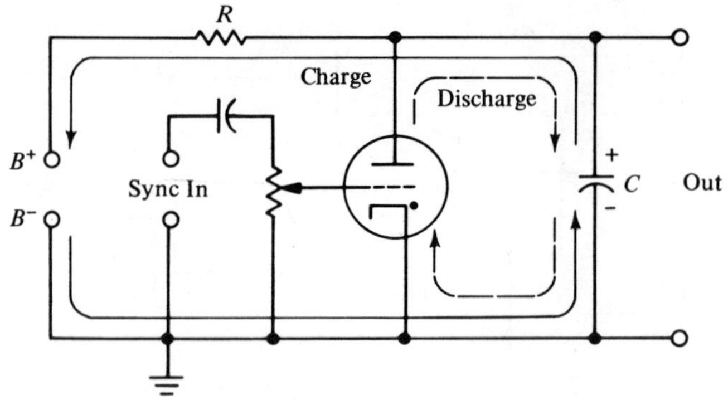

FIGURE 27.7 *Synchronized thyratron sawtooth generator.*

27.3 TRIODE SAWTOOTH GENERATOR

As mentioned, a triode may also be used as the switch in our sawtooth circuit. The sawtooth generator, which uses a vacuum tube triode, is often called a hard tube sweep generator. In the hard tube sweep generator, a signal is applied to the grid in order to allow the capacitor to discharge through the tube. This signal, which causes the triode to conduct, "opens the gate" for electrons from the capacitor. Thus, the term gated generator is often associated with this form of sawtooth generator. (See Figure 27.8.)

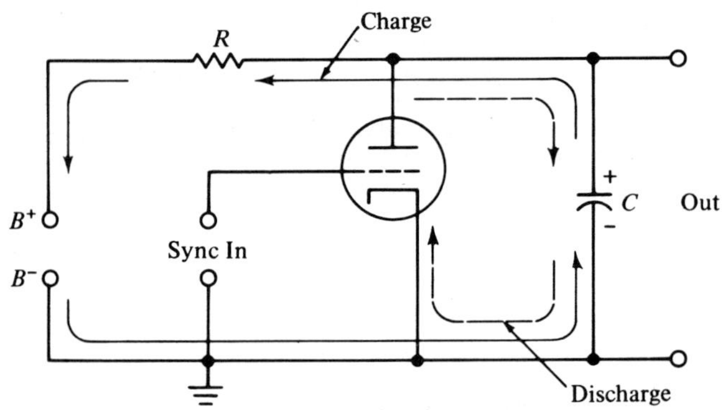

FIGURE 27.8 *Triode (hard tube) gated sawtooth generator.*

When the sync input signal goes negative, driving the grid below cutoff, the tube ceases to conduct, and the capacitor begins charging through R. When the sync input signal, or gate, is positive, the tube conducts, allowing the capacitor to discharge. The resistance of the tube is such that the capacitor does not discharge instantaneously (1 TC = $R_P \times C$). The capacitor begins to charge again when the grid voltage cuts the tube off. (See Figure 27.9.)

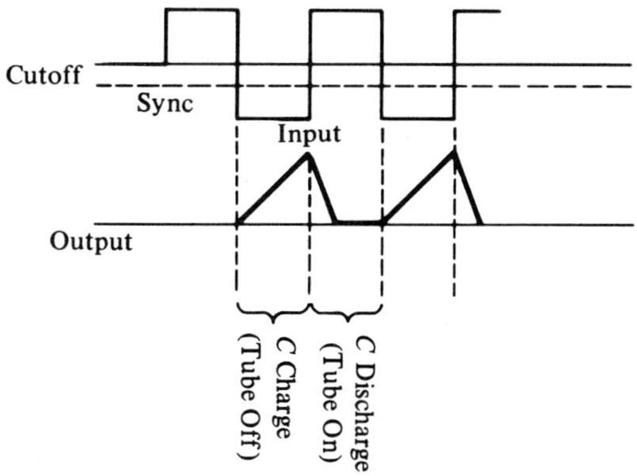

FIGURE 27.9

To reduce the time lapse between the complete discharge of the capacitor and its beginning to charge again, we use an unsymmetrical sync input signal. (See Figure 27.10.)

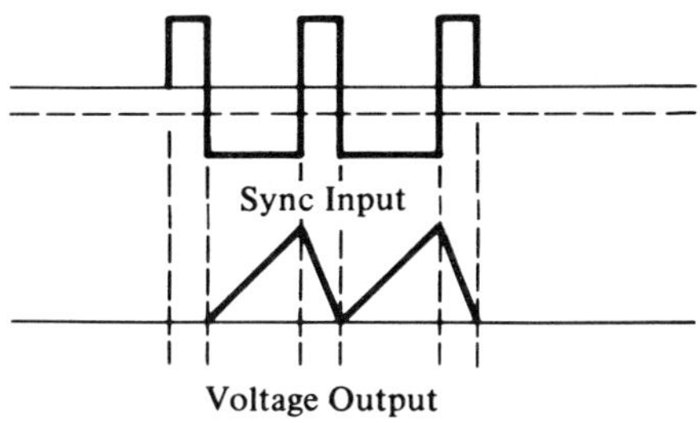

FIGURE 27.10

Questions

(1) The sawtooth generator utilizes only a small portion of the voltage increase which can be built across a capacitor. How may the period of the sawtooth output, that is, the time for this voltage increase, be varied?

CHAPTER 28

Test Instruments

28.1 GALVANOMETER

The instruments that measure voltage, current, resistance, capacitance, inductance, tube and transistor functions, and frequency response enable us to determine whether circuits and individual components are functioning properly. We will conclude our study of electronics by describing how some of these test instruments operate.

First, we will consider the problem of measuring dc voltage and current. Earlier, we described how a coil rotating in a magnetic field has an ac voltage and current induced in it. In fact, this is how we generate ac. Suppose, instead of rotating the coil, we cause a small current to flow in it while it is in a magnetic field. (See Figure 28.1.) The current in the

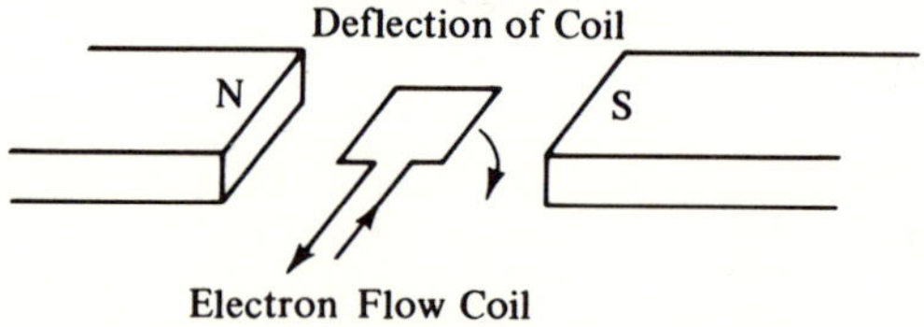

FIGURE 28.1

coil produces a magnetic field, which interacts with the permanent magnetic field and causes the coil to be deflected. The larger the current in the coil, the larger will be the magnetic field about the coil and the greater the deflection of the coil. When a current is used to produce a deflection of a coil instead of the other way around, that is, instead of a mechanical twisting of the coil, producing a current as in a generator, we have a form of motor. Since the "motor" described above does not rotate around and around, but just deflects slightly, it is called a galvanometer.

It is this basic instrument, the galvanometer, which is the heart of voltmeters and ammeters. Thus, to construct these instruments, we must first know something of the characteristics of the galvanometer. For instance, what is the resistance of the coil? How much current or voltage will produce maximum deflection of the meter? To determine the resistance of a galvanometer, construct the circuit shown in Figure 28.2

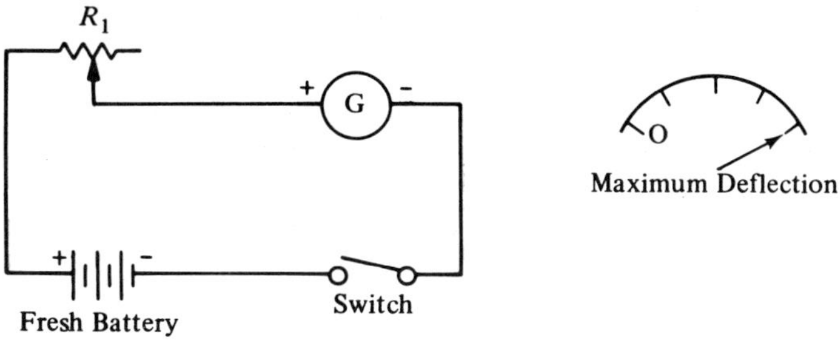

FIGURE 28.2

carefully observing the polarity of the galvanometer. First, close the switch and adjust R_1 for maximum deflection of the pointer. Next open the switch and insert a second variable resistor R_2, being careful not to change the value of R_1. Notice that the meter and R_2 are in parallel. (See Figure 28.3.) Now close the switch again and adjust resistor R_2 for

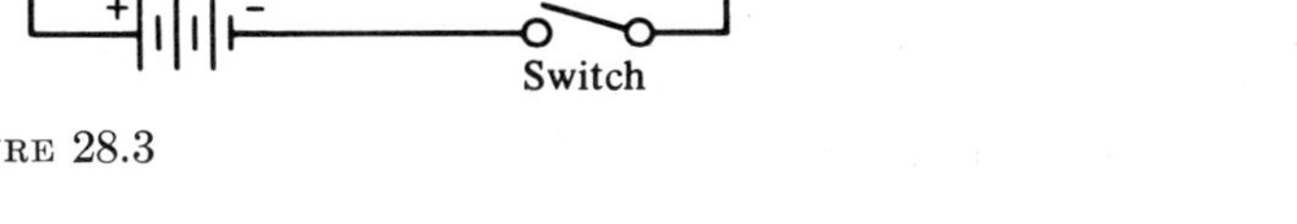

FIGURE 28.3

half scale deflection of the meter. When this is done, half the current in the circuit will be flowing through the coil of the galvanometer and half through resistor R_2. With equal current in each we know, from our study of parallel resistive circuits, that the resistance of the galvanometer and R_2 are equal. Thus, the resistance of the galvanometer is simply the resistance of R_2.

We can now find the total resistance of the first circuit (Figure 28.2), a series resistive circuit, from the resistance of R_1 and the galvanometer. $R_T = R_1 + R_{\text{Galvanometer}}$. Since a fresh battery was used, its voltage is known and hence current for maximum deflection (current in the first circuit) can be found from Ohm's Law. $I_{\text{Meter}} = E/R_T$. The voltage across the meter for maximum deflection (current in the first circuit multiplied by the resistance of the galvanometer) is obtained from the resistance of the meter and the current in the meter. $V_{\text{Meter}} = I_{\text{Meter}} \times R_{\text{Galvanometer}}$.

28.2 DC VOLTMETER

A dc voltmeter consists of a galvanometer and a series resistor. (See Figure 28.4.) The resistance R increases the total opposition to current

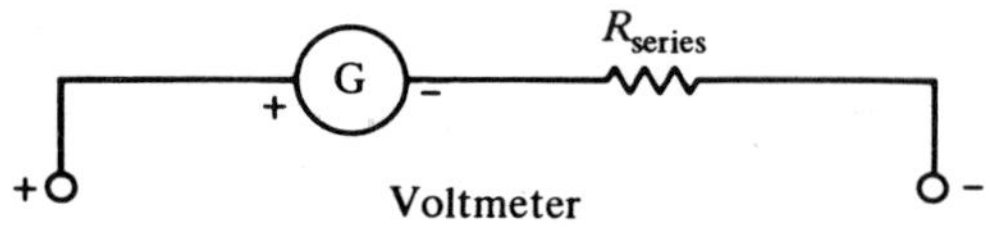

FIGURE 28.4

flow and thus limits the current in the meter to prevent its being damaged. For example, let us assume that the resistance of the galvanometer is $R_{\text{Galvanometer}} = 50$ ohms, and the maximum current the meter can handle (current for full scale deflection) is $I_{\text{Meter}} = 2$ mA $= 0.002$ A. Suppose we want a dc voltmeter which has a maximum scale reading of 100 volts. From the above equation, we know the galvanometer by itself can only handle 0.1 volt. $V_{\text{Meter}} = I_{\text{Meter}} \times R_{\text{Galvanometer}} = 0.002 \text{ A} \times 50 \text{ ohms} = 0.01$ volt. In order to find the series resistance (R_{Series}) required to extend this galvanometer's maximum voltage capability from 0.1 volt to 100 volts, we first determine the total resistance required. $R_T = V_{\text{Meter(desired)}}/I_{\text{Meter}} = 100 \text{ volts}/0.002 \text{ A} = 50{,}000$ ohms. Since the galvanometer and R are in series, $R_T = R_g + R_{\text{Series}}$. $50{,}000 \text{ ohms} = 50 \text{ ohms} + R_{\text{Series}}$ or $R_{\text{Series}} = 49{,}950$ ohms. To construct a 100-volt dc voltmeter from the galvanometer described, we must place a 49,950-ohm resistor in series with the meter.

28.3 DC AMMETER

An ammeter to measure dc current in a circuit must have a means of deviating, or shunting, large currents around a meter movement which can at best manage only a few milliamperes. To accomplish this, an ammeter has a low resistance in parallel with a galvanometer. (See Figure 28.5.) Suppose we design an ammeter that can detect currents up

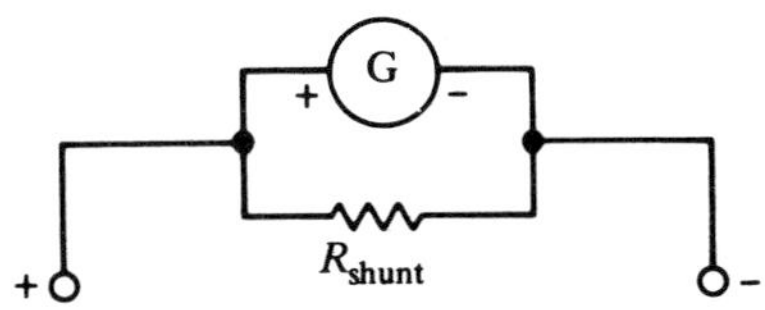

FIGURE 28.5 *Ammeter.*

to 5 A, using the same galvanometer as described above. Since the meter can only handle 0.002 A, most of the current must shunt the meter. $I_{\text{Meter(desired)}} = I_{\text{Meter}} + I_{\text{Shunt}}$; $5\text{ A} = 0.002\text{ A} + I_{\text{Shunt}}$; $I_{\text{Shunt}} = 5\text{ A} - 0.002\text{ A} = 4.998\text{ A}$. The current through the shunt resistance must be 4.998 A. Knowing that the ammeter is a parallel circuit (galvanometer in parallel with shunt resistor), we recall that the maximum voltage across the shunt is the same as the voltage for full scale deflection of the meter, which has already been found to be $V_{\text{Meter}} = 0.1$ volt. Applying Ohm's Law, we can find the resistance of the shunt. $R_{\text{Shunt}} = V_{\text{Shunt}}/I_{\text{Shunt}} = 0.1\text{ volt}/4.998\text{ A} = 0.02\text{ ohm}$. Thus, to construct an ammeter, whose maximum current reading is 5 A using the galvanometer described, we must place a 0.02-ohm shunt resistance in parallel with the meter.

We have stressed the characteristics and construction of dc voltmeters and ammeters because it is frequently necessary to measure dc voltage and current in electronic circuits. Remember that voltage is measured across the circuit component, whereas current is measured within the circuit.

28.4 OHMMETER

We can determine the resistance of a resistor simply by applying Ohm's Law using the voltage and current readings of our meters. (See Figure 28.6.) Thus, the resistance of R is $R = V_{\text{Meter reading}}/I_{\text{Meter reading}}$.

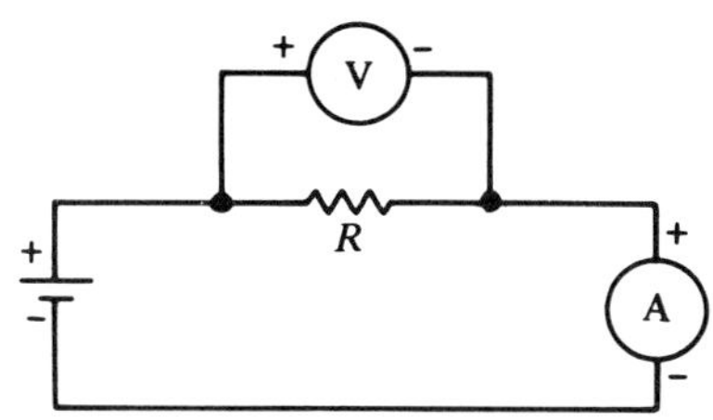

FIGURE 28.6 *Application of dc voltmeter and ammeter to find resistance.*

Suppose the voltage supplied is known and is kept constant. Then increasing the resistance would result in a smaller deflection of the ammeter. Thus, with a known voltage, a galvanometer may be made into an ohmmeter. (See Figure 28.7.) The face of a galvanometer being used

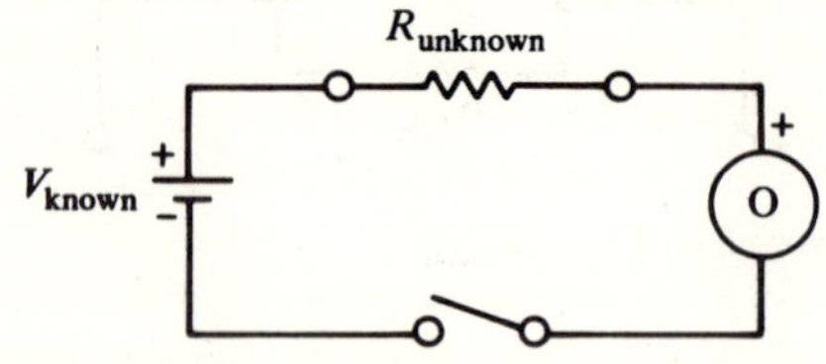

FIGURE 28.7 *Ohmmeter.*

as an ohmmeter is calibrated in ohms. When no resistor is placed in the circuit, no current flows. The pointer on the meter is not deflected, and the resistance is said to be infinite. When an unknown resistance is placed in the circuit, current flows, deflecting the pointer. The resistance, which allows that current, is read directly on the face of the meter.

28.5 WHEATSTONE BRIDGE

An unknown resistance can also be measured using a circuit called a Wheatstone bridge. (See Figure 28.8.) To utilize a Wheatstone bridge for

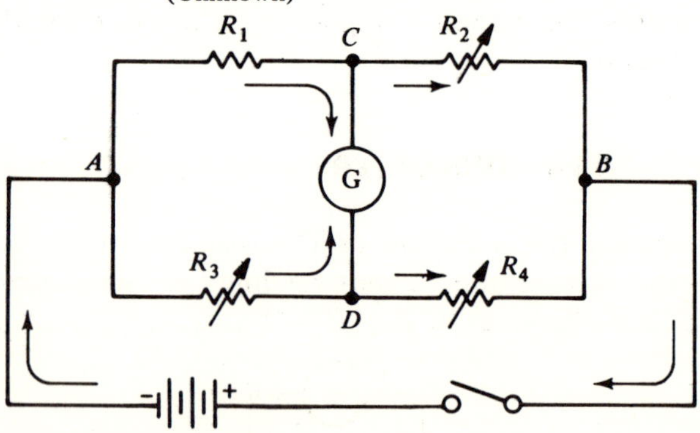

FIGURE 28.8 *Wheatstone bridge.*

this purpose, it is necessary to know the values of the other resistances in the circuit. The "bridge" is, of course, the galvanometer across the parallel branches ACB and ADB. Notice that current flows in both directions in the galvanometer, from AC down through the galvanometer and from AD up through the galvanometer. These opposing currents tend to deflect the galvanometer pointer in opposite directions.

We balance the bridge by adjusting the variable resistors until the galvanometer pointer is not deflected, that is, the current from AC down through the galvanometer equals that from AD up through the galvanometer, or there is no current in the bridge. For there to be no current through the galvanometer, there must be no voltage drop across the galvanometer, or the potential at C and D must be the same. For this to occur, the same voltage must be dropped across R_1 as across R_3. $V_1 = V_3$ or $I_1R_1 = I_3R_3$. In addition, from our knowledge of parallel circuits, the voltage across branches ACB and ADB are equal. That is $V_{ACB} = V_{ADB}$ or $V_1 + V_2 = V_3 + V_4$. Since $V_1 + V_2 = V_3 + V_4$ and $V_1 = V_3$, we see that $V_2 = V_4$ or $I_2R_2 = I_4R_4$. Since there is no current flowing through the galvanometer when the bridge is balanced, the same currents must flow in R_2 as in R_1 or $I_1 = I_2$. And, similarly $I_3 = I_4$. Substituting these values into $I_1R_1 = I_3R_3$, we have $I_2R_1 = I_4R_3$. Dividing the expression $I_2R_2 = I_4R_4$ into the equation just derived, we have $I_2R_1/I_2R_2 = I_4R_3/I_4R_4$ or $R_1/R_2 = R_3/R_4$. Since resistor R_1 was the unknown resistance, we solve the above equation for R_1.

$$R_1 = \frac{R_2R_3}{R_4}$$

Thus, knowing the values of R_2, R_3, and R_4 in a balanced bridge, we could substitute in this equation to find the unknown resistance R_1.

28.6 CAPACITANCE BRIDGE

This same bridge technique can be applied to capacitance and inductance either to determine the value of an unknown or to check a component to ascertain whether it has maintained the characteristic ascribed to it, that is, whether the actual capacitance or inductance is the same as the value printed on the component.

The capacitance bridge is similar to the Wheatstone bridge, but it requires an ac signal rather than dc. (See Figure 28.9.) Normally, an ac signal of 1000 Hz is used, so one of the oscillators already discussed (Hartley, Colpitts, and so on) can be used for the purpose. Since this frequency is in the audible range, we will use a headphone in place of the galvanometer. When the bridge is balanced, there is no sound in the headphone.

When the capacitance bridge is balanced by adjusting variable resistors R_1 and R_2, the voltage across the top branch equals the voltage

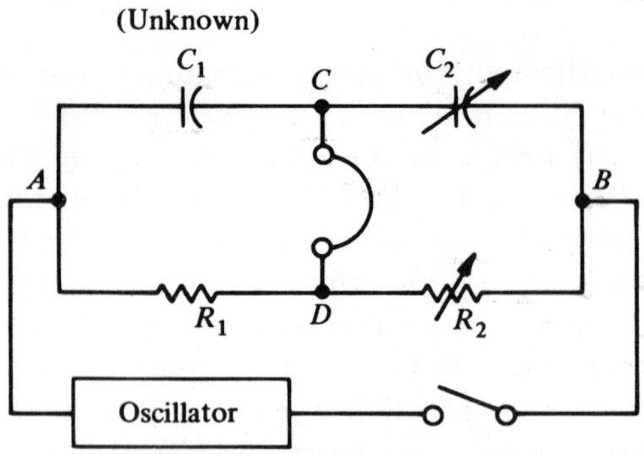

FIGURE 28.9 *Capacitance bridge.*

across the lower branch. $V_{ACB} = V_{ADB}$ or $V_{C1} + V_{C2} = V_{R1} + V_{R2}$. In addition, since there will not be any current flowing through the headphone, the voltage dropped from A to C will equal that dropped from A to D. $V_{AC} = V_{AD}$ or $V_{C1} = V_{R1}$. The voltage drops from C to B and from D to B are also equal. $V_{CB} = V_{DB}$ or $V_{C2} = V_{R2}$. Applying the relation involving capacitance $C = Q/V$ or $V = Q/C$ and Ohm's Law to both equations, we write:

$$V_{C1} = V_{R1} \tag{1}$$

$$\frac{Q_1}{C_1} = I_{R1} \times R_1 \tag{1a}$$

$$\frac{Q_1}{I_{R1}} = C_1R_1 \tag{1b}$$

$$V_{C2} = V_{R2} \tag{2}$$

$$\frac{Q_2}{C_2} = I_{R2} \times R_2 \tag{2a}$$

$$\frac{Q_2}{I_{R2}} = C_2R_2 \tag{2b}$$

With the bridge balanced and hence no current in the headphone, the charge on the two series capacitors must be the same. $Q_1 = Q_2$. No current in the lower branch of the circuit is deviated through the headphone, so the same current flows in R_1 and R_2. $I_{R1} = I_{R2}$. With the charges equal and the currents equal, we know that $Q_1/I_{R1} = Q_2/I_{R2}$. Since the left side of Equation (1b) equals the left side of Equation (2b), then the right sides are also equal. $C_1R_1 = C_2R_2$. Solving for the unknown capacitance, we have

$$C_1 = \frac{C_2R_2}{R_1}$$

28.7 INDUCTANCE BRIDGE

The inductance bridge is also balanced so that no current flows in the headphone and no 1000-Hz tone is heard. (See Figure 28.10.) Again, in

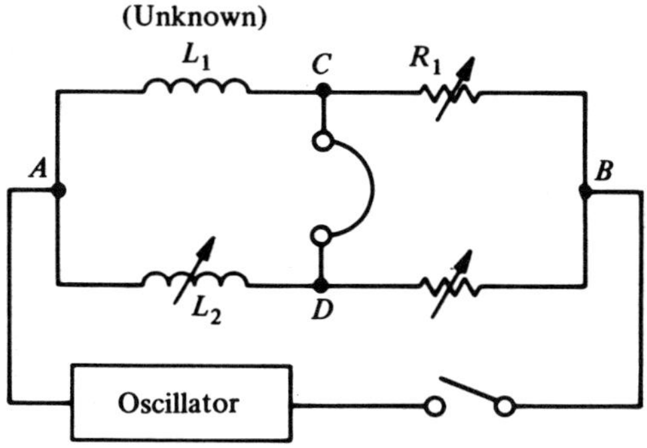

FIGURE 28.10 *Inductance bridge.*

the balanced state, we have $V_{AC} = V_{AD}$ or $V_{L1} = V_{L2}$ and $V_{CB} = V_{DB}$ or $V_{R1} = V_{R2}$. Using Ohm's Law ($V = IR$), we can rewrite these voltages in terms of current and either resistance or inductive reactance.

$$V_{L1} = V_{L2} \tag{1}$$
$$I_1X_{L1} = I_2X_{L2} \tag{1a}$$
$$V_{R1} = V_{R2} \tag{2}$$
$$I_1R_1 = I_2R_2 \tag{2a}$$

Notice we recognize that the current in L_1 is the same as the current in R_1 (both currents are I_1) when the bridge is balanced. Similarly, the same current (I_2) flows in both L_2 and R_2 in the balanced condition. Since inductive reactance is $X_L = 2\pi fL$, we substitute in Equation (1a) and obtain:

$$I_12\pi fL_1 = I_22\pi fL_2 \tag{1b}$$
$$I_1L_1 = I_2L_2 \tag{1c}$$

Dividing Equation (1c) by (2a), we have $I_1L_1/I_1R_1 = I_2L_2/I_2R_2$. Thus, $L_1/R_1 = L_2/R_2$, or

$$L_1 = \frac{L_2R_1}{R_2}$$

which gives the unknown inductance (L_1) in terms of the known quantities L_2, R_1, and R_2.

28.8 AC VOLTMETERS AND AMMETERS

The galvanometer movement, which we have already described and applied to the measurement of dc voltage and current, may also be used to measure ac voltage and current if the ac is first rectified (changed to dc). In the ac meter shown in Figure 28.11, a full-wave bridge rectifier is

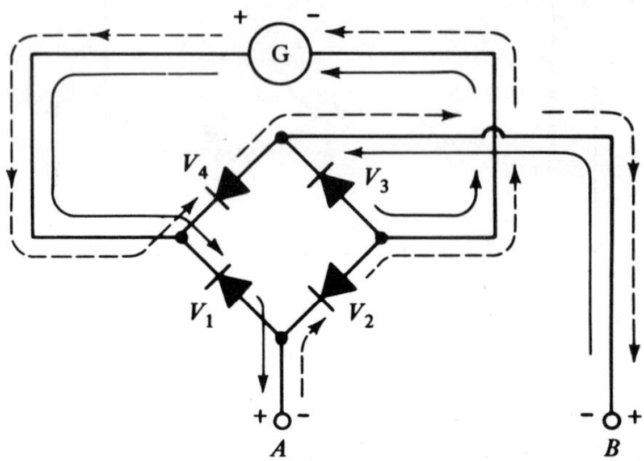

FIGURE 28.11 *ac meter.*

incorporated with the meter movement. Notice that the polarity of the galvanometer is maintained even though the ac input oscillates from positive to negative.

When A is positive and B negative, the anode of V_1 is positive and the cathode of V_3 is negative, so V_1 and V_3 conduct. The flow of current is from B through V_3, the galvanometer, V_4, and back to A. When A is negative and B positive, the anode of V_4 is positive and the cathode of V_2 is negative, so V_2 and V_4 conduct. Current flows from A through V_2, through the galvanometer in the same direction as during the preceding half cycle, through V_4, and back to B.

To construct an ac voltmeter, a series resistor is used with the above ac meter, just as we did for the dc voltmeter. And the ac ammeter consists of the ac meter and a parallel shunt.

28.9 TUBE AND TRANSISTOR TESTERS

Vacuum tube testers are used to determine whether or not a tube is conducting properly. A visual inspection of a vacuum tube, while a voltage is being applied to the filament, will tell us whether the filament is operating satisfactorily, because as it is heated, it turns red. However, since the coated cathode is actually the supplier of electrons, it is possible for the filament to be intact even though few electrons are being given off by the cathode. The test of cathode performance is called the emission test. This test can be used both for diodes or for hot cathode tubes that have grids. (See Figure 28.12.)

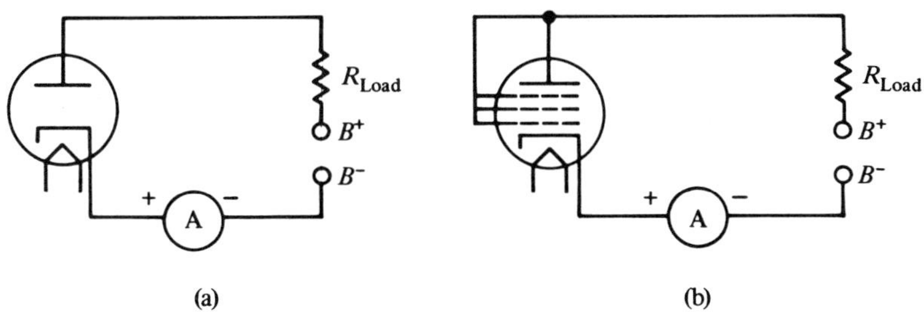

FIGURE 28.12 *(a) Emission test, vacuum diode; (b) emission test, multielement tube.*

The test circuit consists of a dc power supply, a load resistor, and a dc ammeter. The vacuum tube to be tested is placed in the circuit, the filament is given some time to warm up, and the ammeter then indicates the current flowing in the circuit. A weak tube will have a low current reading.

A better test for multielement vacuum tubes is one in which a signal is applied to the grid and then amplified by the tube. This is called a mutual conductance or transconductance test. (See Figure 28.13.) The ac ammeter in the plate circuit shows the performance of the tube when an

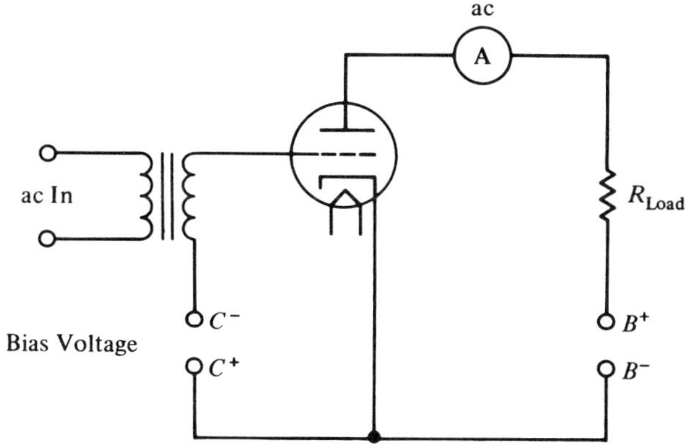

FIGURE 28.13 *Transconductance tube test.*

ac signal of specific amplitude is placed on the grid. This performance is then compared to the tube's rated performance, as indicated by the manufacturer, to determine whether the tube is operating properly.

Similar tests can be used for transistors. (See Figure 28.14.)

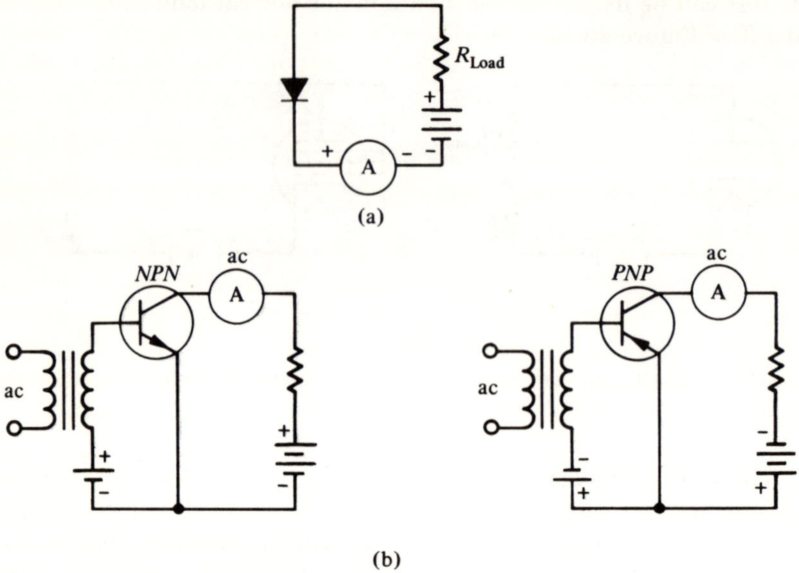

FIGURE 28.14 *(a) Diode test; (b) transistor test.*

28.10 FREQUENCY DETERMINATION BY HETERODYNING

The frequency of an ac signal can be found either by heterodyning or with an oscilloscope. Recalling the operation of an AM superheterodyne receiver, we use a similar circuit to measure an unknown frequency. (See Figure 28.15.) The unknown signal frequency is applied to the mixer

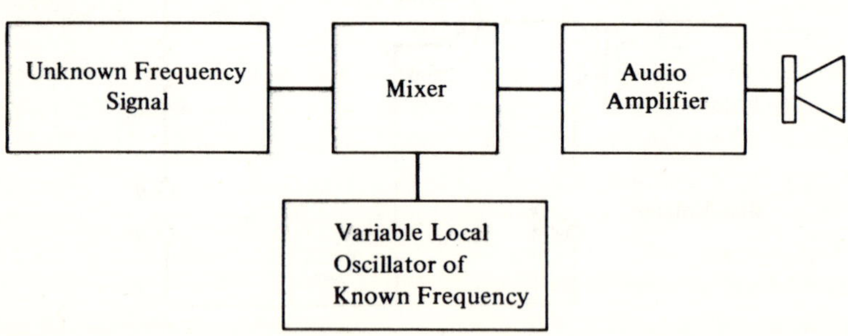

FIGURE 28.15

stage along with a known variable frequency from the local oscillator. The resulting beat frequency, the difference of the two input signals, is within the audio range and is amplified by the audio amplifier, whose output is then applied to a loudspeaker. As the local oscillator frequency comes closer to the unknown frequency, the beat frequency decreases, and the sound from the speaker gets deeper in pitch. When the local oscillator provides exactly the same frequency as the unknown, the beat frequency becomes zero, and there is no output from the speaker.

28.11 OSCILLOSCOPE

Before explaining how the cathode ray oscilloscope is used to determine unknown frequencies, we will have to describe its circuitry and, more especially, the cathode ray tube. The cathode ray tube is a high vacuum, hot cathode tube, which at this time has no semiconductor equivalent. The face, or screen, of the tube provides the "readout" for both the tube and oscilloscope. An indirectly heated cathode supplies electrons, which are accelerated and focused on the screen of the tube by a series of anodes. (See Figure 28.16.)

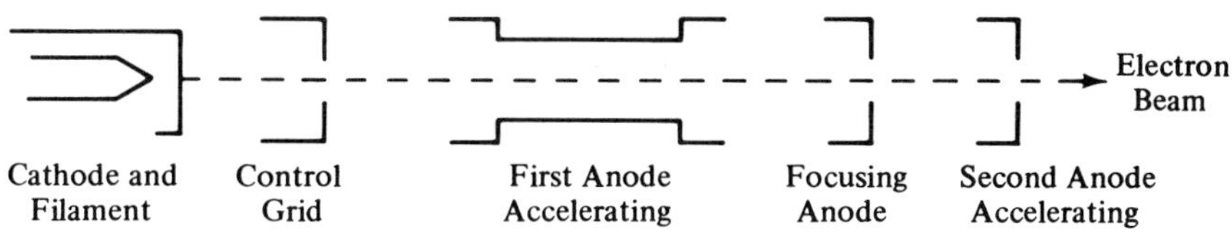

FIGURE 28.16 *Electron gun in cathode ray tube.*

Beyond the electron gun, the electron beam encounters a set of parallel vertical deflection plates. When a voltage is applied across these plates, the electron beam is deflected up or down. For instance, if the upper plate is positive and the lower plate negative, the electrons (negative charges) will be deflected upward. Beyond the vertical deflection plates are the parallel horizontal deflection plates. When a voltage is applied to the horizontal plates, the electron beam is deflected to the right or left. When the electron beam reaches the screen of the cathode ray tube, these fast-moving electrons interact with the phosphorescent material, which coats the inner side of the screen, and cause it to give off visible radiation. (See Figure 28.17.)

The block diagram of Figure 28.18 shows how some of the circuits we have studied are applied in the oscilloscope. The power supply has

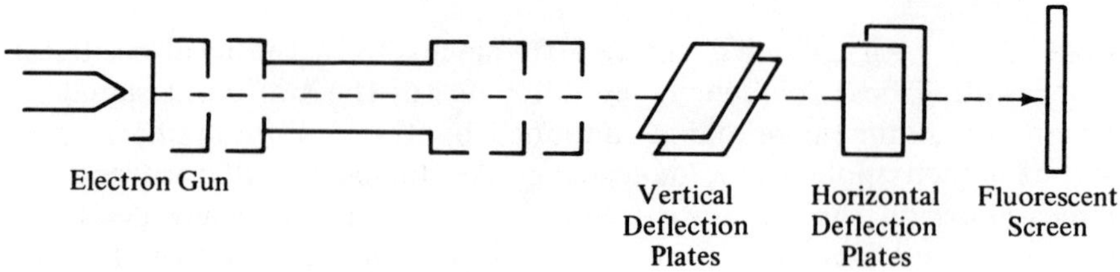

FIGURE 28.17 *Cathode ray tube.*

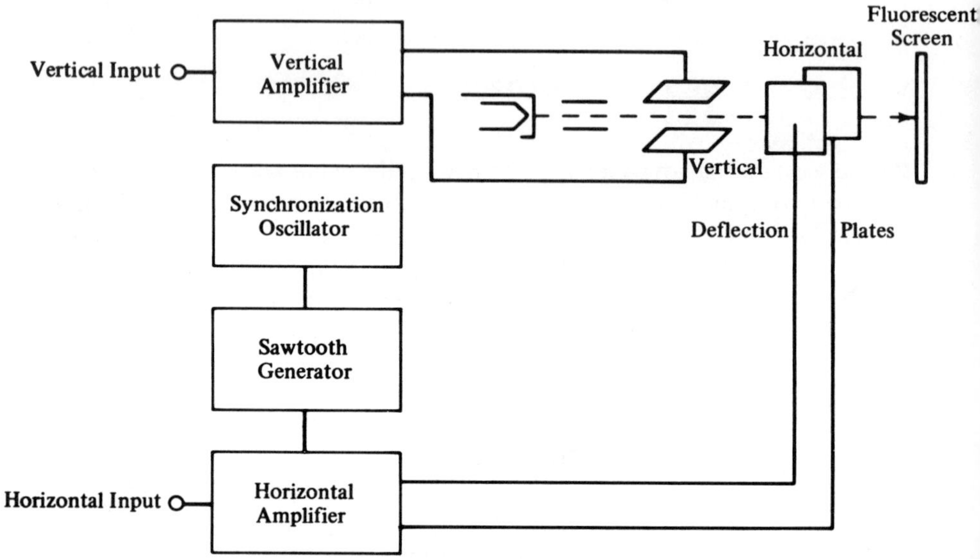

FIGURE 28.18 *Block diagram of oscilloscope.*

been omitted to simplify the diagram; however, each circuit does require dc voltage for operation. For the electron beam to be deflected vertically by the vertical deflection plates, the potential difference, or voltage, applied across the two plates must be large. If the vertical input signal is too weak to noticeably deflect the beam, an amplifier is employed to increase the vertical signal strength before it is placed on the deflection plates. The horizontal deflection plates not only have a horizontal input signal applied to them but also a continually changing voltage from the sawtooth generator. The steady rise of voltage output from the sawtooth generator, when applied to the horizontal deflection plates, moves the electron beam from left to right on the screen. The sudden decrease in sawtooth voltage produces a rapid return of the electron beam to the left side of the screen. (See Figures 28.19 and 28.20.)

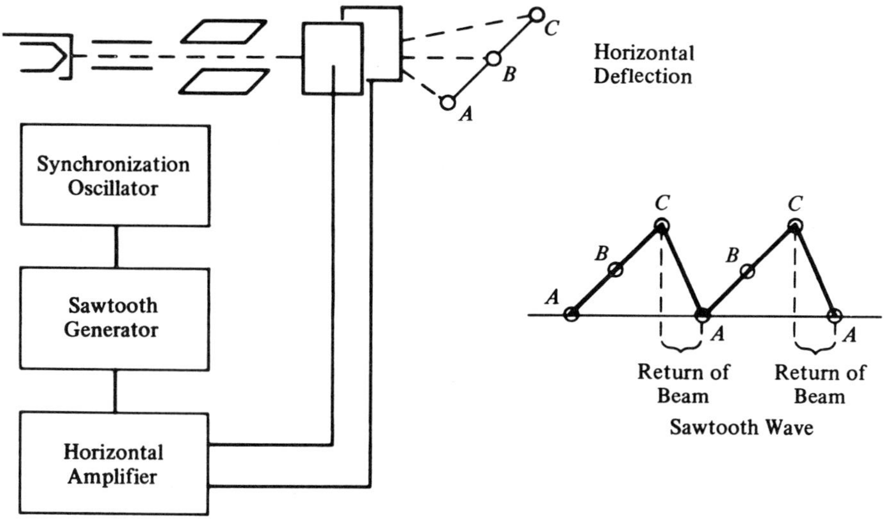

FIGURE 28.19 *Deflection left to right due to sawtooth generator.*

28.12 FREQUENCY DETERMINATION BY LISSAJOUS FIGURES

The oscilloscope is used with a calibrated signal generator to determine the frequency of an unknown signal. The unknown signal is placed on the vertical input and the signal generator on the horizontal input. Then the two signals are beaten together, and the result can be seen on the screen of the oscilloscope. (See Figure 28.21.) If the two frequencies are equal, the picture on the face of the oscilloscope will be a circle. If the known frequency from the signal generator, which is applied to the horizontal deflection plates, is twice that of the unknown signal applied to the vertical deflection plates, the wave pattern will be a figure eight "8." Similarly, if the signal generator frequency is one half that of the unknown signal, the trace on the screen will be an eight on its side "∞." These wave patterns are called Lissajous figures.

The unknown frequency is found by observing the number of times the Lissajous figure touches a horizontal line along the edge of the figure and the number of times it touches a vertical line next to the figure. For example, when the frequencies of the known signal and the unknown signal are equal, the wave pattern (circle) touches the horizontal and vertical lines once each. (See Figure 28.22.) When the signal generator puts out a signal whose frequency is twice the unknown frequency, the figure touches the horizontal line once and the vertical line twice. (See

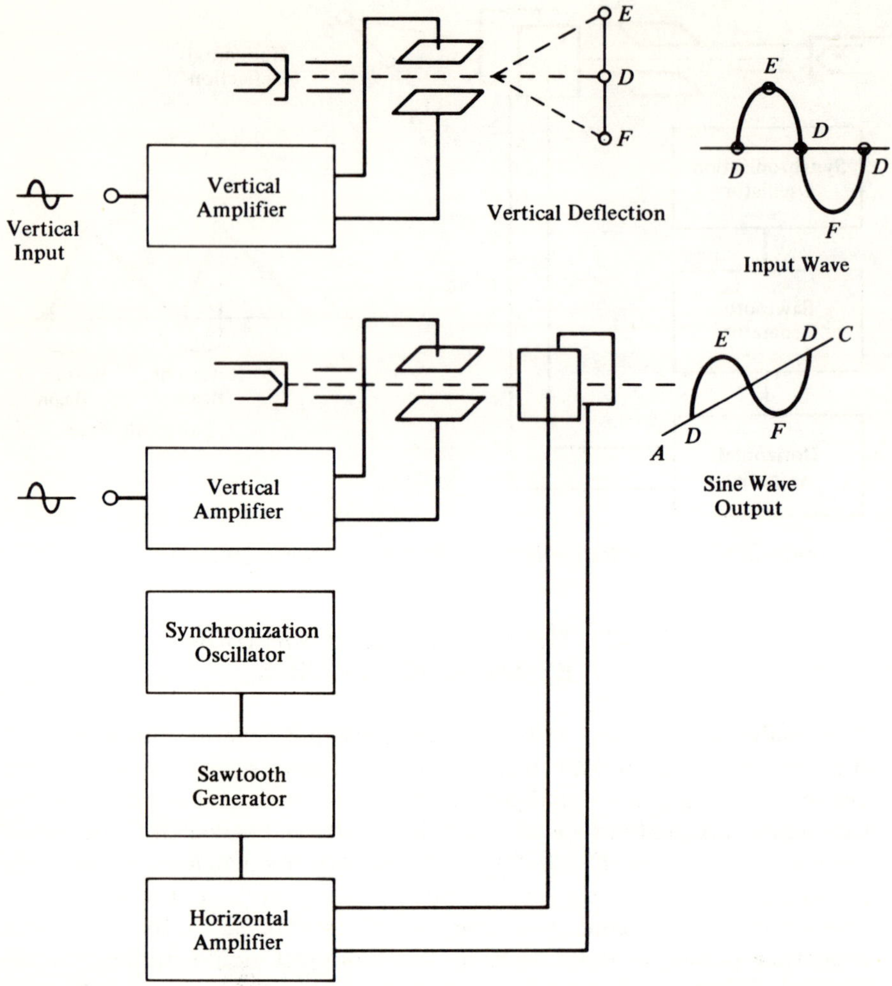

FIGURE 28.20 *Vertical deflection while electron beam is moved across screen by sawtooth input.*

Figure 28.23.) Observe that if we multiply the number of times the figure touches a vertical side by the frequency applied to the vertical input, this product is equal to the product of the number of times the figure touches the horizontal side times the frequency of the horizontal input.

number of vertical touches × vertical input frequency
= number of horizontal touches × known horizontal frequency

That is, if the horizontal input from the signal generator is 200 Hz and the figure is the figure eight "8" shown in Figure 28.23, then the

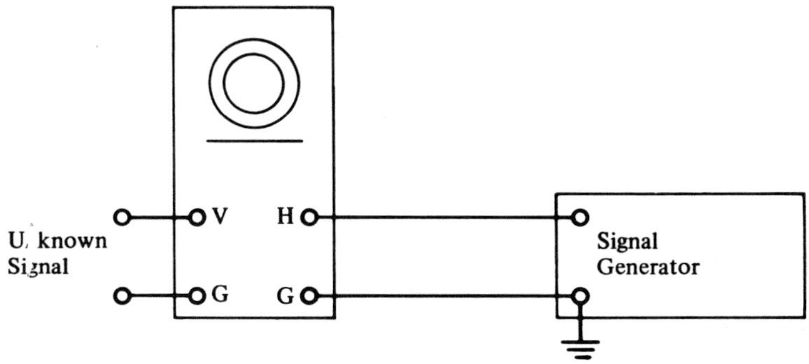

FIGURE 28.21

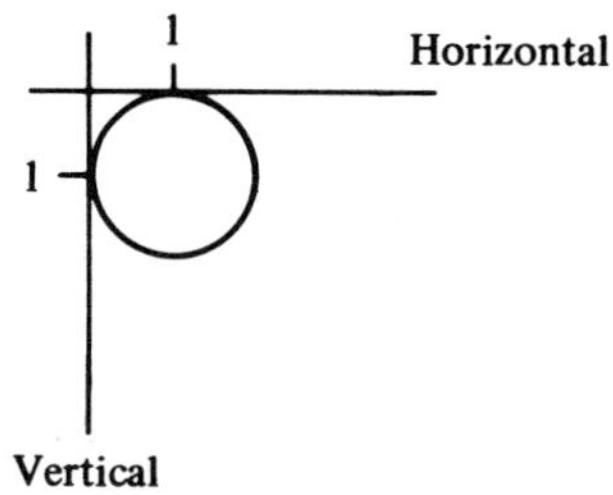

FIGURE 28.22 (*Note: Unknown frequency is equal to known frequency from signal generator.*)

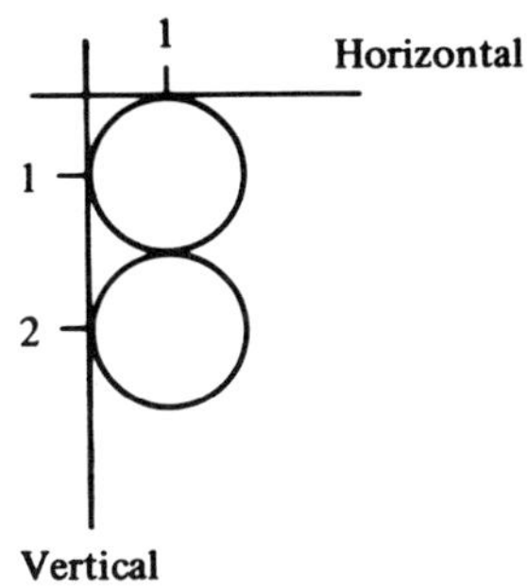

FIGURE 28.23 (*Note:* 2 × *unknown frequency is equal to the known frequency from the signal generator.*)

unknown vertical frequency is:

$$2 \times \text{vertical frequency} = 1 \times \text{horizontal frequency}$$
$$2 \times \text{vertical frequency} = 1 \times 200 \text{ Hz}$$
$$\text{vertical frequency} = 100 \text{ Hz}$$

Consider one more example. Suppose the horizontal input is 60 Hz and the oscilloscope wave pattern is as shown in Figure 28.24. Then the

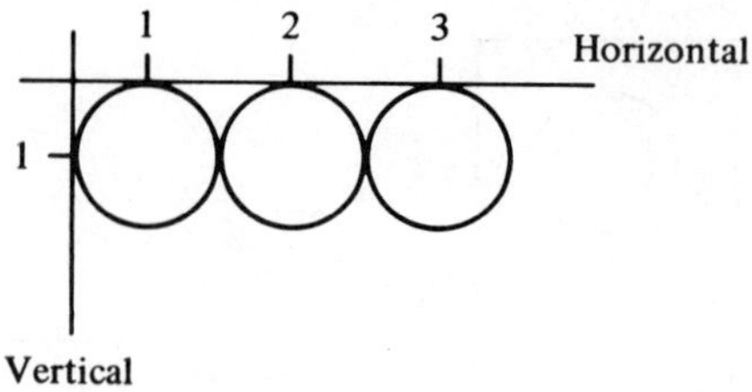

FIGURE 28.24

equations above become:

$$\begin{aligned} 1 \times \text{vertical frequency} &= 3 \times \text{horizontal frequency} \\ 1 \times \text{vertical frequency} &= 3 \times 60 \text{ Hz} \\ \text{vertical frequency} &= 180 \text{ Hz} \end{aligned}$$

This value of 180 Hz for the unknown vertical frequency may be checked by increasing the horizontal input frequency from the signal generator to 180 Hz. The Lissajous figure should then be a circle (unknown frequency = known frequency). (See Figure 28.25.)

$$\begin{aligned} 1 \times \text{vertical frequency} &= 1 \times \text{horizontal frequency} \\ 1 \times \text{vertical frequency} &= 1 \times 180 \text{ Hz} \\ \text{vertical frequency} &= 180 \text{ Hz} \end{aligned}$$

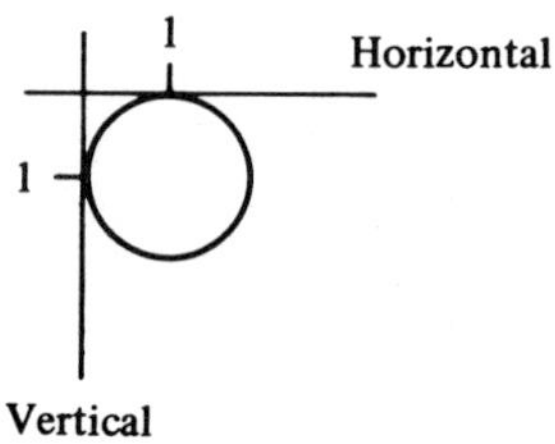

FIGURE 28.25

28.13 VOLTAGE MEASUREMENT WITH THE OSCILLOSCOPE

Another useful feature of the oscilloscope is its ability to measure peak-to-peak ac voltages. In order to do this, the screen of the oscilloscope must be calibrated. (See Figure 28.26.) The ac input voltage is first measured with an ac voltmeter and then applied to the vertical deflection plates of an oscilloscope. If the sawtooth generator is operating, a sine wave will be observed on the screen; if the sweep generator is turned off, a vertical line appears on the screen. When the ac input voltage is varied, the height of the sine wave, or of the vertical line, also varies. These variations

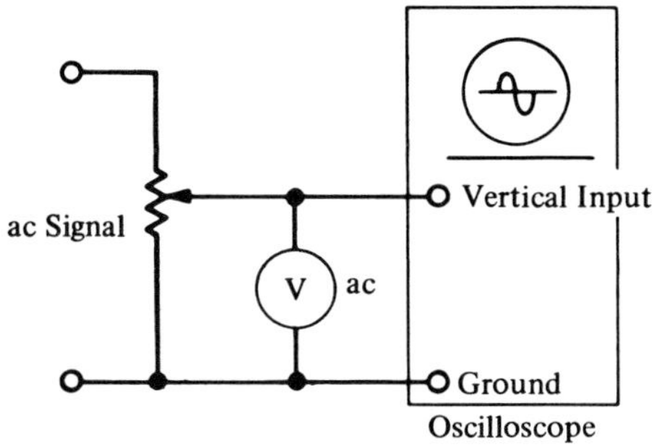

FIGURE 28.26 *Calibration of oscilloscope screen to be used as an ac peak-to-peak voltmeter.*

in ac input are also observed on the ac voltmeter. Remember that an ac voltmeter measures effective voltage, not peak-to-peak voltage. $V_{\text{Effective}} = 0.707 \times V_{\text{Maximum (or peak)}}$; $V_{\text{Maximum (or peak)}} = 1.414 \times V_{\text{Effective}}$. The peak-to-peak voltage will be twice the maximum ac voltage. Thus, $V_{\text{Peak to peak}} = 2 \times V_{\text{Maximum}} = 2 \times 1.414\ V_{\text{Effective}} = 2.828 \times V_{\text{Effective}}$. For instance, if the ac voltmeter reads 10 volts, the peak-to-peak voltage is $V_{\text{Peak to peak}} = 2.828\ V_{\text{Effective}} = 2.828 \times 10$ volts $= 28.28$ volts. The height of the sine wave, or of the vertical line, would thus be equivalent to 28.28 volts, not 10 volts.

The vertical amplifier has a gain adjustment on the front of the oscilloscope, which enables us to vary the height of the sine wave or vertical line. With a known ac voltage input to the oscilloscope, we adjust the vertical gain control to obtain the desired height. In our example, the known voltage input is 28.2 volts peak to peak. Most cathode ray tubes have a grid on the front of the screen marked off in squares. If we adjust the vertical gain control to make the height of the vertical input 28.2 squares, then each square would represent 1 volt. We can now remove the ac voltmeter from the circuit. As long as the vertical gain control is not disturbed, the oscilloscope remains calibrated to measure peak-to-peak voltages of the vertical input. For example, if an unknown ac vertical input causes a vertical deflection of 15 squares on the oscilloscope screen "readout," its peak-to-peak voltage is 15 volts (1 volt for each vertical square).

28.14 INPUT AND OUTPUT DEVICES

The input to an electrical circuit may have one of two forms. It may produce a current as does a battery, generator, photoelectric cell, or piezoelectric crystal; it may cause a change in the current of the electronic circuit by producing a change in resistance, capacitance, or inductance.

Those input devices that cause a change in current either by producing a current themselves or by altering an existing current are called transducers. We will consider several examples of transducers.

A common type of transducer is the phonograph pickup, which utilizes either the piezoelectric or generator principle to produce a small current when the stylus moves back and forth. The dynamic and ribbon microphones also utilize the generator principle. Sound waves move either a diaphragm or a metal ribbon which is between the poles of a magnet. As these metal parts move through the magnetic field, a small current is produced.

The photoelectric generation of current is often used to turn street lights on and off and in burglar alarm systems.

A transducer that produces changes in an existing current flow is the capacitance microphone. The capacitance of the microphone is varied when sound waves striking a flexible plate change the distance between the two plates. The carbon microphone also produces changes in current flow. In this microphone, sound waves compress carbon granules between one fixed and one movable plate. When the granules are compressed, the resistance between the two plates varies; thus, the carbon microphone acts as a variable resistance.

The strain gauge is another transducer which produces a change in current flow by varying resistance. In the case of the strain gauge, a very fine wire is stretched causing its length and diameter to change. We recall that resistance of a wire is directly proportional to length and inversely proportional to cross-sectional area. Thus, the resistance of the strain gauge is varied, and the gauge is sensitive to mechanical deformations.

Probably the most common transducer that produces a change in current flow is the simple switch. When the switch is open, or off, its resistance is infinite; when the switch is closed, or on, its resistance to current flow is zero. This switching action, causing an extreme change in resistance, may be brought about by changes in temperature, as for example in a thermostat which turns a furnace or air conditioner on and off automatically, or by changes in pressure.

Another rather interesting example of a transducer that changes resistance is an ionization detector, such as a Geiger–Mueller tube, which is used to detect the presence of certain types of radiation. The tube essentially consists of two electrodes immersed in a gas which is easily ionized (stripped of one or more electrons). A high potential is placed across the electrodes. Initially, the gas offers a high resistance to current flow. But when it is ionized by some form of radiation (gamma radiation), current flows between the two electrodes.

At the other end of an electronic circuit, changes in electrical current may be transformed into mechanical motion. Unlike input transducers, output devices do not necessarily produce any current or change the rate of flow of current. They do, however, change the form of energy. A simple output device is the Edison light bulb which changes electrical energy to light and heat. In the galvanometer, an electrical current moves a needle, thus changing electrical energy into mechanical energy. And a loud speaker, as we know, changes the energy of an amplified electrical current into sound energy.

Questions

(1) How may one be assured that the voltage reading of a voltmeter is correct and that the meter itself did not reduce the actual voltage?

(2) Why is a headphone used in the capacitance and inductance bridge instead of a galvanometer as in the Wheatstone bridge?

(3) Many tube testers provide a "short" test, a test which indicates if any of the internal components of a vacuum tube are in contact with each other. For example, if the grid came loose and was touching the cathode, this would cause a "short." The short test usually consists of a small bulb. If the bulb glows brightly, there is a short and the tube must be replaced. Where in the basic transconductance tube test circuit would you place a small light bulb to test for a short?

(4) By what other means besides charged metal plates could the electron beam from the cathode ray tube be deflected vertically and horizontally?

(5) Find the ratio of horizontal to vertical frequency using the following Lissajous figure. (See Figure 28.27.)

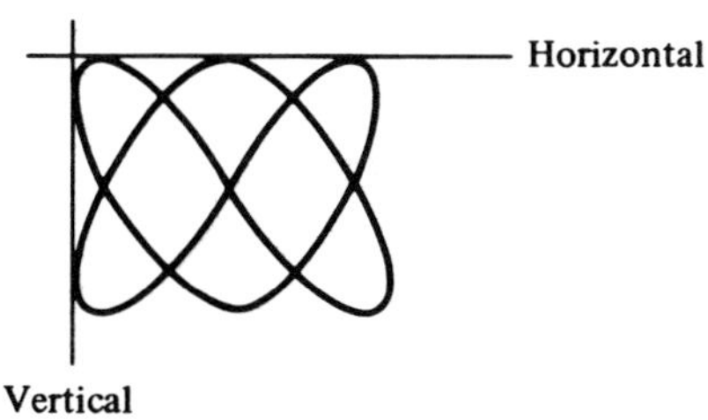

FIGURE 28.27

Appendix I

I.1 EQUATIONS

An equation is just a mathematical sentence, which tells us that two values are the same, or equivalent. In order to state the sentence as simply as possible, symbols are used in place of words. Thus, when we first discuss Ohm's Law, we see that current is equal to voltage divided by resistance. This sentence is then shortened to $I = E/R$ with I representing current, E voltage, and R resistance.

We can think of the two values on either side of an equation as being on a balance. Since the values are equal, the balance is at its zero rest position. (See Figure AI.1.) You know that if you add equal amounts

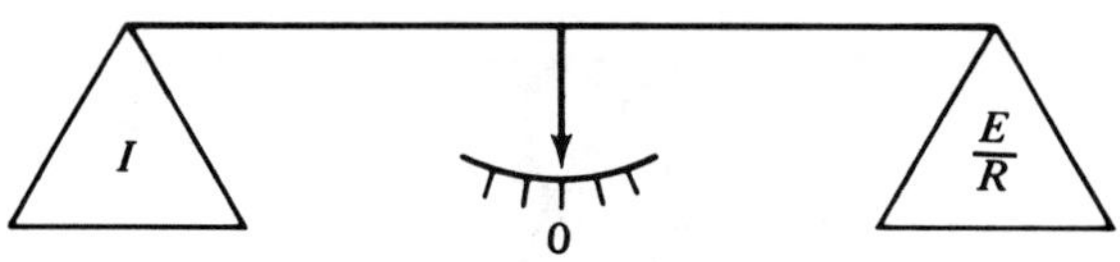

FIGURE AI.1

to both sides of a balance, you do not disturb its equilibrium, that is, it remains balanced. We can say much the same thing about an equation. We can do the same thing, or perform the same operation, on both sides of an equation without changing the balance, or the equivalence.

In our work, we want to perform the following operations on equations: (1) add or subtract the same quantity on both sides of the equation; (2) multiply or divide by the same quantity on both sides of the equation; (3) take the square root of both sides of the equation.

Let us consider the equation $I = E/R$. Suppose we multiply both sides of this equation by R. $I \times R = (E/R) \times R$. Look at the right side: $(E/R) \times R$. Since dividing by a number (R) and then multiplying by the same number (R) leaves us with the value with which we started, then $E/R \times R$ is equal to E. Thus, $I \times R = E$, which you recognize as another form of Ohm's Law.

If we take this equation ($E = IR$) and divide both sides by I, we have $E/I = IR/I$. Again, on the right we divide and multiply by the same number (I), so $E/I = R$, the third form of Ohm's Law.

Suppose we look at another example. In a footnote, it is observed that the expression for the total resistance of two resistors in parallel, which we usually find by the "product divided by the sum" method $R_T = R_1R_2/(R_1 + R_2)$, could also be written as $1/R_T = 1/R_1 + 1/R_2$. Let us show that these two equations are the same, that is, that we can transform one equation into the other by using the accepted operations on equations. Start with $1/R_T = 1/R_1 + 1/R_2$. Multiply both sides by R_1 and R_2, or by $R_1 \times R_2$.

$$R_1 \times R_2 \times \frac{1}{R_T} = R_1 \times R_2 \left(\frac{1}{R_1} + \frac{1}{R_2}\right)$$

When we multiply $R_1 \times R_2$ by a sum, we must multiply it by each one of the terms.

$$R_1 \times R_2 \times \frac{1}{R_T} = R_1 \times R_2 \times \frac{1}{R_1} + R_1 \times R_2 \times \frac{1}{R_2}$$

$$R_1 \times R_2 \times \frac{1}{R_T} = R_2 + R_1$$

Now multiply each side by R_T.

$$R_1 \times R_2 \times \frac{1}{R_T} \times R_T = (R_2 + R_1) \times R_T$$

$$R_1 \times R_2 = (R_1 + R_2) \times R_T$$

Finally, divide each side by $(R_1 + R_2)$.

$$\frac{(R_1 \times R_2)}{(R_1 + R_2)} = \frac{[(R_1 + R_2) \times R_T]}{(R_1 + R_2)}$$

$$\frac{(R_1 \times R_2)}{(R_1 + R_2)} = R_T$$

which is the same as the first equation.

Let us take one more example. Remember when we discussed resonance, we said that $X_L = X_C$ at the resonant frequency. And we know $X_L = 2\pi fL$ and $X_C = 1/2\pi fC$. Thus, $2\pi fL = 1/2\pi fC$ at resonance. If we solve this equation for f, we have a "sentence," which tells us the resonant frequency of an LC circuit. First multiply both sides by $2\pi fC$.

$$2\pi fL \times 2\pi fC = (1/2\pi fC) \times 2\pi fC$$
$$4\pi^2 f^2 LC = 1$$

Remember π^2 just means $\pi \times \pi$ and $f^2 = f \times f$. Divide both sides by $4\pi^2 LC$.

$$\frac{4\pi^2 f^2 LC}{4\pi^2 LC} = \frac{1}{4\pi^2 LC}$$
$$f^2 = \frac{1}{4\pi^2 LC}$$

To find f, recall that the square root of a number N, written $\sqrt{N}$, is that value, which, when multiplied by itself, gives the number N. Thus, the square root of f^2 is f ($\sqrt{f^2} = f$), because $f \times f = f^2$. Similarly, $\sqrt{4} = 2$ ($2 \times 2 = 4$), $\sqrt{\pi^2} = \pi$, and $\sqrt{1} = 1$. Taking the square root of both sides of the above equation, we have

$$\sqrt{f^2} = \sqrt{\frac{1}{4\pi^2 LC}}$$
$$f = \frac{1}{2\pi\sqrt{LC}}$$

I.2 SCIENTIFIC NOTATION

Scientists frequently must work with numbers which are extremely large, such as 6,250,000,000,000,000,000 electrons each second, or quite small, such as 0.000000000002 farad. Since these numbers are awkward to write, a shorthand notation has been devised for them.

You have no doubt noticed that multiplying or dividing a number by 10 or 100 or 1000 or any power of 10 does not change the digits of the number but just changes its magnitude. By a power of 10, we simply mean the number of times 10 is multiplied by itself. Thus, $10^5 = 10 \times 10 \times 10 \times 10 \times 10 = 100{,}000$. For example,

$$5.13 \times 10{,}000 = 51{,}300$$

and

$$\frac{5.13}{1000} = 0.00513$$

Thus, any number can be written as a value between 1 and 10 either multiplied or divided by a power of 10. In our example,

$$51{,}300 = 5.13 \times 10^4$$

and

$$0.00513 = \frac{5.13}{10^3}$$

Notice that the power, or exponent, tells how many places the decimal point is moved. If we multiply by a power of 10, the decimal point is moved to the right; if we divide, the decimal point is moved to the left. Thus,

$$6.7 \times 10^8 = 670{,}000{,}000$$

8 places to right

and

$$\frac{4.3}{10^5} = 0.000043$$

5 places to left

Usually, division is written as multiplication by a negative power of 10, that is,

$$\frac{4.3}{10^5} = 4.3 \times 10^{-5}$$

and

$$\frac{5.13}{10^3} = 5.13 \times 10^{-3}$$

We can now express the two numbers mentioned earlier in scientific notation. $6{,}250{,}000{,}000{,}000{,}000{,}000 = 6.25 \times 10^{18}$ and

$$0.000000000002 = 2 \times 10^{-12}$$

Some prefixes used to represent powers of ten are as follows:

mega	10^6	M
kilo	10^3	k
centi	10^2	c
milli	10^{-3}	m
micro	10^{-6}	μ
pico	10^{-12}	p

Thus, 2×10^{-12} farad = 2 picofarads (2 pF).

Appendix II

II.1 RESISTOR COLOR CODE

The resistance of a resistor may be read if you know the color code. A typical resistor has three or four colored bands around it. The first band is at one end of the resistor with spaces between the following bands. (See Figure AII.1.) The first band represents the first number in the value

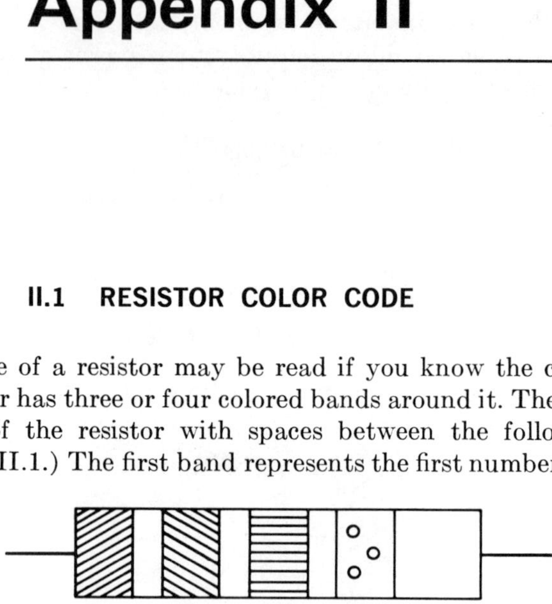

FIGURE AII.1

TABLE AII.1

Color	Number	Number of Zeros	Color	Tolerance
Black	0	None	Gold	5 percent
Brown	1	0	Silver	10 percent
Red	2	00	No 4th Band	20 percent
Orange	3	000		
Yellow	4	0000		
Green	5	00000		
Blue	6	000000		
Violet	7	0000000		
Gray	8	00000000		
White	9	000000000		

of the resistance in ohms. The second band represents the second number in the value of the resistance in ohms. The third band indicates how many zeros follow the second number. The fourth band represents the tolerance for that specific resistor, that is, the percentage by which the color band value of resistance may differ from the actual resistance. A simple mnemonic device to remember the order of colors is:

*B*lack	*B*lack
*B*rown	*B*ruins
*R*ed	*R*aid
*O*range	*O*ur
*Y*ellow	*Y*ellow
*G*reen	*G*rain
*B*lue	*B*lue
*V*iolet	*V*iolets
*G*ray	*G*row
*W*hite	*W*ild

Suppose, for example, you had a resistor with the following color bands: Red White Blue Gold. (See Figure AII.2.) The resistance would be:

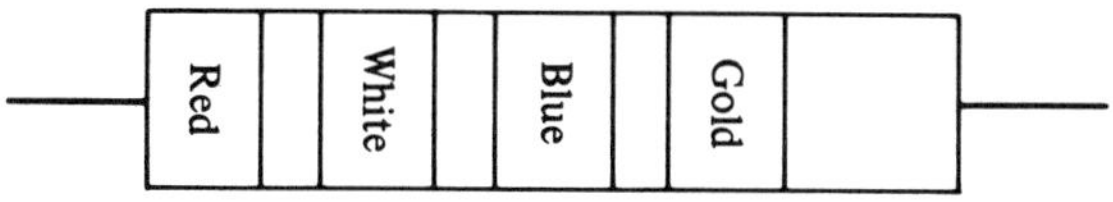

FIGURE AII.2

Red = 2, White = 9, Blue = 000000, Gold = 5 percent, or 29,000,000 ohms ± 5 percent (plus or minus 5 percent).

Appendix III

Appendix III Symbols used in schematic diagrams.

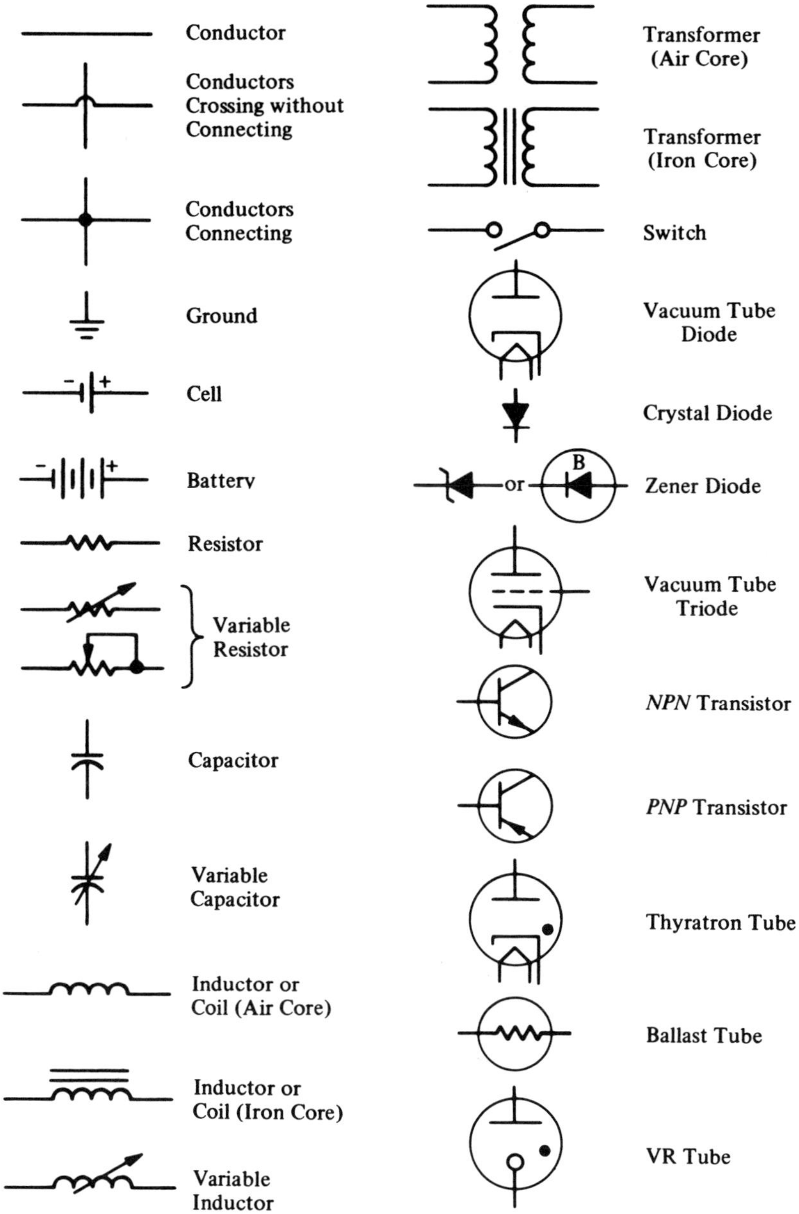

Appendix III Symbols used in schematic diagrams.

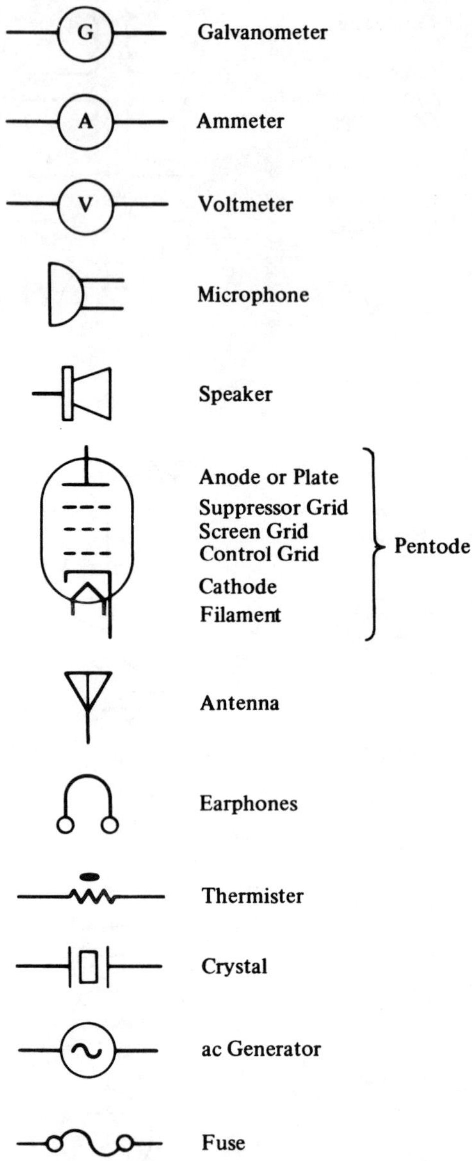

Index

Index

Q

R

S

T

U

V